D0146254

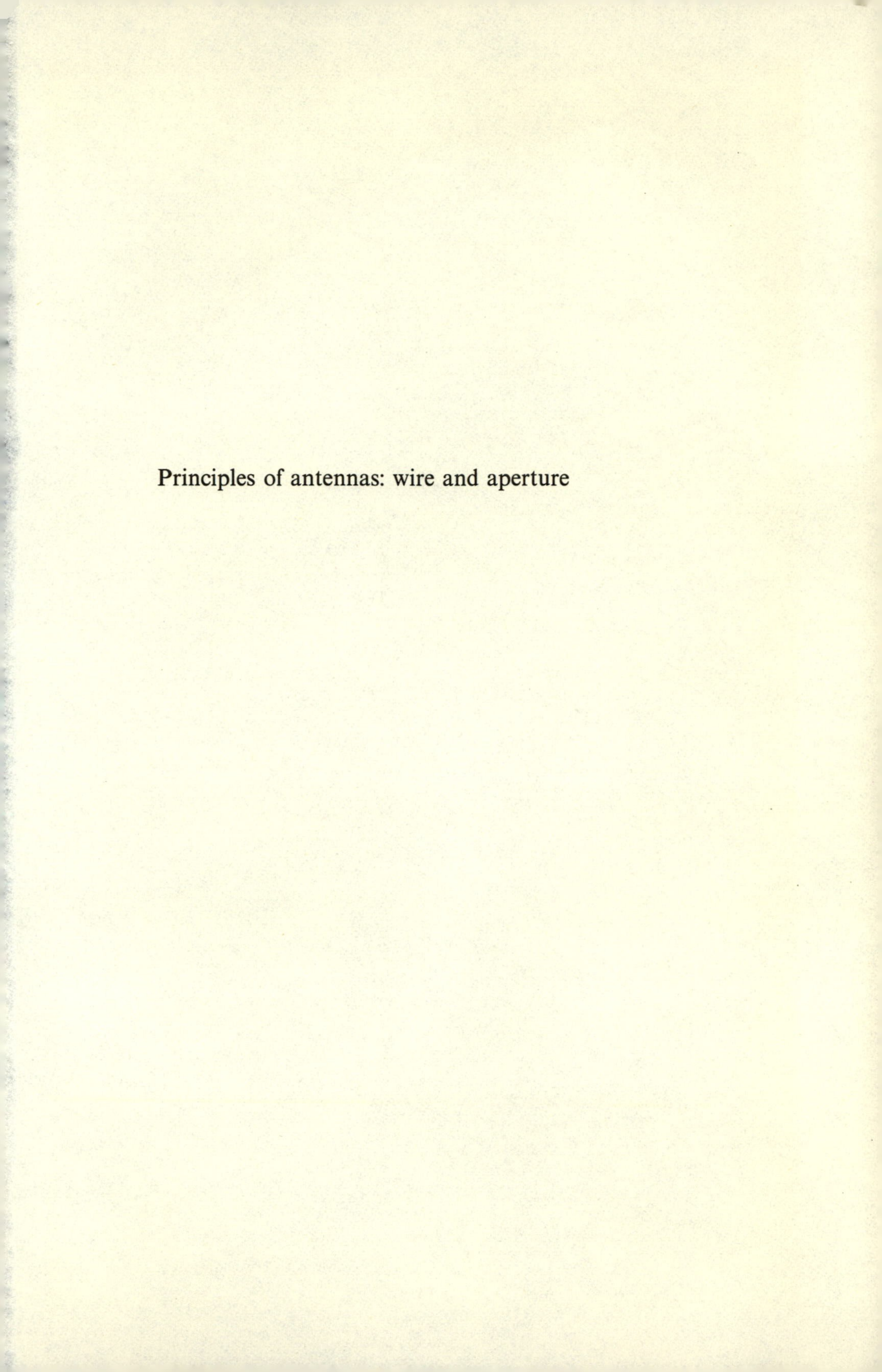

Principles of antennas: wire and aperture

Principles of antennas

wire and aperture

T.S.M. MACLEAN
Department of Electronic and Electrical Engineering
The University of Birmingham

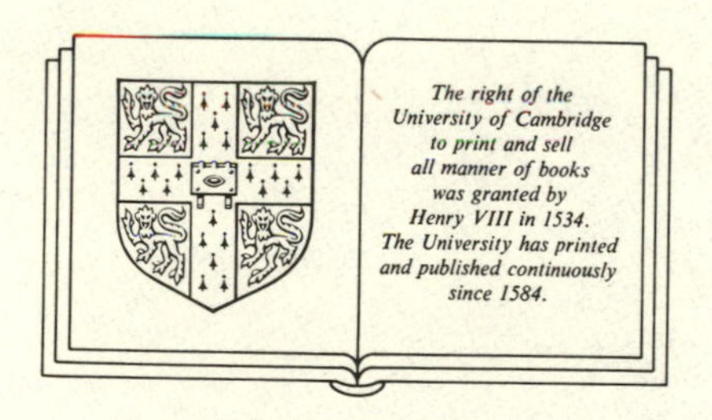

CAMBRIDGE UNIVERSITY PRESS
Cambridge
London New York New Rochelle
Melbourne Sydney

Published by the Press Syndicate of the University of Cambridge
The Pitt Building, Trumpington Street, Cambridge CB2 1RP
32 East 57th Street, New York, NY 10022, USA
10 Stamford Road, Oakleigh, Melbourne 3166, Australia

First published 1986

Printed in Great Britain at the University Press, Cambridge

British Library cataloguing in publication data

Maclean, T.S.M.
Principles of antennas: wire and aperture.
1. Antennas (Electronics)
I. Title
621.38'028'3 TK7871.6

Library of Congress cataloging-in-publication data

Maclean, T.S.M. (Thomas Stewart Mackenzie), 1926-
Principles of antennas.
1. Antennas (Electronics) I. Title.
~~TK7871.6.M33 1986~~ 621.38'028'3 85-21267

ISBN 0 521 30668 X

TP

Contents

	Preface	xi
1	**Planar sources of uniform plane waves**	**1**
1.1	Introduction	1
1.2	First law of electricity – Ampère's law	2
1.3	Second law of electricity – Faraday's law	4
1.4	Application of Maxwell's equations to generation of uniform plane waves by electric currents	5
1.5	Generation of uniform plane waves by magnetic current sheet	6
1.6	Distinction between generating sheet and perfect conductor	8
2	**Current element sources**	**9**
2.1	Line current elements	9
2.2	Electric line current element **Idl**	9
2.3	Fields of current element **Idl** along z-axis calculated using current I alone	10
2.4	Fields of current element calculated using current I and charge Q	12
2.5	Cylindrical current element	15
2.6	Short current element of finite length	17
2.7	Analysis of current element with constant current and uniformly distributed charges	18
2.8	Self impedance of element	20
2.9	Mutual impedance between collinear elements	21
2.10	Mutual impedance between orthogonal elements	23
2.11	General expression for mutual impedance between finite current elements in a plane	25
2.12	Current elements along orthogonal axes	27

2.13 Power flow from current element 29
2.14 Magnetic current element $\mathbf{I}_m\mathbf{dl}$ 30
2.15 Fields of magnetic current element $\mathbf{I}_m\mathbf{dl}$ along z-axis calculated using magnetic current alone 30
2.16 Fields of magnetic current element calculated using magnetic current I_m and pole strength M 32
2.17 Power flow from magnetic current element 33
2.18 Filamentary and cylindrical dipoles and monopoles 34
2.19 Infinitesimal Hertzian dipole 34
2.20 Radiation resistance and input resistance of Hertzian dipole 35
2.21 Input impedance of Hertzian dipole 36
2.22 Short dipole of finite radius 36

3 Dipole and monopole antennas 39
3.1 Filamentary dipole of finite length 39
3.2 Centre-fed transmitting dipole of length $2H$ 39
3.3 Short-circuited reflecting dipole of length $2H$ 43
3.4 Open-circuited linear dipole of length $2H$ 47
3.5 Output impedance of linear dipole of length $2H$ 49
3.6 Impedance loaded linear dipole of length $2H$ 50
3.7 Travelling wave current on filamentary dipole 53
3.8 Radiation pattern of linear dipole of length $2H$ with standing wave of current 54
3.9 General expression for fields of linear dipole of length $2H$ with standing wave of current 56
3.10 Transmitting and receiving power gains of dipole and monopole antennas 58
3.11 Filamentary half-wave dipole in corner reflector 62

4 Computer solutions of dipole and monopole antennas 65
4.1 Computer solutions of linear dipoles 65
4.2 Transmitting operation 66
4.3 Receiving operation 68
4.4 Monopole antenna 69
4.5 Computer solution of half-wave dipole in corner reflector 78

5 Loop antennas 84
5.1 Flux linking approach to rectangular receiving loop antenna 85
5.2 Flux cutting approach to rectangular receiving loop antenna 86
5.3 Electric field approach to rectangular receiving loop antenna 87
5.4 Input impedance of rectangular loop 96

5.5	Electric field approach to circular receiving loop antenna	99
5.6	Rectangular loop antenna as magnetic field probe	105
5.7	Loop antenna as direction finder	106
5.8	Screened loop antenna	108
5.9	Transmitting loop antenna	111
5.10	Radiation pattern of large rectangular loop	113
6	**Helical antennas**	**118**
6.1	Normal mode helical antennas	118
6.2	Analysis of small circular helical antenna – open helix	120
6.3	Analysis of small circular helical antenna – closed helix	122
6.4	Radiation resistance of open and closed small helical antennas	122
6.5	Axial mode circular helical antenna	123
6.6	Analysis of axial mode helical antenna	125
6.7	Linearly polarised helical antennas	138
7	**Yagi–Uda antennas**	**140**
7.1	Two dipole array	141
7.2	Long Yagi–Uda array with continuous distribution of short director elements	145
7.3	Experimental approach to Yagi–Uda antenna design	151
8	**Frequency independent and logarithmically periodic antennas**	**157**
8.1	Planar equiangular spiral antenna	159
8.2	Practical equiangular spiral antenna	162
8.3	Linearly polarised frequency independent antenna	164
8.4	Unidirectional linearly polarised frequency independent antenna	166
8.5	Triangular logarithmically periodic linearly polarised antenna	166
8.6	Logarithmically periodic dipole array	168
8.7	Network analysis of uniformly loaded periodic structure	171
8.8	Application of uniformly loaded periodic structure to logarithmically periodic dipole array	175
9	**Noise power delivered by wire antennas**	**176**
9.1	Noise power delivered to matched load of infinitesimal dipole in isotropic noise field	176
9.2	Noise power delivered to matched load of any antenna in isotropic noise field	178

9.3 Network representation of noise power delivered by wire antennas 181

10 Aperture antennas **189**
10.1 Field equivalence theorem 190
10.2 Equivalent currents over spherical surface of radius *a* for original current element 191
10.3 Application and simplification of the equivalence theorem applied to aperture antennas 199
10.4 Further comparison of three aperture field techniques 201
10.5 Radiation due to elemental area of magnetic current in infinite ground plane 202
10.6 Radiation fields due to finite distribution of magnetic current in infinite ground plane 203
10.7 Radiation due to elemental area of electric current in infinite magnetic ground plane 205
10.8 Radiation fields due to finite distribution of electric current in infinite magnetic ground plane 206
10.9 Radiation fields due to combined magnetic and electric currents 207

11 Angular spectrum of plane waves **210**
11.1 Angular spectrum for one-dimensional slot 210
11.2 Infinite slot of width 2*a* in planar conducting sheet 214
11.3 Angular spectrum for two-dimensional aperture 223
11.4 Radiation field from two-dimensional aperture in conducting plane 227
11.5 Circular aperture of radius *a* in planar conducting sheet 229
11.6 Radiation fields of uniformly excited circular aperture 237
11.7 Power gain of uniformly excited circular aperture 238
11.8 Radiation fields of uniformly excited rectangular aperture 239
11.9 Power gain of uniformly excited rectangular aperture 241
11.10 Measured radiation patterns 242

12 Waveguide radiators **244**
12.1 Radiation pattern of rectangular waveguide carrying TE_{10} mode 245
12.2 *H*-plane sectoral horn 248
12.3 Power gain of *E*-plane sectoral horn 253
12.4 Power gain of pyramidal horn 253
12.5 Cylindrical waveguide: derivation of wave equation 254

12.6 Cylindrical waveguide: solution of wave equation for TE modes 255
12.7 Tangential electric and magnetic fields in aperture of cylindrical waveguide carrying TE_{11} mode 257
12.8 Radiation pattern of cylindrical waveguide carrying TE_{11} mode 259

13 Paraboloidal reflector 266
13.1 Equations of parabola 266
13.2 Paraboloid excited by normally incident uniform plane wave 269
13.3 Analysis of fields produced by paraboloid irradiated by normally incident plane wave 271
13.4 Scattered magnetic fields on axis 273
13.5 Scattered magnetic field at focus 274
13.6 Scattered electric fields on axis including focus 275
13.7 Asymmetry of scattered fields on axis about focus 276
13.8 Numerical values of scattered fields on axis for normally irradiated paraboloid 277
13.9 Focal plane magnetic fields for paraboloid with normally incident plane wave 279
13.10 Focal plane electric fields for paraboloid with normally incident plane wave 286
13.11 Power flow across focal plane 291

14 Receiving paraboloidal reflector with feed 296
14.1 Infinitesimal dipole feed 296
14.2 Receiving radiation pattern of paraboloid with infinitesimal dipole feed 298
14.3 Power coupling between apertures 300
14.4 Coupling between paraboloid and TE_{11} circular cylindrical wave-guide feed 303
14.5 Computed aperture efficiency for cylindrical waveguide carrying TE_{11} mode placed at focus 305
14.6 Coupling between paraboloid and HE_{11} mode corrugated waveguide feed 307
14.7 Signal to noise ratio of receiving paraboloidal reflector plus feed 309

15 Analysis of transmitting paraboloids 312
15.1 Radiation pattern of paraboloid with current element feed – aperture field approach 312

15.2 Radiation pattern of paraboloid with current element feed – current distribution approach 318
15.3 Power gain on axis of paraboloid excited by y-directed current element 324
15.4 Radiation pattern of paraboloid excited by current element feed 325
15.5 Reflected fields from paraboloid with incident field $(E_\theta(\theta, \phi) + E_\phi(\theta, \phi))$ 326
15.6 Radiation from paraboloid using TE_{11} mode cylindrical waveguide feed 327
15.7 Power gain of paraboloid excited by TE_{11} mode cylindrical waveguide feed 330

16 Cassegrain and offset reflector analysis 333
16.1 Cassegrain antennas 333
16.2 Hyperboloidal reflector 334
16.3 Radiation pattern of hyperboloid with current element feed – current distribution approach 338
16.4 Forward scattered field of hyperboloid with cylindrical waveguide feed 346
16.5 Offset reflector antennas 348
16.6 Tangential electric fields in focal aperture plane of offset paraboloid 353
16.7 Offset paraboloid radiation pattern 354

Index 357

Preface

This work is intended both as a students' textbook at final year undergraduate and postgraduate level, and for use by engineers in industry who wish to ask the question 'why', rather than simply 'how', a particular approach is used in considering different antennas. The book is not intended, however, to teach commercial antenna design.

A deliberate choice has been made to devote approximately equal amounts of space to wire antennas and aperture antennas. Most other texts tend to concentrate rather heavily on one or the other, but here the author has attempted to cover both fields from as unified a standpoint as possible.

The book is quantitative rather than descriptive because it is economic in terms of a student's time that it should be so. It is in no sense exhaustive for any of the antennas which are considered, and many matters are not considered at all. But the aim has been to combine a physical understanding with a quantitative outlook for the antennas which are considered.

The author wishes to express his appreciation to the research students with whom he has worked at the University of Birmingham. The titles of theses written by some of these are listed in the References, but all of them have contributed at least indirectly to the form of the book, although the responsibility for the contents of the book remains with the author.

TSM Maclean

Fig. 1.1. Uniform plane wave with E_x, H_y, travelling in z-direction.

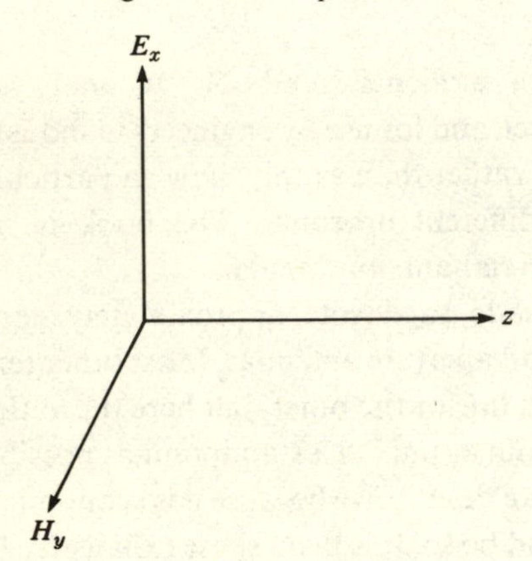

Fig. 1.2. Two uniform plane waves with E_x, $\pm H_y$ travelling in $\pm z$-directions.

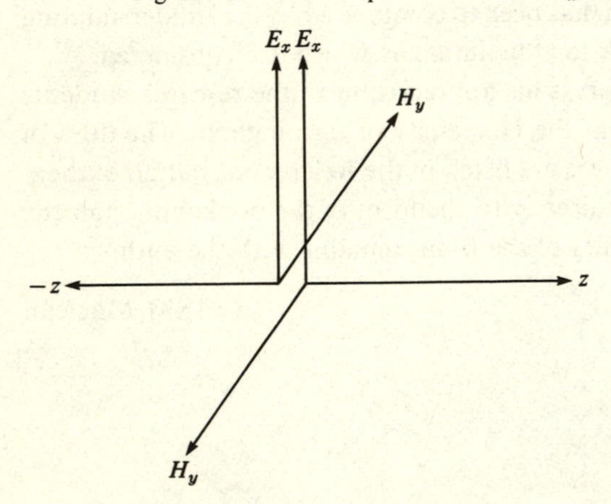

1

+−+

Planar sources of uniform plane waves

1.1 Introduction

The simplest type of radio wave is a uniform plane wave. This we define to be a wave which, in the direction orthogonal to the direction of propagation, is uniform in amplitude, and which also possesses a planar phase front. Such a wave would travel through free space with a velocity equal to the velocity of light, with two orthogonal field components **E** and **H**, both of which are perpendicular to the direction of propagation. The electric field **E** is measured in volts/metre and it is convenient to regard this as being directed along the x-axis of a right-handed coordinate system as shown in Fig. 1.1. The magnetic field **H** is therefore directed along the y-axis, so that in rotating a unit vector which is initially along the electric axis x, to be along the magnetic axis y, the direction of travel of a right-handed screw would be along the direction of propagation z.

The question is naturally asked, what mechanism is able to produce such a simple type of wave. To answer this question we will consider, as in Fig. 1.2, that there are two such uniform plane waves travelling in opposite directions, namely the positive and negative z-directions. Retaining the directions E_x and H_y for the electric and magnetic field components of the wave travelling to the right along the positive z-axis, we have to decide what the corresponding directions are for the wave travelling to the left. We will assume, quite arbitrarily initially, that the same x-direction is retained for the electric field, and therefore its magnetic field must be in the opposite, i.e. negative y-direction, in order to satisfy the right-handed orthogonality condition between these two fields and the direction of propagation.

It follows that there must be some source of energy responsible for producing these two outgoing waves travelling in the positive and negative z-directions. To find out what form this source of energy takes, we must use the known laws of electricity which describe quantitatively how electric and magnetic fields are related to each other, and to the flow of currents which are responsible for producing them.

1.2 First law of electricity – Ampère's law

Every electric current is accompanied by the presence of a magnetic field and the relation between the current and this field is given by Ampère's law. Where the current does not vary with time, the law takes on its simplest form, and says that the integral of $\mathbf{H} \cdot \mathbf{dl}$ round any closed path is equal to the current I enclosed by that path, i.e.

$$\oint \mathbf{H} \cdot \mathbf{dl} = I \tag{1.1}$$

For example, let eqn (1.1) be applied to the case of an infinitely long filament, which is assumed to be carrying a constant direct current as shown in Fig. 1.3. The path of integration will be chosen to be a circle of radius r centred on the axis of the wire, so that by symmetry the magnitude of $\mathbf{H}$ can be taken to be constant, and this then gives for the circumferential component of magnetic field

$$H_\phi = \frac{I}{2\pi r}$$

Fig. 1.3. Infinitely long filament carrying direct current.

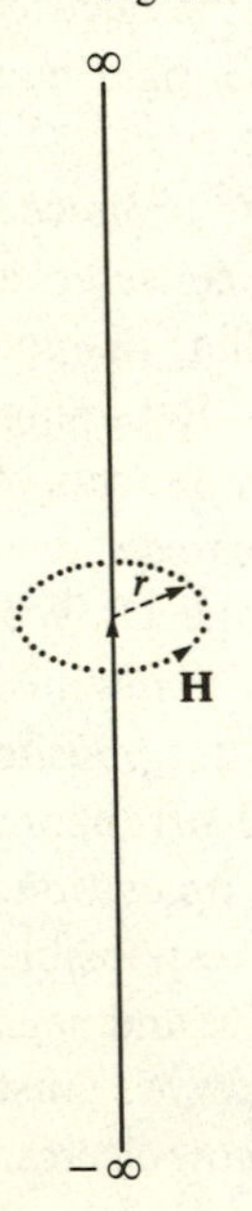

Where the current does vary with time, Ampère's law becomes

$$\oint \mathbf{H}\cdot d\mathbf{l} = I + \frac{\partial \psi}{\partial t} \tag{1.2}$$

where ψ is the total electric flux which is enclosed by the path of integration. In this case for the infinite filament the axial current I sets up an axially directed electric field inside every circle of radius r, which is in opposition to the direction of current flow. But if the filament is imagined to be expanded radially from zero radius up to a radius a, and if in addition it is constructed from material of infinite conductivity, so that no electric field can exist within it, then if the path of integration is taken circumferentially round the path of radius a only, then again

$$H_{\phi_{r=a}} = \frac{I}{2\pi a} \tag{1.3}$$

This argument and result for an alternating current apply also to a linear conductor of any length, unlike the direct current case, which had to be restricted to one of infinite length so that the magnetic effect of the return path could be ignored. The magnetic field given by eqn (1.3) is thus the only component of magnetic field which exists for a linear conductor of any length.

The relation between the value of **H** given above, and the current I which produces it, is clearly a vector relationship. The circumferential direction of **H** at the surface of the conductor is orthogonal to the axial direction of the current. Considering now the surface current density and the tangential magnetic field at the surface of any perfectly conducting circular cylinder, we have from eqn (1.3) that the linear current density **J** in amperes per metre is equal to the surface magnetic field $\mathbf{H}_s$, through the relation

$$\mathbf{J} = \mathbf{n} \times \mathbf{H}_s \tag{1.4}$$

where **n** is a unit vector normal to the conducting surface.

Similarly for a planar perfect conductor which carries a uniform current density **J** over its entire surface, application of eqn (1.2) to the closed path $ABCD$ of length $2l$ shown in Fig. 1.4, gives for the surface magnetic field $\mathbf{H}_s$

$$H_s l = Jl \tag{1.5}$$

where the path l is orthogonal to the current density **J**. This type of current density would exist, for example, if the perfectly conducting plane were irradiated by a normally incident uniform plane wave, and eqn (1.5) is simply the scalar form of the vector eqn (1.4).

Thus for both cylindrical and planar surfaces the relationship between surface tangential magnetic field and linear current density in the form of

Fig. 1.4. $\mathbf{H}_s$ at surface of perfect planar conductor.

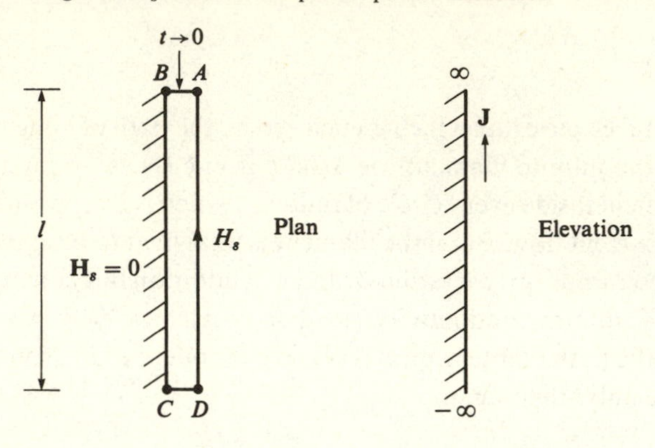

eqn (1.4) has been established by the application of the generalised form of Ampère's law, given by eqn (1.2), to perfect conductors. The conductors do not have to be infinite in extent, and the relationships are valid at any point on the surface.

1.3 Second law of electricity – Faraday's law

Every time-varying magnetic flux sets up an electric field, whose line integral round any closed path surrounding the flux is equal to minus the time rate of change of the enclosed flux. This may be expressed algebraically as

$$\oint \mathbf{E}\cdot d\mathbf{l} = -\frac{\partial \Phi}{\partial t} \tag{1.6}$$

However although this equation is known to be exact in all experimental situations, it turns out that it is valuable for later work to generalise it into a form similar to eqn (1.2), by postulating an additional term on the right-hand side analogous to the electric current in that equation. Such a term would be a magnetic current consisting of a movement of magnetic poles if such entities existed as single units. Thus the generalised form of Faraday's law is taken to be

$$\oint \mathbf{E}\cdot d\mathbf{l} = -I_m - \frac{\partial \Phi}{\partial t} \tag{1.7}$$

where the minus sign associated with the I_m is used simply to maintain symmetry in the right-hand side of the equation.

Were such a magnetic current able to exist in practice and to flow in a perfect magnetic conductor, the relation between the magnetic current density on the surface of this conductor and its associated tangential electric field, would be

$$\mathbf{J}_m = -\mathbf{n} \times \mathbf{E}_s \tag{1.8}$$

by applying eqn (1.7) to, say, a cylindrical linear magnetic conductor of any length, or a planar perfect magnetic conductor.

It is appropriate to refer to the general eqns (1.2) and (1.7) as the generalised forms of Maxwell's equations, and to eqns (1.2) and (1.6) simply as Maxwell's equations.

1.4 Application of Maxwell's equations to generation of uniform plane waves by electric currents

We return now to the situation shown in Fig. 1.2 in order to find out how two oppositely directed uniform plane waves may be generated by an infinite planar generator. The two waves are shown in both plan and elevation in Fig. 1.5, and it is necessary that the application of Maxwell's equations to the generating plane in these figures should result in self consistent fields and currents.

Consider first the path *ABCD* in Fig. 1.5(*a*). Applying the line integral of magnetic field intensity round this closed circuit we obtain for a time variation $e^{j\omega t}$, where ω is the angular frequency,

$$H_{y1}l + H_{z1}t + H_{y2}l + H_{z2}t = I_x + j\omega\varepsilon E_x lt \tag{1.9}$$

Now let the width t of the closed circuit tend to zero so that since H_{y1} is equal to H_{y2}, then

$$\left.\begin{aligned} 2H_{y1}l &= J_x l \\ \text{or} \qquad |J_x| &= |2H_y| \end{aligned}\right\} \tag{1.10}$$

Fig. 1.5. Generation of uniform plane waves by electric currents.

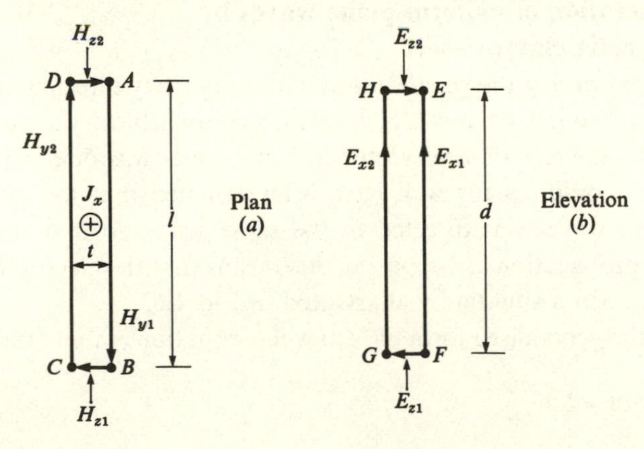

i.e. the linear current density is equal to the discontinuity in tangential magnetic field. It will be noted that the direction of the current flow J_x is in the negative x-direction, i.e. opposite to the radiated electric field E_x. This is in contradistinction to the relative directions of electric field and current in a dissipative network where both are directed in the same sense. In this case it indicates that the power associated with the flow of current is an outgoing or radiated power, as opposed to an inflowing or dissipative power in network devices.

The rate of work done by the current in driving against the electric field is given by

$$P = \tfrac{1}{2}E_x J_x$$

where E_x, J_x are both peak values of fields. From eqn (1.10) this can be written

$$P = E_x H_y \tag{1.11}$$

i.e. the radiated power density P is the sum of the power densities $\tfrac{1}{2}EH$ radiated by the current sheet to right and left.

The above information has been derived from the application of Ampère's law to the current sheet. When the integral form of Faraday's law is applied to the path $EFGH$ in Fig. 1.5(*b*), the result is

$$-E_{x1}d + E_{z1}t + E_{x2}d + E_{z2}t = -j\omega\mu Hd\cdot t$$

Again letting the thickness t of the path of integration tend to zero, confirms that the tangential electric field is continuous across the sheet. Hence the fields which have been postulated satisfy both of Maxwell's equations, and show that uniform plane waves in both directions normal to the sheet may be generated by a uniform, in-phase electric current density in the infinite, planar, conducting sheet considered.

1.5 Generation of uniform plane waves by magnetic current sheet

In considering the generation of two oppositely directed uniform plane waves in Section 1.4, it will be recalled that an arbitrary assumption was made that the electric field vector in both waves was directed in the same sense. It could equally well have been assumed that the magnetic vector in both waves was directed in the same sense, so that since the directions of propagation are opposite, this means that the electric vectors are opposed. Such a situation is illustrated in Fig. 1.6.

Recalling the generalised form of Maxwell's equations in integral form,

$$\oint \mathbf{H}\cdot\mathbf{dl} = I + \frac{\partial\psi}{\partial t} \tag{1.2}$$

and

$$\oint \mathbf{E}\cdot\mathbf{dl} = -I_m - \frac{\partial \Phi}{\partial t} \tag{1.7}$$

where I_m is a magnetic conduction current, these equations will now be applied to the paths *ABCD* and *EFGH* in Fig. 1.6(*a*) and (*b*). Integrating **H·dl** first round *ABCD* gives

$$H_{y1}l + H_{z1}t - H_{y2}l + H_{z2}t = I + j\omega\varepsilon Elt$$

Since H_{y1} is equal to H_{y2} this confirms, when the thickness t tends to zero, that there is no conduction current I. Likewise integrating **E·dl** round *EFGH* gives

$$\begin{aligned} -E_{x1}d + E_{z1}t - E_{x2}d + E_{z2}t &= -I_m - j\omega\mu Hdt \\ &= -J_m d - j\omega\mu Hdt \end{aligned}$$

Again letting t tend to zero gives

$$|J_m| = (E_{x1} + E_{x2}) = |2E_x| \tag{1.12}$$

i.e. the linear magnetic current density is equal to the discontinuity in tangential electric field.

The direction of flow of this magnetic current density will be seen to be opposite to the magnetic force **H**. Hence power must be supplied to drive the magnetic poles constituting this current against the force **H**, and this power density then becomes the power density in the outgoing waves. i.e.

$$P_m = \tfrac{1}{2}H_y J_m$$

where H, J_m are both peak values. From eqn (1.12) this becomes

$$P_m = E_x H_y \tag{1.13}$$

Fig. 1.6. Generation of uniform plane waves by magnetic currents.

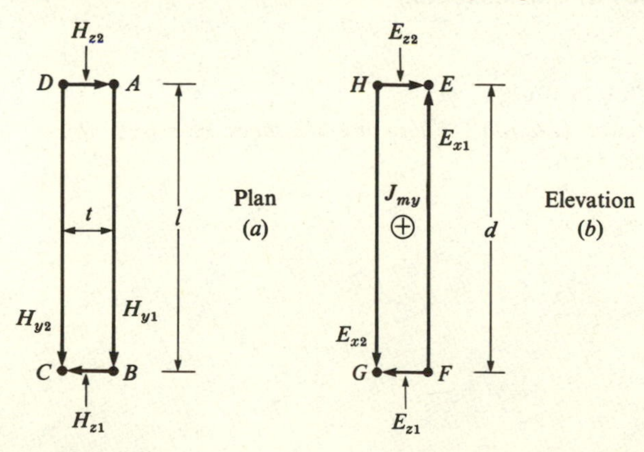

1.6 Distinction between generating sheet and perfect conductor

At an electrically perfectly conducting planar sheet the boundary conditions to be satisfied are that the total tangential electric field must be zero, and that the linear current density is in magnitude equal to, and in direction orthogonal to, the tangential magnetic field. Expressed mathematically these conditions are

$$\mathbf{E}_t = 0$$

and

$$\mathbf{J} = \mathbf{n} \times \mathbf{H}$$

By contrast, in a generating plane, a tangential E must both exist and be in antiphase to the driving linear current density, in order to provide the power density which is radiated into both half-spaces on each side of the plane. It is, however, legitimate to consider that this driving tangential electric field can be nullified by an incoming wave from an external source. In such a case the resultant tangential electric field would be zero, and the generating plane could then be replaced by a perfect conductor without altering the radiated field.

Care must be taken, however, in evaluating the Poynting vector when a generator consisting of a perfectly conducting sheet is being considered. The Poynting vector parallel to the sheet will necessarily be zero since the magnetic field normal to the sheet is zero. The Poynting vector along the outward normal to the sheet will be equal to that of the incoming wave since although the magnetic fields are codirectional the electric fields are reversed.

Likewise the impedance looking parallel to the sheet will be infinite since the magnetic field normal to the sheet is zero, and that looking normal to the sheet will vary between zero and infinity because of the standing wave pattern of fields in that direction.

Further reading

J.R. Pierce: *Electrons, Waves and Messages*: Hanover House, New York, 1956.

2

+−+

Current element sources

2.1 Line current elements

The radiated waves considered in the previous chapter were produced by the summation of the radiation from each infinitesimal element which makes up the planar radiating sheet. Because all antennas consist of a grouping of such infinitesimal elements, the fields of a single line element of length, dl, carrying a constant electric current I will now be considered.

2.2 Electric line current element Idl

Since the current in the element dl is of constant magnitude, this means that there are no electric charges along its length. Charges only exist where the current changes; negative charges if the current increases, since current is assumed to be associated with positive charge, and an increase of current therefore means that positive charge has been taken away from the point of observation. Likewise positive charge accumulation is associated with a decrease of electric current. But although there are no charges distributed along the length dl, there are point charges located at its ends. If the current is flowing in the positive z direction in Fig. 2.1, then positive charge of magnitude $\int I dt$ exists at the top of dl, and an equal negative charge at the bottom.

The current element **Idl** as shown in Fig. 2.1, is placed along the z-axis at the origin of a right-handed coordinate system. Both a rectangular and a spherical coordinate system are shown, because it is frequently simpler to calculate the fields initially in rectangular coordinates and then transform to spherical coordinates, in which the point of observation is often expressed. Because the element **Idl** has both current and electric charge associated with it, it is obvious that there will be a magnetic field set up by the current and an electric field set up by the charge. But it should not be thought that the electric field exists only because of the existence of the electric charges at the ends of the element. Such an electric field will exist even in the absence of charges, simply by virtue of the time-varying nature of the current I. This follows from the use of Maxwell's equations, and the complete fields of the element **Idl** will initially be worked out, without considering the charge explicitly. The alternative solution which does consider the effect of the charges explicitly, will then be used to confirm the validity of the first solution.

Very frequently the term line current element is abbreviated simply to current element and this practice will henceforth be followed.

2.3 Fields of current element Idl along z-axis calculated using current I alone

Because the current I flows in the z-direction only, there is only a z-directed component of magnetic vector potential, which is given by

$$A_z = \frac{\mu I dl}{4\pi} \frac{e^{-jkr}}{r} \tag{2.1}$$

where μ is the permeability of the medium surrounding the element and r is the distance from the centre of the element to the point of observation. This vector potential by itself provides a complete description of the electrical effect of the current element anywhere in space, both in magnitude and

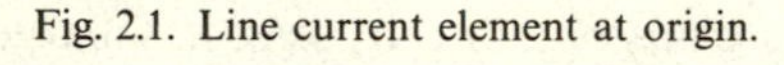
Fig. 2.1. Line current element at origin.

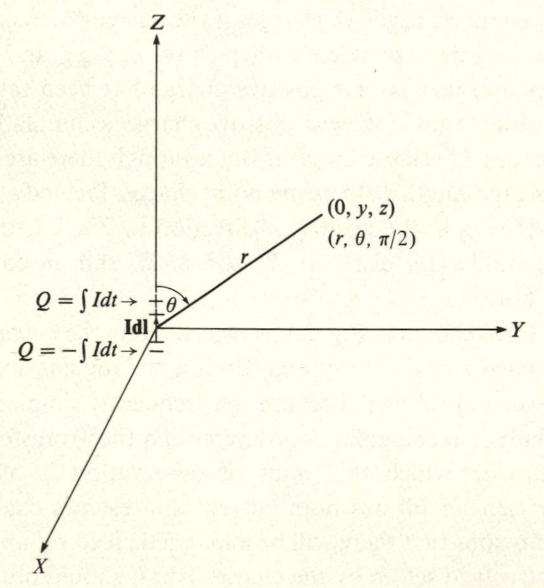

direction. But although it is complete it is deficient from an experimental viewpoint, in that there is no means of measuring it directly. Because of this it is customary to operate on it, so as to produce, first, the corresponding magnetic field **H** anywhere in space, and then, secondly, operate on **H** to produce the electric field at the same point. Both **H** and **E** are field quantities which can be measured, the first with a loop probe, and the second with a probe dipole.

The magnetic field **B** is obtained from the vector potential **A** by performing the curl operation on **A**. In rectangular coordinates this then gives

$$H_x = \frac{1}{\mu}\left(\frac{\partial A_z}{\partial y} - \frac{\partial A_y}{\partial z}\right) = \frac{1}{\mu}\frac{\partial A_z}{\partial y} \tag{2.2}$$

$$H_y = \frac{1}{\mu}\left(\frac{\partial A_x}{\partial z} - \frac{\partial A_z}{\partial x}\right) = -\frac{1}{\mu}\frac{\partial A_z}{\partial x} \tag{2.3}$$

$$H_z = \frac{1}{\mu}\left(\frac{\partial A_y}{\partial x} - \frac{\partial A_x}{\partial y}\right) = 0 \tag{2.4}$$

Now the distance r from the source element dl at the origin to a field point of observation (x, y, z) is

$$r = [x^2 + y^2 + z^2]^{\frac{1}{2}}$$

Considering an observation point in the yz-plane for simplicity, so that x is equal to zero, gives

$$\frac{\partial A_z}{\partial y} = \frac{\mu I dl}{4\pi}\frac{[re^{-jkr}(-jk) - e^{-jkr}]}{r^2}\frac{y}{r}$$

i.e.

$$H_x = -\frac{I dl \sin\theta}{4\pi}\left(\frac{1}{r^2} + \frac{jk}{r}\right)e^{-jkr} \tag{2.5}$$

This field in the x-direction is also the circumferential field for the observation point in the yz-plane, and for this same point H_y is zero.

Similarly for an observation point in the xz-plane it will be found that

$$H_y = +\frac{I dl \sin\theta}{4\pi}\left(\frac{1}{r^2} + \frac{jk}{r}\right)e^{-jkr}$$

This again is the circumferential field, and in general at any point in space

$$H_\phi = \frac{I dl \sin\theta}{4\pi}\left(\frac{1}{r^2} + \frac{jk}{r}\right)e^{-jkr} \tag{2.6}$$

and there are no other components orthogonal to this.

Because of this time-varying component of magnetic field, an electric field will be generated whose magnitude is proportional to the curl of **H**.

The components of $\mathbf{E}$ are more readily derived in spherical coordinates giving

$$E_r = \frac{1}{j\omega\varepsilon}\left[\frac{1}{r}\frac{\partial H_\phi}{\partial \theta} + \frac{H_\phi \cot\theta}{r}\right]$$

$$E_\theta = \frac{1}{j\omega\varepsilon}\left[-\frac{\partial H_\phi}{\partial r} - \frac{H_\phi}{r}\right]$$

and

$$E_\phi = 0$$

Hence

$$E_r = \frac{Idl\cos\theta}{2\pi\omega\varepsilon}\left(-\frac{j}{r^3} + \frac{k}{r^2}\right)e^{-jkr} \tag{2.7}$$

and

$$E_\theta = \frac{Idl\sin\theta}{4\pi\omega\varepsilon}\left(-\frac{j}{r^3} + \frac{k}{r^2} + \frac{jk^2}{r}\right)e^{-jkr} \tag{2.8}$$

It is of interest also to express the component of electric field along the axis of the current element, to give its longitudinal component. This is given by

$$E_z = E_r\cos\theta - E_\theta\sin\theta$$

For points on the surface of the current element, θ equals zero or π and e^{-jkr} can be expanded in a power series to give

$$E_{zs} \approx \frac{Idl}{2\pi\omega\varepsilon}\left[\left(\frac{k}{r^2} - \frac{j}{r^3}\right)\left(1 - jkr - \frac{k^2r^2}{2} + j\frac{k^3r^3}{6}\right)\right]$$

$$= \frac{Idl}{2\pi\omega\varepsilon}\left[-\frac{k^3}{3} - j\left(\frac{1}{r^3} + \frac{k^2}{2r} - \frac{k^4 r}{6}\right)\right] \tag{2.9}$$

Thus there is a constant component of E_{zs} on the surface, which is opposed to the current flow and independent of distance r from the origin, equal in magnitude to

$$E_{zs} = -\frac{k^3 Idl}{6\pi\omega\varepsilon} \tag{2.10}$$

2.4 Fields of current element calculated using current *I* and charge *Q*

The calculation of the magnetic field $\mathbf{H}$, since it depends on I only, is the same as in the preceding section. But instead of calculating $\mathbf{E}$ from Maxwell's equation it will be calculated instead directly from the equation

$$\mathbf{E} = -\frac{\partial \mathbf{A}}{\partial t} - \nabla V$$

where V is the scalar potential due to the point charges at the ends of the current element.

Then the components of electric field due to the vector potential are

$$E_{r1} = -j\omega A_z \cos\theta$$
$$= -\frac{j\omega\mu I dl \cos\theta}{4\pi}\frac{e^{-jkr}}{r} \tag{2.11}$$

and

$$E_{\theta 1} = +j\omega A_z \sin\theta$$
$$= \frac{j\omega\mu I dl \sin\theta}{4\pi}\frac{e^{-jkr}}{r} \tag{2.12}$$

Similarly the components of electric field due to the scalar potential are

$$E_{r2} = -\frac{\partial V}{\partial r}$$

and

$$E_{\theta 2} = -\frac{1}{r}\frac{\partial V}{\partial \theta}$$

where

$$V = \frac{Q}{4\pi\varepsilon}\frac{e^{-jkr_1}}{r_1} - \frac{Q}{4\pi}\frac{e^{-jkr_2}}{r_2} \tag{2.13}$$

Q is the charge at the top of the current element, of magnitude $I/j\omega$ for a

Fig. 2.2. Point charges at ends of current element.

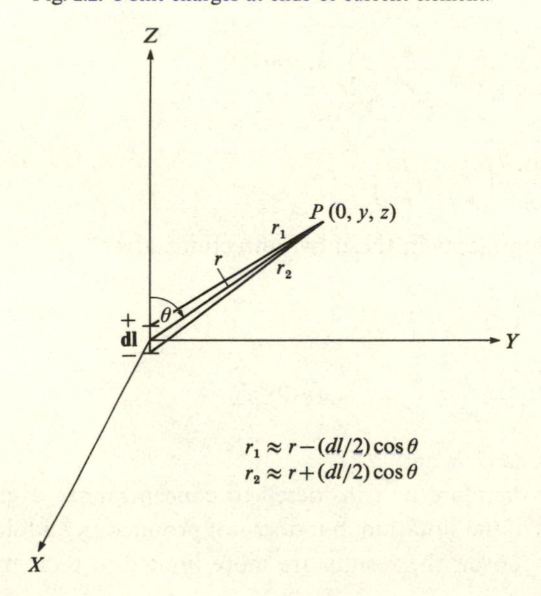

time varying current $Ie^{j\omega t}$, while r_1 and r_2 are the distances from the positive and negative charges to the point of observation P. If these distances are much greater than the element length dl, then they are both related to the distance r from the origin to P, through the equations

$$r_1 \approx r - \frac{dl}{2}\cos\theta \tag{2.14(a)}$$

$$r_2 \approx r + \frac{dl}{2}\cos\theta \tag{2.14(b)}$$

as shown in Fig. 2.2.

This then enables the scalar potential to be written as

$$V \approx \frac{I}{j\omega 4\pi\varepsilon}\frac{e^{-jkr}}{r_1 r_2}\left(r_2 e^{jk(dl/2)\cos\theta} - r_1 e^{-jk(dl/2)\cos\theta}\right)$$

Since $r_1 r_2$ is approximately equal to r^2 for the case considered, and the exponential terms can be expanded in their power series forms for small arguments this gives

$$\begin{aligned} V &\approx \frac{I}{j\omega 4\pi\varepsilon}\frac{e^{-jkr}}{r^2}\left[dl\cos\theta\cos\left(k\frac{dl}{2}\cos\theta\right)\right. \\ &\qquad \left. + j2r\sin\left(k\frac{dl}{2}\cos\theta\right)\right] \\ &\approx -\frac{jIdl\cos\theta}{4\pi\omega\varepsilon}(1+jkr)\frac{e^{-jkr}}{r^2} \end{aligned} \tag{2.15}$$

Hence

$$E_{r2} = -\frac{jIdl\cos\theta}{2\pi\omega\varepsilon}\left(\frac{1}{r^3}+\frac{jk}{r^2}-\frac{k^2}{2r}\right)e^{-jkr} \tag{2.16}$$

and

$$E_{\theta 2} = -\frac{jIdl\sin\theta}{4\pi\omega\varepsilon}\left(\frac{1}{r^3}+j\frac{k}{r^2}\right)e^{-jkr} \tag{2.17}$$

Then adding the components in these two directions gives

$$E_r = \frac{Idl\cos\theta}{2\pi\omega\varepsilon}\left(\frac{k}{r^2}-\frac{j}{r^3}\right)e^{-jkr} \tag{2.18}$$

$$E_\theta = \frac{Idl\sin\theta}{4\pi\omega\varepsilon}\left(-\frac{j}{r^3}+\frac{k}{r^2}+\frac{jk^2}{r}\right)e^{-jkr} \tag{2.19}$$

in agreement with eqns (2.7) and (2.8).

This method may therefore be considered to concentrate to a greater extent on the physics of the situation, but does not produce any additional information and, moreover, the results are more limited in their proved region of applicability, through the restriction imposed by eqns (2.14(a)) and (2.14(b)).

2.5 Cylindrical current element

In many practical situations a current element, as shown in Fig. 2.3, consists of an areal current density **i** flowing along a short cylinder of length $2b$ and of radius a. It is important to know in such cases how the fields set up by the cylinder compare with those calculated on the assumption that all the current flows along the axis. In both cases the current is assumed to be z-directed, so that all the fields at a point (a, ϕ, z) can be derived from a z-directed vector potential given by

$$A_{z1} = \frac{\mu}{4\pi} \iiint i_z(\rho', \phi', z') \frac{e^{-jkr}}{r} \rho' \, d\phi' \, d\rho' \, dz' \tag{2.20}$$

where the source coordinates are (ρ', ϕ', z') and

$$\begin{aligned} r &= [(a \cos \phi - \rho' \cos \phi')^2 + (a \sin \phi - \rho' \sin \phi')^2 + (z - z')^2]^{\frac{1}{2}} \\ &= [a^2 + \rho'^2 - 2a\rho' \cos(\phi - \phi') + (z - z')^2]^{\frac{1}{2}} \end{aligned} \tag{2.21}$$

For an isolated filament $i_z(\rho', \phi', z')$ will not be a function of ϕ' and so becomes $i_z(\rho', z')$. Moreover for a short current element the current will not be a function of z' either, and so may be further simplified to $i_z(\rho')$.

Fig. 2.3. Cylindrical current element.

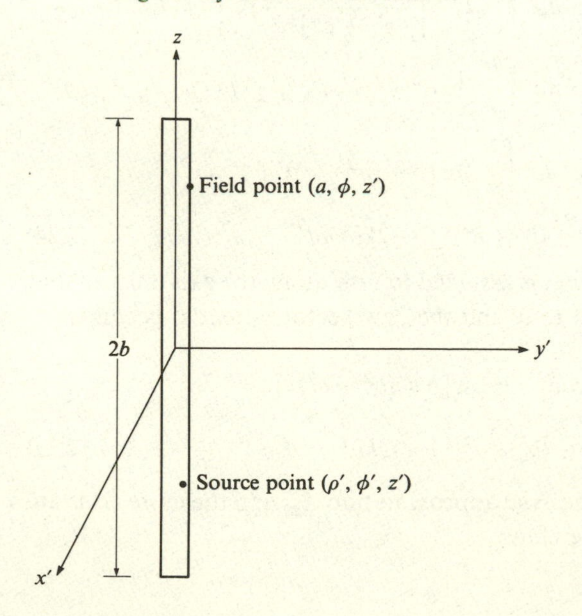

Then

$$A_{z1} = \frac{\mu}{4\pi}\int i_z(\rho')\rho'\,d\rho'$$
$$\times \iint \frac{e^{-jk[a^2+\rho^{12}-2a\rho'\cos(\phi-\phi')+(z-z')^2]^{\frac{1}{2}}}}{[a^2+\rho'^2-2a\rho'\cos(\phi-\phi')+(z-z')^2]^{\frac{1}{2}}}\,d\phi'\,dz' \tag{2.22}$$

For points of observation close to the short cylindrical element the exponential term can be expanded by the first two terms in its power series to give

$$A_{z1} = \frac{\mu}{4\pi}\int i_z(\rho')\rho'\,d\rho'\int_0^{2\pi}\int$$
$$\times\left[\frac{1}{[a^2+\rho'^2-2a\rho'\cos(\phi-\phi')+(z-z')^2]^{\frac{1}{2}}}-jk\right]d\phi'\,dz'$$
$$= \frac{\mu}{4\pi}\int i_z(\rho')\rho'\,d\rho'\int_0^{2\pi}\int\left[\frac{1}{[c^2+(z-z')^2]^{\frac{1}{2}}}-jk\right]d\phi'\,dz' \tag{2.23}$$

where

$$c^2 = a^2+\rho^{12}-2a\rho'\cos(\phi-\phi')$$

In particular at a point on the surface of the cylinder $(a, \phi, 0)$, where the cylinder extends in the z-direction from $-b$ to $+b$,

$$A_{z1} = \frac{\mu}{4\pi}\int i_z(\rho')\rho'\,d\rho'\int_0^{2\pi}\int_{-b}^{+b}\left[\frac{1}{[z'^2+c^2]^{\frac{1}{2}}}-jk\right]d\phi'\,dz'$$
$$= \frac{\mu}{4\pi}\int i_z(\rho')\rho'\,d\rho'\int_0^{2\pi}\{\ln[z'+(z'^2+c^2)^{\frac{1}{2}}]_{-b}^{+b}-jk[z']_{-b}^{+b}\}\,d\phi'$$
$$= \frac{\mu}{4\pi}\int i_z(\rho')\rho'\,d\rho'\int_0^{2\pi}\{\ln[b+(b^2+c^2)^{\frac{1}{2}}]$$
$$-\ln[-b+(b^2+c^2)^{\frac{1}{2}}]-j2kb\}\,d\phi' \tag{2.24}$$

In the case when the current is assumed to flow along the axis only, so that ρ' is zero, then c is equal to a, and the new vector potential becomes

$$A_{z2} = \frac{\mu}{4\pi}\int i_z(\rho')\rho'\,d\rho'\cdot 2\pi\{\ln[b+(b^2+a^2)^{\frac{1}{2}}]$$
$$-\ln[-b+(b^2+a^2)^{\frac{1}{2}}]-j2kb\} \tag{2.25}$$

The difference between the axial approximation A_{z2} and the more accurate approximation A_{z1} is therefore

$$\Delta A_z = A_{z1} - A_{z2}$$

Consider now the integral,

$$A_{z1} = \frac{\mu}{4\pi} \int i_z(\rho')\rho'\,d\rho' \int_0^{2\pi} \{\ln[b + (b^2 + c^2)^{\frac{1}{2}}] - \ln[-b + (b^2 + c^2)^{\frac{1}{2}}] - j2kb\}\,d\phi'$$

This can readily be evaluated numerically for a fixed point of observation on the surface, say $(a, \pi, 0)$, when the current flows in a thin skin at ρ' equal to a, to give

$$A_{z1}(a, \pi, 0) = \frac{\mu}{4\pi} F_1\left(\frac{b}{a}\right) \int_0^a i(\rho')\rho'\,d\rho' - j\frac{\mu}{4\pi} 2kb \cdot 2\pi \int_0^a i(\rho')\rho'\,d\rho' \tag{2.26}$$

where

$$F_1\left(\frac{b}{a}\right) = \int_0^{2\pi} \left\{ \ln a\left[\frac{b}{a} + \left(\frac{b^2}{a^2} + 2(1 + \cos\phi')\right)^{\frac{1}{2}}\right] - \ln a\left[-\frac{b}{a} + \left(\frac{b^2}{a^2} + 2(1 + \cos\phi')\right)^{\frac{1}{2}}\right]\right\} d\phi' \tag{2.27}$$

Likewise

$$A_{z2}(a, \pi, 0) = \frac{\mu}{4\pi} F_2\left(\frac{b}{a}\right) I - \frac{j\mu}{4\pi}(2kb)I \tag{2.28}$$

where

$$F_2\left(\frac{b}{a}\right) = \ln[b + (a^2 + b^2)^{\frac{1}{2}}] - \ln[-b + (a^2 + b^2)^{\frac{1}{2}}] \tag{2.29}$$

and

$$I = 2\pi \int_0^a i(\rho')\rho'\,d\rho'$$

There is thus no error in the imaginary part of the vector potential, but the real part has a percentage error which may be defined as

$$\%\,\text{error} = \frac{F_1\left(\frac{b}{a}\right) - 2\pi F_2\left(\frac{b}{a}\right)}{2\pi F_2\left(\frac{b}{a}\right)} \tag{2.30}$$

This is less than 1% for b/a equal to approximately 4, and hence the line current approximation is very good when the length/diameter ratio is > 4.

2.6 Short current element of finite length

The short element considered in Section 2.2 was of infinitesimal length dl, and it was shown in the preceding section that for negligible error in calculating the vector potential due to a cylindrical element of radius a, the element length could be taken as greater than, say, eight times this radius. But in real situations not even this radius is strictly infinitesimal, much less the length of the cylindrical element. For example, in analysing linear dipoles, it is customary in dealing with computer solutions to consider element lengths which are as large as $\lambda/10$ in length. The reason for this relatively large length is to reduce the size of matrix which the computer has to handle, this matrix being a square matrix of size equal to the number of segments into which the dipole is divided.

Thus the question has to be asked, whether a current element of length $\lambda/10$, with its current flowing along the axis, and its associated point charges at the ends of the element, can be used to describe adequately the vector potential at the surface of the element. It is the value of this vector potential which is going to determine, for example, the self impedance of the element, which depends on time and space derivatives of the vector potential. The space derivative in the form of the divergence is proportional to the scalar potential, and the scalar potential depends on the charge distribution of the element. Where the element length is short, i.e. $dl \ll \lambda$, such a model must be perfectly satisfactory, but where the length is of the order of $\lambda/10$, the point discontinuities in charge distribution give erroneous answers for the element self impedance.

A model which makes use of uniformly distributed charges in conjunction with a constant axial current is clearly a much better physical representation of a cylindrical conductor than a point charge model. To produce this model the previously considered point charges are each considered to be spread uniformly on both sides of their initial position over a length equal to the element length. Thus a uniform positive charge distribution exists along the axis from the centre of the element for a length of $\lambda/10$, say, and likewise a negative charge distribution in the opposite direction along an equal length. An analysis of this model will therefore be carried out.

2.7 Analysis of current element with constant current and uniformly distributed charges

The model to be analysed is shown in Fig. 2.4. A constant current I flows along the z-axis of the element so that the total charges associated with this current are $\pm I/j\omega$. The positive charge is distributed uniformly along the axis from the origin at the centre of the element, to a point l along the axis, where l is the element length. Likewise the negative charge is distributed from the origin to $-l$ along the axis.

Expressions have previously been derived in eqns (2.6)–(2.8) for the magnetic and electric fields set up by an element of length dl carrying a constant current I. Hence the corresponding fields for this finite sized element of length l can, in principle, be derived by integrating the previous expression for vector potential along the length of the element and then performing curl operations as before. In particular, at an observation point (a, z') on the surface of the element the vector potential is

$$A_z(a,z') = \frac{\mu I}{4\pi} \int_{-l/2}^{+l/2} \frac{e^{-jk[a^2+(z-z')^2]^{\frac{1}{2}}}}{[a^2+(z-z')^2]^{\frac{1}{2}}} dz \tag{2.31}$$

The integration may either be performed numerically for particular dimensions, and for a given frequency, or the integrand may be expanded to give

$$A_z(a,z') \approx \frac{\mu I}{4\pi} \int_{-l/2}^{+l/2} \left\{ \frac{1}{[a^2+(z-z')^2]^{\frac{1}{2}}} - jk \right\} dz$$

By making a change of variable

$$u = (z - z')$$

the integration may be performed using standard integrals to give

$$A_z(a,z') = \frac{\mu I}{4\pi} \left\{ \ln \frac{\left(\frac{l}{2} - z'\right) + \left[a^2 + \left(\frac{l}{2} - z'\right)^2\right]^{\frac{1}{2}}}{\left(-\frac{l}{2} - z'\right) + \left[a^2 + \left(-\frac{l}{2} - z'\right)^2\right]^{\frac{1}{2}}} - jkl \right\} \tag{2.32}$$

Fig. 2.4. Constant current segment with uniformly distributed charge.

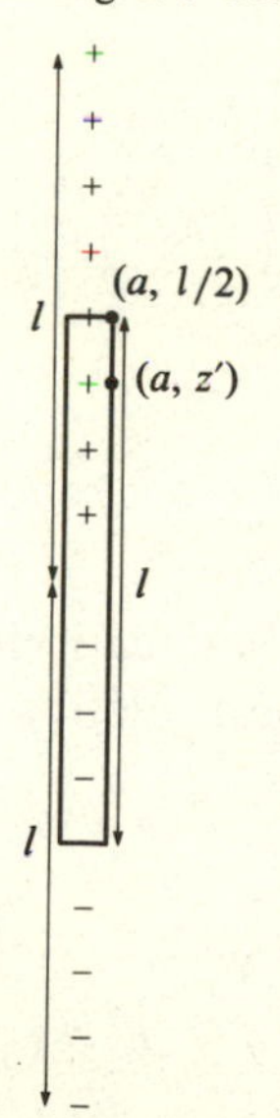

2.8 Self impedance of element

The self impedance of this element is defined as

$$Z_{11} = -\frac{\int_{-l/2}^{+l/2} E_{z'}\,dz'}{I} \tag{2.33}$$

i.e. it is the line integral of electric field along the element divided by the current through it. There are two components of electric field, one associated with the electric current and the other with the charges. There is thus a component of impedance due to the current alone, given by

$$Z_{11_{current}} = -\frac{\int_{-l/2}^{-l/2} j\omega A_z\,dz'}{I} \tag{2.34}$$

This is a complex impedance whose real component is

$$Z_{11_{real}} = \frac{\int_{-l/2}^{+l/2} 30k^2 lI\,dz'}{I} = 30k^2 l^2 \tag{2.35}$$

Its quadrature component can be evaluated by integrating the real component of $A_z(a, z')$ which is of the form $\int \ln(x+r)\,dx$. Although this is readily integrable the numerical contribution which it makes is negligible compared with the corresponding contribution from the electric charges, which are next considered.

The electric field produced by the distributed electric charges, tangential to the surface of the element can be integrated from $-l/2$ to $+l/2$. It is, however, simpler analytically to calculate the difference of potential between these two end points directly. Thus at the top surface of the element the potential is

$$V\left(a, \frac{l}{2}\right) = V_A = \frac{I}{j\omega l 4\pi\varepsilon}\left\{\int_0^l \frac{e^{-jk\left[a^2+\left(z-\frac{l}{2}\right)^2\right]^{\frac{1}{2}}}}{\left[a^2+\left(z-\frac{l}{2}\right)^2\right]^{\frac{1}{2}}} - \int_{-l}^{0} \frac{e^{-jk\left[a^2+\left(z-\frac{l}{2}\right)^2\right]^{\frac{1}{2}}}}{\left[a^2+\left(z-\frac{l}{2}\right)^2\right]^{\frac{1}{2}}}\right\}dz \tag{2.36}$$

These integrals may again be evaluated numerically, or alternatively may be expanded in power series form to obtain an analytical solution

$$V_A = \frac{I}{j\omega l 4\pi\varepsilon}\left\{\left(1-\frac{k^2a^2}{4}\right)\ln \frac{\left(\frac{l}{2}+\left(a^2+\frac{l^2}{4}\right)^{\frac{1}{2}}\right)\left(-\frac{3l}{2}+\left(a^2+\frac{9l^2}{4}\right)^{\frac{1}{2}}\right)}{\left(-\frac{l}{2}+\left(a^2+\frac{l^2}{4}\right)^{\frac{1}{2}}\right)^2} - j\frac{k^3l^3}{6}\right\} \tag{2.37}$$

Hence the impedance contribution from the charges is

$$Z_{11_{charges}} = \frac{V_A - V_B}{I} = \frac{2V_A}{I}$$

$$\approx \frac{2}{\omega l 4\pi\varepsilon}\left[-\frac{k^3 l^3}{6} - j\left(1 - \frac{k^2 a^2}{4}\right)\ln\left(\frac{l^2}{3a^2}\right)\right] \quad (2.38)$$

The total self impedance of the element of length l and radius a then becomes

$$Z_{11} \approx 20k^2 l^2 - j\frac{\ln\left(\dfrac{l^2}{3a^2}\right)}{\omega l 2\pi\varepsilon} \quad (2.39)$$

2.9 Mutual impedance between collinear elements

Consider two adjacent collinear elements, each of length l and radius a as shown in Fig. 2.5. The mutual impedance between them is defined as

$$Z_{12} = Z_{21} = -\frac{\int_{l/2}^{3(l/2)} E_{z'}\, dz'}{I} \quad (2.40)$$

where $E_{z'}$ is the tangential electric field on the surface of the upper element due to a constant current I in the lower element. But this field can be divided into two components, one of which is due to the current and the other to the charges associated with the lower element. Thus the component due to the

Fig. 2.5. Two collinear elements each of length l and radius a.

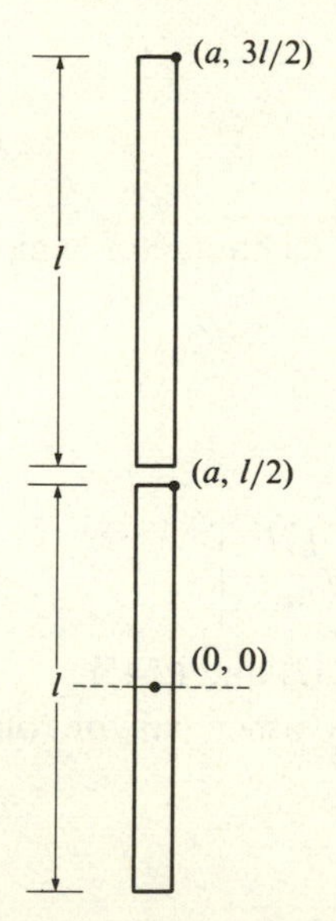

current is

$$E_{z'} = -\frac{j\omega\mu I}{4\pi}\int_{-l/2}^{+l/2}\frac{e^{-jk[a^2+(z-z')^2]^{\frac{1}{2}}}}{[a^2+(z-z')^2]^{\frac{1}{2}}}dz \tag{2.41}$$

This is an identical expression to that used for calculating the self impedance Z_{11}, but the range of values of z' is of course different. Nevertheless to a first approximation the real part of the input impedance due to the current is the same, i.e.

$$Z_{11_{real}} = 30k^2l^2 \tag{2.42}$$

and the quadrature component can again be neglected by comparison with the quadrature component due to the charges.

To find the impedance component due to the charges, the potential at the top of the upper element, produced by the current in the lower segment is expressed as

$$V\left(a,\frac{3l}{2}\right) = \frac{I}{j\omega l 4\pi\varepsilon}\left\{\int_0^l \frac{e^{-jk\left[a^2+\left(z-\frac{3l}{2}\right)^2\right]^{\frac{1}{2}}}}{\left[a^2+\left(z-\frac{3l}{2}\right)^2\right]^{\frac{1}{2}}} - \int_{-l}^{0}\frac{e^{-jk\left[a^2+\left(z-\frac{3l}{2}\right)^2\right]^{\frac{1}{2}}}}{\left[a^2+\left(z-\frac{3l}{2}\right)^2\right]^{\frac{1}{2}}}dz\right\} \tag{2.43}$$

From this has to be subtracted the potential at the bottom of this element, which is identical with the potential at the top of the lower element, viz. $V(a, l/2)$, given in eqn (2.37). Then

$$Z_{12_{charges}} = \frac{V\left(a,\frac{3l}{2}\right) - V\left(a,\frac{l}{2}\right)}{I}$$

$V(a, 3l/2)$ can be readily evaluated by the normal technique of expanding the exponential terms to give

$$V\left(a,\frac{3l}{2}\right) \approx \frac{I}{j\omega l 4\pi\varepsilon}\left\{\ln\frac{9}{5}+\frac{k^2l^2}{2}-j\frac{k^3l^3}{2}\right\} \tag{2.44}$$

Hence

$$Z_{12_{charges}} \approx \frac{1}{j\omega l 4\pi\varepsilon}\left[\ln\left(5.4\frac{a^2}{l^2}\right)+\frac{k^2l^2}{2}-j\frac{k^3l^3}{6}\right] \tag{2.45}$$

and the total Z_{12} is given by the sum of eqns (2.42) and (2.45).

Similarly, the mutual impedances between any other pair of collinear elements can readily be evaluated.

2.10 Mutual impedance between orthogonal elements

When two elements are placed in the same plane, but orthogonal to each other as shown in Fig. 2.6, there is no component of mutual impedance associated with the current in either element, since both vector potentials are orthogonal to the other coupled element. But there remains a component of mutual impedance due to the charges. For example, taking a source charge distribution along the axis of element one, gives a potential at point A on element two.

$$V\left(x_2+\frac{l}{2}, z_2-a\right) = \frac{I}{j\omega l 4\pi\varepsilon}\left\{\int_0^l \frac{e^{-jk\left[\left(x_2+\frac{l}{2}\right)^2+(z_2-z-a)^2\right]^{\frac{1}{2}}}}{\left[\left(x_2+\frac{l}{2}\right)^2+(z_2-z-a)^2\right]^{\frac{1}{2}}}dz - \int_{-l}^0 \frac{e^{-jk\left[\left(x_2+\frac{l}{2}\right)^2+(z_2-z-a)^2\right]^{\frac{1}{2}}}}{\left[\left(x_2+\frac{l}{2}\right)^2+(z_2-z-a)^2\right]^{\frac{1}{2}}}dz\right\} \tag{2.46}$$

Likewise, the potential at point B is

$$V\left(x_2-\frac{l}{2}, z_2-a\right) = \frac{I}{j\omega l 4\pi\varepsilon}\left\{\int_0^l \frac{e^{-jk\left[\left(x_2+\frac{l}{2}\right)^2+(z_2-z-a)^2\right]^{\frac{1}{2}}}}{\left[\left(x_2-\frac{l}{2}\right)^2+(z_2-z-a)^2\right]^{\frac{1}{2}}}dz - \int_{-l}^0 \frac{e^{-jk\left[\left(x_2-\frac{l}{2}\right)^2+(z_2-z-a)^2\right]^{\frac{1}{2}}}}{\left[\left(x_2-\frac{l}{2}\right)^2+(z_2-z-a)^2\right]^{\frac{1}{2}}}dz\right\} \tag{2.47}$$

The mutual impedance due to the charges is then

$$Z_{12} = \frac{V\left(x_2+\frac{l}{2}, z_2-a\right) - V\left(x_2-\frac{l}{2}, z_2-a\right)}{I} \tag{2.48}$$

Consider now the reverse situation where the source charges are on element two, and it is required to find the mutual impedance with element one. Then, using the current directions shown,

$$Z_{21} = \frac{V\left(x_2 - a, z_2 - \frac{l}{2}\right) - V\left(x_2 - a, z_2 + \frac{l}{2}\right)}{I} \tag{2.49}$$

where

$$V\left(x_2 - a, z_2 - \frac{l}{2}\right) = \frac{I}{j\omega l 4\pi\varepsilon}\left\{\int_0^l \frac{e^{-jk\left[(x_2 - a - x)^2 + \left(z_2 - \frac{l}{2}\right)^2\right]^{\frac{1}{2}}}}{\left[(x_2 - a - x)^2 + \left(z_2 - \frac{l}{2}\right)^2\right]^{\frac{1}{2}}} dx - \int_{-l}^0 \frac{e^{-jk\left[(x_2 - a - x)^2 + \left(z_2 - \frac{l}{2}\right)^2\right]^{\frac{1}{2}}}}{\left[(x_2 - a - x)^2 + \left(z_2 - \frac{l}{2}\right)^2\right]^{\frac{1}{2}}} dx\right\} \tag{2.50}$$

and

$$V\left(x_2 - a, z_2 + \frac{l}{2}\right) = \frac{I}{j\omega l 4\pi\varepsilon}\left\{\int_0^l \frac{e^{-jk\left[(x_2 - a - x)^2 + \left(z_2 + \frac{l}{2}\right)^2\right]^{\frac{1}{2}}}}{\left[(x_2 - a - x)^2 + \left(z_2 + \frac{l}{2}\right)^2\right]^{\frac{1}{2}}} dx - \int_{-l}^0 \frac{e^{-jk\left[(x_2 - a - x)^2 + \left(z_2 + \frac{l}{2}\right)^2\right]^{\frac{1}{2}}}}{\left[(x_2 - a - x)^2 + \left(z_2 + \frac{l}{2}\right)^2\right]^{\frac{1}{2}}} dx\right\} \tag{2.51}$$

Fig. 2.6. Two orthogonal elements each of length l and radius a.

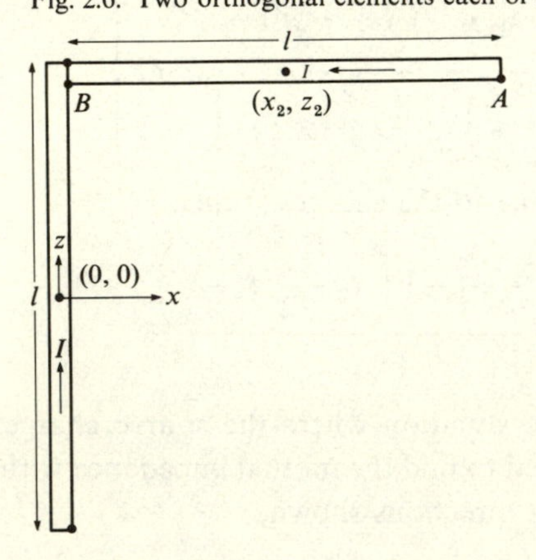

It is clear that, for the particular case of x_2 and z_2 being equal, so that the two elements are then symmetrically placed with regard to the corner, the two mutual impedances are then equal. However, for other relative positions this equality does not hold exactly, which is an indication that the model adopted for these finite length elements is not completely precise. Nevertheless it is found in practice that the small inequality which exists does not significantly affect calculated results for the input impedance and current distributions of antennas comprising such elements.

2.11 General expression for mutual impedance between finite current elements in a plane

The two preceding results for the mutual impedances between collinear and orthogonal elements will now be generalised to a form which is directly applicable to any two current elements in the same plane, with an arbitrary orientation between them. Consider the elements m and n shown in Fig. 2.7. A current I is assumed to flow along the axis of element n from $\bar{n}$ to $\overset{+}{n}$ spaced a distance l apart. The mid-point of element n is denoted by $\hat{n}$, so that a positive charge density extends a distance l from $\hat{n}$, and a negative charge density a like distance l in the opposite direction from $\hat{n}$. Similarly for element m, but in this case the middle point $\hat{m}$ and the end points $\bar{m}$ and $\overset{+}{m}$ are all considered to lie on the nearer surface of m and not on its axis.

Then the vector potential set up at any point m on the surface of element m by the axial current I flowing in element n is

$$\mathbf{A} = \frac{\mu I}{4\pi} \int_{-l/2}^{+l/2} \frac{e^{-jkr_{mn}}}{r_{mn}} \mathbf{dl} \tag{2.52}$$

Similarly the scalar potentials set up at the end point $\overset{+}{m}$ by the positive and negative line segments of charge on the axis of element n is

$$V(\overset{+}{m}) = \frac{I(n)}{j\omega l 4\pi\varepsilon} \left[\int_{0}^{l} \frac{e^{-jkr_{\overset{+}{m}\overset{+}{n}}}}{r_{\overset{+}{m}\overset{+}{n}}} - \int_{-l}^{0} \frac{e^{-jkr_{\overset{+}{m}\bar{n}}}}{r_{\overset{+}{m}\bar{n}}} \right] dl \tag{2.53}$$

Fig. 2.7. Two current elements with arbitrary orientation.

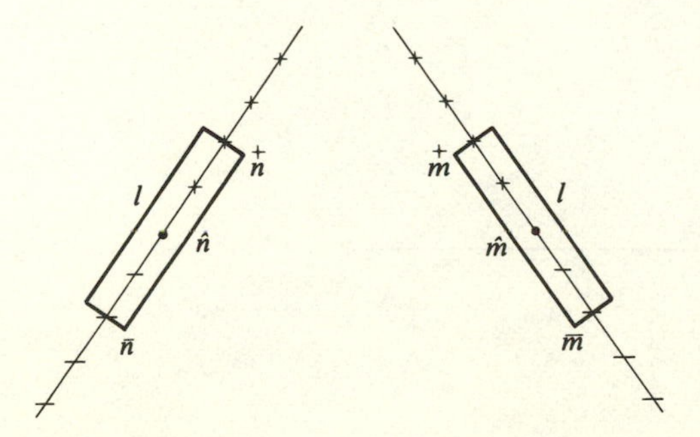

where the positive and negative signs associated with the symbol n indicate the polarities of the line charge densities. Likewise the potential at the end point $\bar{m}$ is

$$V(\bar{m}) = \frac{I(n)}{j\omega l 4\pi\varepsilon}\left[\int_0^l \frac{e^{-jkr_{\bar{m}\overset{+}{n}}}}{r_{\bar{m}\overset{+}{n}}} - \int_{-l}^0 \frac{e^{-jkr_{\bar{m}\bar{n}}}}{r_{\bar{m}\bar{n}}}\right] dl \tag{2.54}$$

These expressions may be written more concisely as

$$V(\overset{+}{m}) = \frac{I(n)}{j\omega l 4\pi\varepsilon}[\psi(\overset{+}{m}\overset{+}{n}) - \psi(\overset{+}{m}\bar{n})] \tag{2.55}$$

and

$$V(\bar{m}) = \frac{I(n)}{j\omega l 4\pi\varepsilon}[\psi(\bar{m}\overset{+}{n}) - \psi(\bar{m}\bar{n})] \tag{2.56}$$

where

$$\psi(\overset{+}{m}\overset{+}{n}) = \int_0^l \frac{e^{-jkr_{\overset{+}{m}\overset{+}{n}}}}{r_{\overset{+}{m}\overset{+}{n}}} dl \qquad \text{etc...} \tag{2.57}$$

Then the mutual impedance between the elements may be written generally as

$$Z_{mn} = j\omega\mu \mathbf{l}_m \cdot \mathbf{l}_n \psi(m, n) + \frac{1}{j\omega\varepsilon}[\psi(\overset{+}{m}, \overset{+}{n}) - \psi(\overset{+}{m}, \bar{n}) - \psi(\bar{m}, \overset{+}{n}) + \psi(\bar{m}, \bar{n})] \tag{2.58}$$

where the scalar product in the first term takes account of the orientations of the two elements. This result also includes the self impedance case when m is equal to n.

Fig. 2.8. Current element **Idl** at origin along x-axis.

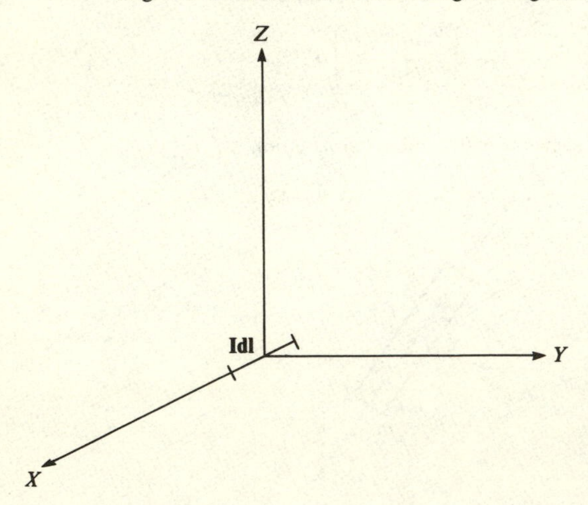

2.12 Current elements along orthogonal axes

The current element **Idl** previously positioned along the z-axis is rotated through 90° to lie along the x-axis as shown in Fig. 2.8. The vector potential is therefore directed along the x-axis and is of magnitude, as before,

$$A_x = \frac{\mu I dl}{4\pi} \frac{e^{-jkr}}{r} \tag{2.59}$$

It is desired to know the magnetic and electric fields at the observation point $P(r, \theta, \phi)$ so that their form can be compared with the previous results when the dipole was directed along the z-axis.

Using the rectangular–spherical transformation table

	x	y	z
R	$\sin\theta\cos\phi$	$\sin\theta\sin\phi$	$\cos\theta$
θ	$\cos\theta\cos\phi$	$\cos\theta\sin\phi$	$-\sin\theta$
ϕ	$-\sin\phi$	$\cos\phi$	0

it follows that the spherical components of the vector potential A_x are

$$A_r = \frac{\mu I dl}{4\pi} \frac{e^{-jkr}}{r} \sin\theta\cos\phi \tag{2.60}$$

$$A_\theta = \frac{\mu I dl}{4\pi} \frac{e^{-jkr}}{r} \cos\theta\cos\phi$$

$$A_\phi = -\frac{\mu I dl}{4\pi} \frac{e^{-jkr}}{r} \sin\phi$$

Hence from the relation

$$\mathbf{B} = \nabla \times \mathbf{A}$$

we obtain

$$H_r = \frac{1}{r\sin\theta}\left[\frac{\partial}{\partial\theta}(A_\phi \sin\theta) - \frac{\partial A_\theta}{\partial\phi}\right]$$

$$H_\theta = \frac{1}{r\sin\theta}\frac{\partial A_r}{\partial\phi} - \frac{1}{r}\frac{\partial}{\partial r}(rA_\phi)$$

$$H_\phi = \frac{1}{r}\left[\frac{\partial}{\partial r}(rA_\theta) - \frac{\partial A_r}{\partial\theta}\right]$$

Hence

$$H_r = \frac{1}{r\sin\theta}\frac{I dl}{4\pi}\frac{e^{-jkr}}{r}[-\sin\phi\cos\theta + \cos\theta\sin\phi] = 0$$

$$H_\theta = \frac{I dl}{4\pi}\frac{e^{-jkr}}{r}\left[-\frac{\sin\phi}{r} - jk\sin\phi\right]$$

$$= -\frac{Idl}{4\pi}\frac{e^{-jkr}}{r}\sin\phi\left(\frac{1}{r}+jk\right) \tag{2.61}$$

$$H_\phi = -\frac{Idl}{4\pi}\frac{e^{-jkr}}{r}\cos\theta\cos\phi\left(\frac{1}{r}+jk\right)$$

Likewise using Maxwell's equations the corresponding values of electric field are found from

$$\mathbf{E} = \frac{1}{j\omega\varepsilon}\nabla\times\mathbf{H}$$

giving

$$\left.\begin{aligned} E_r &= +\frac{Idl}{2\pi j\omega\varepsilon}\frac{e^{-jkr}}{r^2}\left(\frac{1}{r}+jk\right)\cos\phi\sin\theta \\ E_\theta &= -\frac{Idl}{j\omega\varepsilon 4\pi}\frac{e^{-jkr}}{r^2}\left(\frac{1}{r}+jk-k^2r\right)\cos\theta\cos\phi \\ \text{and}\quad E_\phi &= \frac{Idl}{j\omega\varepsilon 4\pi}\frac{e^{-jkr}}{r^2}\left(\frac{1}{r}+jk-k^2r\right)\sin\phi \end{aligned}\right\} \tag{2.62}$$

The E_ϕ and H_θ patterns in the yz-plane are independent of θ as this is the equatorial plane for a dipole directed along the x-axis. Likewise for a y-axis directed dipole

$$\left.\begin{aligned} A_r &= \frac{\mu Idl}{4\pi}\frac{e^{-jkr}}{r}\sin\theta\sin\phi \\ A_\theta &= \frac{\mu Idl}{4\pi}\frac{e^{-jkr}}{r}\cos\theta\sin\phi \\ A_\phi &= \frac{\mu Idl}{4\pi}\frac{e^{-jkr}}{r}\cos\phi \end{aligned}\right\} \tag{2.63}$$

Hence

$$\left.\begin{aligned} H_r &= \frac{Idl}{r\sin\theta}\frac{e^{-jkr}}{r}\frac{1}{4\pi}(\cos\phi\cos\theta-\cos\theta\cos\phi)=0 \\ H_\theta &= \frac{Idl}{4\pi}\frac{e^{-jkr}}{r}\left(\frac{1}{r}+jk\right)\cos\phi \\ H_\phi &= -\frac{Idl}{4\pi}\frac{e^{-jkr}}{r}\left(\frac{1}{r}+jk\right)\cos\theta\sin\phi \end{aligned}\right\} \tag{2.64}$$

and

$$\left.\begin{aligned} E_r &= \frac{Idl}{j\omega\varepsilon 2\pi}\frac{e^{-jkr}}{r^2}\left(\frac{1}{r}+jk\right)\sin\theta\sin\phi \\ E_\theta &= -\frac{Idl}{j\omega\varepsilon 4\pi}\frac{e^{-jkr}}{r^2}\left(\frac{1}{r}+jk-k^2r\right)\cos\theta\sin\phi \end{aligned}\right\} \tag{2.65}$$

$$E_\phi = -\frac{Idl}{j\omega\varepsilon 4\pi}\frac{e^{-jkr}}{r^2}\left(\frac{1}{r}+jk-k^2r\right)\cos\phi \Bigg]$$

Again the H_θ and E_ϕ patterns in the xz-plane are independent of θ as this is the equatorial plane for a current element along the y-axis.

The complete fields for a current element along each of the three orthogonal axis have thus been developed. For an element oriented in any direction (θ, ϕ) in space the resultant fields can be found from

$$A_x = A\sin\theta\cos\phi$$

$$A_y = A\sin\theta\sin\phi$$

$$A_z = A\cos\theta$$

where $A = (\mu Idl/4\pi)(e^{-jkr}/r)$, and the fields produced by A_x, A_y, A_z have already been given above and in Section 2.3.

2.13 Power flow from current element

It has previously been pointed out that the direction of current flow in an infinitesimal current element is opposite to that of the tangential electric field at its surface. Power has therefore to be supplied to drive the current against this field, and this power goes into the electromagnetic field which is being radiated, since all losses are at present being ignored. The magnitude of the power is

$$W = -\tfrac{1}{2}E_{zs}Idl \qquad (2.66)$$

which from eqn (2.10) gives

$$W = \frac{1}{2}\frac{k^3I^2dl^2}{6\pi\omega\varepsilon} = 10k^2dl^2I^2 \qquad (2.67)$$

It is customary to express this power in a form which suggests that the current in the element is being passed through a resistance R_r, called the radiation resistance of the current element, and that all the radiated power is being dissipated in this resistance. Then

$$\tfrac{1}{2}I^2R_r = \frac{k^3I^2dl^2}{12\pi\omega\varepsilon}$$

so that the magnitude of resistance is

$$R_r = \frac{k^2dl^2\sqrt{(\mu\varepsilon)}}{6\pi\varepsilon}$$

$$= 20k^2dl^2 = 80\pi^2\frac{dl^2}{\lambda^2} \qquad (2.68)$$

To confirm that this power is actually radiated into space the radial Poynting vector will be integrated over the surface area of a large sphere at a radius at which only the inverse distance fields are significant, i.e. the electric and magnetic fields will be taken to be

$$E_\theta = j\frac{k^2 I dl \sin\theta}{4\pi\omega\varepsilon}\frac{e^{-jkr}}{r} \tag{2.69}$$

$$H_\phi = j\frac{kIdl \sin\theta}{4\pi}\frac{e^{-jkr}}{r} \tag{2.70}$$

Then the Poynting vector in the radial direction is

$$P_r = \tfrac{1}{2}\mathrm{Re}(E_\theta H_\phi^*) \tag{2.71}$$

where H_ϕ^* is the complex conjugate of H_ϕ. This gives

$$P_r = \frac{k^3 I^2 dl^2 \sin^2\theta}{32\pi^2 \omega\varepsilon r^2} \tag{2.72}$$

Integrating this power density, which exists over an elemental area of a sphere $r^2 \sin\theta\, d\theta\, d\phi$, to obtain the total radiated power gives

$$\begin{aligned} W &= \int_{\theta=0}^{\pi}\int_{\phi=0}^{2\pi} P_r r^2 \sin\theta\, d\theta\, d\phi \\ &= \frac{k^3 I^2 dl^2}{16\pi\omega\varepsilon}\int_0^{\pi} \sin^3\theta\, d\theta \\ &= \frac{k^3 I^2 dl^2}{12\pi\omega\varepsilon} \end{aligned} \tag{2.73}$$

which is identical with the result obtained in eqn (2.67).

It may readily be confirmed that the Poynting vector components in the elevational and azimuthal directions, P_θ and P_ϕ, are identically zero.

2.14 Magnetic current element $\mathbf{I_m dl}$

A magnetic current element refers to an element of length dl which is carrying a constant magnetic current, i.e. a uniform flow of isolated magnetic poles. Although such isolated poles do not exist, many aperture antennas can have their operation explained most simply if the conceptual radiating properties of magnetic current elements are understood.

2.15 Fields of magnetic current element $\mathbf{I_m dl}$ along *z*-axis calculated using magnetic current alone

Just as an electric current I sets up a magnetic vector potential $\mathbf{A}$, so a magnetic current I_m can be postulated to set up an electric vector potential $\mathbf{F}$ in the same direction as I_m. Hence for $\mathbf{I_m dl}$ along the z-axis

$$F_z = \frac{\varepsilon I_m dl}{4\pi}\frac{e^{-jkr}}{r} \tag{2.74}$$

The electric displacement **D** is obtained from the electric vector potential **F** through the equation

$$\mathbf{D} = -\nabla \times \mathbf{F} \tag{2.75}$$

In rectangular coordinates this gives

$$\left.\begin{aligned} E_x &= -\frac{1}{\varepsilon}\frac{\partial F_z}{\partial y} \\ E_y &= \frac{1}{\varepsilon}\frac{\partial F_z}{\partial x} \\ E_z &= 0 \end{aligned}\right\} \tag{2.76}$$

For an observation point in the yz-plane, for which

$$r = [y^2 + z^2]^{\frac{1}{2}}$$

the derivatives are readily obtained to give

$$\left.\begin{aligned} E_x &= \frac{I_m dl \sin\theta}{4\pi}\left(\frac{1}{r^2} + \frac{jk}{r}\right)e^{-jkr} \\ E_y &= E_z = 0 \end{aligned}\right\} \tag{2.77}$$

The E_x-field is perpendicular to the plane in which the observation point is located and this is true for every observation point, so that the electric field **E** due to the magnetic current is a circumferential field E_ϕ only, given by

$$E_\phi = -\frac{I_m dl \sin\theta}{4\pi}\left(\frac{1}{r^2} + \frac{jk}{r}\right)e^{-jkr} \tag{2.78}$$

If this magnetic current is time varying, the electric field E_ϕ will set up magnetic field components given by

$$H_r = -\frac{1}{j\omega\mu}\left[\frac{1}{r}\frac{\partial E_\phi}{\partial\theta} + \frac{E_\phi \cot\theta}{r}\right]$$

$$H_\theta = -\frac{1}{j\omega\mu}\left[-\frac{\partial E_\phi}{\partial r} - \frac{E_\phi}{r}\right] \tag{2.79}$$

and

$$H_\phi = 0$$

Hence

$$H_r = \frac{I_m dl \cos\theta}{2\pi\omega\mu}\left(\frac{k}{r^2} - \frac{j}{r^3}\right)e^{-jkr} \tag{2.80}$$

and

$$H_\theta = \frac{I_m dl \sin\theta}{4\pi\omega\mu}\left(\frac{jk^2}{r} + \frac{k}{r^2} - \frac{j}{r^3}\right)e^{-jkr} \tag{2.81}$$

The component of magnetic force **H** along the axis of the magnetic current element is

$$H_z = H_r \cos\theta - H_\theta \sin\theta$$

so that for points on the axis where θ is zero, and where e^{-jkr} can be expressed in a power series

$$H_{za} = \frac{I_m dl}{2\pi\omega u}\left(\frac{k}{r^2} - \frac{j}{r^3}\right)\left(1 - jkr - \frac{k^2r^2}{2} + j\frac{k^3r^3}{6}\right)$$

$$= \frac{I_m dl}{2\pi\omega\mu}\left[-\frac{k^3}{3} - j\left(\frac{1}{r^3} + \frac{k^2}{2r} - \frac{k^4 r}{6}\right)\right] \tag{2.82}$$

Thus an antiphase component of magnetic force H_{za} exists along the full length of the element dl, in opposition to the magnetic current I_m and of magnitude

$$H_{za} = -\frac{k^3 I_m dl}{6\pi\omega\mu} \tag{2.83}$$

2.16 Fields of magnetic element calculated using magnetic current I_m and pole strength M

The calculation of the electric field **E** is carried out as in the preceding section. But instead of deriving **H** from Maxwell's equations it will be calculated instead from

$$\mathbf{H} = -\frac{\partial \mathbf{F}}{\partial \mathbf{t}} - \nabla U \tag{2.84}$$

where U is the magnetic scalar potential due to the magnetic poles at the ends of the current element.

The components of H due to the electric vector potential are

$$H_{r1} = -j\omega F_z \cos\theta$$

$$= -\frac{j\omega\varepsilon I_m dl \cos\theta}{4\pi}\frac{e^{-jkr}}{r} \tag{2.85}$$

and

$$H_{\theta 1} = +j\omega F_z \sin\theta$$

$$= \frac{j\omega\varepsilon I_m dl \sin\theta}{4\pi}\frac{e^{-jkr}}{r} \tag{2.86}$$

Similarly the components of H due to the scalar potential are

$$H_{r2} = -\frac{\partial U}{\partial r} \tag{2.87}$$

and

$$H_{\theta 2} = -\frac{1}{r}\frac{\partial U}{\partial \theta} \tag{2.88}$$

where

$$U = \frac{M}{4\pi\mu}\frac{e^{-jkr_1}}{r_1} - \frac{M}{4\pi\mu}\frac{e^{-jkr_2}}{r_2} \tag{2.89}$$

and M is the pole strength in webers at the two ends of the magnetic element. For a time-varying current $I_m e^{j\omega t}$

$$M = \frac{I_m}{j\omega} e^{j\omega t} \tag{2.90}$$

The distances r_1, r_2 from the positive and negative poles to the point of observation $P(r, \theta, \phi)$ are assumed to be much greater than the element length dl, so that they are approximately related to the distance r from the origin to P in the usual way by the equations

$$\left.\begin{aligned} r_1 &\approx r - \frac{dl}{2}\cos\theta \\ r_2 &\approx r + \frac{dl}{2}\cos\theta \end{aligned}\right\} \tag{2.91}$$

This then enables the magnetic scalar potential to be written in the same way as for the electric scalar potential in Section 2.4, viz.

$$U \approx -\frac{jI_m dl\cos\theta}{4\pi\omega\mu}(1 + jkr)\frac{e^{-jkr}}{r^2} \tag{2.92}$$

Hence

$$H_{r2} = -\frac{jI_m dl\cos\theta}{2\pi\omega\mu}\left(\frac{1}{r^3} + \frac{jk}{r^2} - \frac{k^2}{2r}\right)e^{-jkr} \tag{2.93}$$

and

$$H_{\theta 2} = -\frac{jI_m dl\sin\theta}{4\pi\omega\mu}\left(\frac{1}{r^3} + \frac{jk}{r^2}\right)e^{-jkr} \tag{2.94}$$

Adding the components of fields due to the current and pole strength gives the total fields as

$$H_r = \frac{I_m dl\cos\theta}{2\pi\omega\mu}\left(\frac{k}{r^2} - \frac{j}{r^3}\right)e^{-jkr} \tag{2.95}$$

and

$$H_\theta = \frac{I_m dl\sin\theta}{4\pi\omega\mu}\left(j\frac{k^2}{r} + \frac{k}{r^2} - \frac{j}{r^3}\right)e^{-jkr} \tag{2.96}$$

which agrees with the results previously obtained using the magnetic current alone.

2.17 Power flow from magnetic current element

As with the electric current element the power flow may be calculated either by integration of the Poynting vector over a large sphere of radius r, or more simply from multiplying the magnetic current by the constant tangential magnetic intensity integrated over the length dl. Thus, by inspection from eqn (2.83), the radiated power is

$$W = -\tfrac{1}{2}H_{za}I_m dl = \frac{k^3 I_m^2 dl^2}{12\pi\omega\mu} \tag{2.97}$$

2.18 Filamentary and cylindrical dipoles and monopoles

An electric current element in isolation does not constitute a practical radio antenna, both because the power radiated by it is proportional to the square of an infinitesimal length dl, and also because the constant current assumed for the current element must fall to zero at its ends in a practical case. If the radiating length dl is retained while the current is assumed to fall linearly from a value I at its mid-point to zero current at both ends, the antenna so formed will be referred to as a Hertzian dipole. Since in addition the thickness of the element will be assumed to be zero, the antenna will strictly be a Hertzian dipole of filamentary type.

2.19 Infinitesimal Hertzian dipole

The distribution of current assumed for this antenna is of the triangular form shown in Fig. 2.9. The antenna is positioned at the origin and is directed along the z-axis, with the point of observation at $P(r, \theta, \phi)$. Because of the variation of current along the element length there will a charge density distribution along the antenna, given by

$$q = -\frac{1}{j\omega}\frac{\partial I}{\partial z} = \frac{2}{j\omega}\frac{I}{dl} \tag{2.98}$$

Fig. 2.9. Infinitesimal Hertzian dipole with triangular current distribution.

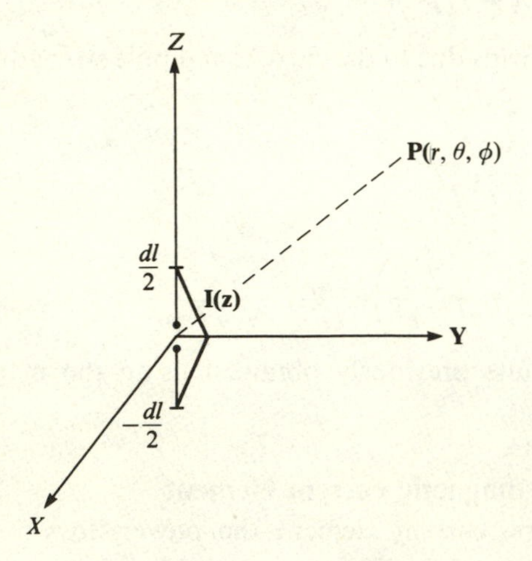

This means that the charge density is positive and constant between the origin and the top of the antenna at $dl/2$, and likewise negative and of equal magnitude between the bottom of the antenna at $-dl/2$ and the origin.

Because the magnetic intensity **H** depends only on the current, and the electric field can be calculated from Maxwell's equation using the field **H** only, it is not necessary to take note analytically of the above charge distribution. As with the current element example given earlier, all field quantities can be obtained from the vector potential set up by this Hertzian dipole. Since the current is cophasal over the infinitesimal length dl, the vector potential at any point will be exactly half that of the previously considered current element value, i.e.

$$A_z = \frac{\mu I dl}{8\pi} \frac{e^{-jkr}}{r} \tag{2.99}$$

Consequently the magnetic intensity H_ϕ and the electric field components E_r, E_θ are likewise one-half that of each of the values calculated previously for the current element. The radiated power will therefore be reduced by a factor of four.

2.20 Radiation resistance and input resistance of Hertzian dipole

Because the power radiated by the Hertzian dipole is one-fourth of that radiated by a current element carrying the same current I, given by eqn (2.67), the Hertzian dipole radiated power is

$$W = \tfrac{5}{2} k^2 dl^2 I^2$$

where I is the maximum value of the current in the dipole, which occurs at its mid-point. Hence the radiation resistance R_r, defined as that resistance which when multiplied by one-half the square of the maximum current gives the radiated power, is

$$R_r = 5k^2 dl^2 \tag{2.100}$$

If there are no losses in the Hertzian dipole then the radiated power is equal to the input power. Since this input power can be written as

$$W_i = \tfrac{1}{2} I_i^2 R_i$$

where the suffix i refers to the input terminals, and since I_i is equal to the maximum current I on the dipole, then it follows that

$$R_i = R_r \tag{2.101}$$

The input resistance is in this case equal to the radiation resistance, and this will always be the case when the maximum current on the antenna is at the input terminals. There are some antennas in which this equality does not hold, but where the input current is less than the maximum current on the antenna. In such cases the input resistance will be greater than the radiation resistance.

2.21 Input impedance of Hertzian dipole

The input resistance has already been found from the calculation of the real power radiated by the dipole. Since the quadrature as well as the antiphase electric field close to the surface of the dipole can be readily found from the solution given for the current element, this means that an expression for the reactive power there can also be found. Unfortunately, because the dipole has been considered to be filamentary, this reactive power becomes infinite at the origin, and a more realistic model for a short dipole has therefore to be considered.

2.22 Short dipole of finite radius

A short dipole of length l and radius a is assumed to be fed at its centre, so that the current distribution along its surface may be taken to be linear, falling from I at the feeding point to zero at its two ends. For the purpose of all field calculations this current will be assumed to flow along the axis of the dipole, since it has previously been shown that provided $1 \gg a$ this results in negligible error for a current element.

Since the current can be expressed as

$$I = I_0\left(1 \mp \frac{2z}{l}\right), \quad z \gtrless 0 \tag{2.102}$$

the charge density distribution is

$$\begin{aligned} q &= -\frac{1}{j\omega}\frac{\partial I}{\partial z} \\ &= \pm\frac{2}{j\omega}\frac{I_0}{l} \end{aligned} \tag{2.103}$$

A uniform positive charge density thus exists over the top half of the dipole, and an equal negative one over the bottom half.

The self impedance of the dipole will be defined as

$$Z_{11} = \frac{\int_{-l/2}^{l/2} E(z')\left(1 \mp \frac{2z}{l}\right) dz'}{I_0} \tag{2.104}$$

where $E_{z'}$ is the total tangential electric field at the surface of the dipole, made up of components due to both the electric current and the distributed charges over the length l. The component of $E_{z'}$ due to the current is

$$E_{z'_{current}} = -\frac{j\omega\mu}{4\pi}\frac{\int_{-l/2}^{l/2} I_0\left(1 \mp \frac{2z}{l}\right)e^{-jk[a^2+(z-z')^2]^{\frac{1}{2}}}}{[a^2+(z-z')^2]^{\frac{1}{2}}}dz \tag{2.105}$$

i.e.

$$E_{z'_{current}} \approx -\frac{j\omega\mu I_0}{4\pi}\int_{-l/2}^{l/2}\left(1 \mp \frac{2z}{l}\right)\left[\frac{1}{[a^2+(z-z')^2]^{\frac{1}{2}}} - jk\right]dz \tag{2.106}$$

As in the case of the current element previously considered the quadrature component will be negligible compared with the quadrature component due to the charges, and so will be neglected. The real component becomes

$$E_{z'_{current}} = -30k^2\left[z \mp \frac{z^2}{l}\right]_{-l/2}^{+l/2} I_0$$
$$= -15k^2 l I_0 \tag{2.107}$$

This is a constant and substitution into eqn (2.104) for Z_{11} gives

$$Z_{11_{current}} = 7.5k^2 l^2 \tag{2.108}$$

which, as could have been expected, is one-quarter of the value for the current element.

Instead of dealing with the component of $E_{z'}$, due to the charges the potential difference across the dipole length l will be found. This can be derived from the result for the current element, where the integrations were performed over a length l instead of the required $l/2$ in the case of the dipole. The result is

$$Z_{11_{charges}} \approx -\tfrac{35}{18}k^2 l^2 + \frac{2}{j\omega l 4\pi\varepsilon}\ln\frac{a[l+(a^2+l^2)^{\frac{1}{2}}]}{\left[-\frac{l}{2}+\left(a^2+\frac{l^2}{4}\right)^{\frac{1}{2}}\right]^2}$$

$$\approx -\tfrac{35}{18}k^2 l^2 - j\frac{60}{kl}\ln\left(\frac{l}{2a}\right) \tag{2.109}$$

Hence the total impedance due to currents and charges is

$$Z_{11} \approx 5.5k^2 l^2 - j\frac{60}{kl}\ln\left(\frac{l}{2a}\right) \tag{2.110}$$

It will be noted that the effect of increasing the diameter of the Hertzian dipole is to increase the input resistance slightly, and that it also provides a finite value for the capacitive input reactance.

Further reading

C.A. Balanis: *Antenna Theory: Analysis and Design*: Harper and Row, New York, 1982.

R.F. Harrington: *Field Computation by Moment Methods*: Macmillan, New York, 1968.

J.D. Kraus: *Antennas*: McGraw-Hill, New York, 1950.

E.B. Moullin: *Radio Aerials*: Clarendon Press, Oxford, 1949.

S.A. Schelkunoff and H.T. Friis: *Antennas, Theory and Practice*: Wiley and Sons, New York, 1952.

3

+−+

Dipole and monopole antennas

3.1 Filamentary dipole of finite length

The discussion of the filamentary Hertzian dipole was based on the assumption that the current fell linearly from a maximum value at its mid-point to zero at both ends. This is a good overall approximation when the dipole is short, and the nature of separate approximations involved in it will become clearer in the following analysis. The case to be considered is that of a filamentary dipole of length $2H$, which is not necessarily short. Both transmitting and receiving operation of the dipole will be considered in separate analyses.

3.2 Centre-fed transmitting dipole of length $2H$

The linear dipole, extending from $-H$ to $+H$ as shown in Fig. 3.1, is energised at its centre. As a result currents are constrained to flow along it, and wherever the current varies charges will accumulate along the filament. The distribution of this current and associated charge must be such as to satisfy the two following conditions:

(a) the current must fall to zero at the end points $\pm H$;

(b) the tangential electric field at every point along the filament due to these current and charge distributions must be identically zero if the filament is a perfect conductor.

Since the tangential electric field at the surface is zero,

$$0 = -j\omega A_z - \nabla_z V \tag{3.1}$$

Fig. 3.1. Centre-fed transmitting dipole of length $2H$.

$z = -H$ $z = 0$ $z = z'$ $z = +H$ $2a$

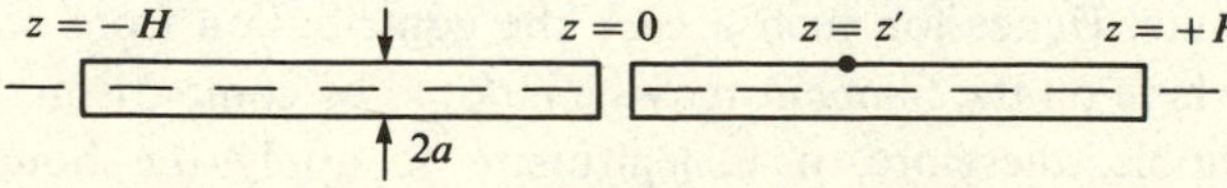

But the vector and scalar potentials are related by

$$\nabla \cdot \mathbf{A} = -j\omega\mu\varepsilon V \tag{3.2}$$

so that eqn (3.1) may be written

$$0 = -j\omega A_z - \frac{\partial}{\partial z}\left(-\frac{1}{j\omega\mu\varepsilon}\frac{\partial A_z}{\partial z}\right)$$

for a vector potential which, like the current, exists in the z-direction only. i.e.

$$-\frac{j\omega}{k^2}\left(\frac{\partial^2 A_z}{\partial z^2} + k^2 A_z\right) = 0 \tag{3.3}$$

The solution to eqn (3.3) being a second order differential equation with constant coefficients, is

$$A_z = B \sin kz + C \cos kz \tag{3.4}$$

where B, C are constants. This solution is exact, and from eqn (3.2) leads to another exact solution for the scalar potential on the surface of the filamentary dipole, viz.

$$V = \frac{jk}{\omega\mu\varepsilon}(B \cos kz - C \sin kz) \tag{3.5}$$

The advantage of having two exact solutions for the scalar and vector potentials on the surface of this linear dipole is, however, greatly reduced by the fact that it is not possible to apply boundary conditions to either of these equations in order to find the constants B and C. This is because neither the scalar nor the vector potential are known exactly at any point on the surface. The only boundary conditions which are known are those which relate to current, which say, as stated previously, that the current must be zero at the two ends of the filament.

If, however, it were assumed that the vector potential at any point on the surface is directly proportional to the current at the same point, then it immediately becomes possible to find the constants B and C. To see the nature of this assumption consider the expression for the vector potential at a point z' on the surface

$$A_z = \frac{\mu}{4\pi}\int_{-H}^{+H} \frac{I(z)e^{-jk[a^2+(z-z')^2]^{\frac{1}{2}}}}{[a^2+(z-z')^2]^{\frac{1}{2}}}dz$$

where for a filament $a \to 0$. It is immediately seen that when a is zero and z' is equal to z the contribution to the vector potential at z' from the current at point z is infinite. Hence for such a case the contribution from current elements elsewhere on the filament may be ignored by comparison. For a filamentary dipole, therefore, it is legitimate to apply the boundary condition that the vector potential is zero at its two ends. Thus from eqn (3.4),

$$0 = B \sin kH + C \cos kH \qquad z > 0$$
$$0 = -B_1 \sin kH + C_1 \cos kH \quad z < 0$$

giving

$$A_{z1} = \frac{C}{\sin kH} \sin k(H - z) \quad z > 0 \tag{3.6}$$

$$A_{z2} = \frac{C_1}{\sin kH} \sin k(H + z) \quad z < 0 \tag{3.7}$$

But from continuity of the vector potential at $z = 0, C = C_1$. Similarly, from eqn (3.2),

$$V_1 = -\frac{jkC}{\omega\mu\varepsilon \sin kH} \cos k(H - z) \quad z > 0 \tag{3.8}$$

$$V_2 = \frac{jkC}{\omega\mu\varepsilon \sin kH} \cos k(H + z) \qquad z < 0 \tag{3.9}$$

It will be noted that the charges on the upper and lower halves of the dipole are equal in magnitude and opposite in sign.

The equations for A_z and V still retain one unknown constant, but this may be found by applying boundary conditions at the generator terminals. For example, for a generator balanced with respect to its two terminals so that the potentials at these terminals are, say, $\pm V$, where z is zero, eqns (3.8) and (3.9) both give

$$C = \frac{j\omega\mu\varepsilon V}{k} \tan kH \tag{3.10}$$

Moreover since for an infinitesimally thin dipole the current at any point is proportional to the vector potential at the same distance z' on the surface through the equation

$$A_z(z') = L(z')I(z) \tag{3.11}$$

where

$$L(z') = \frac{\mu}{4\pi} \int_{\substack{-H \\ a \to 0}}^{+H} \frac{e^{-jk[a^2 + (z - z')^2]^{\frac{1}{2}}}}{[a^2 + (z - z')^2]^{\frac{1}{2}}} dz \tag{3.12}$$

the solution for the current may be written

$$I(z) = \frac{j\omega\mu\varepsilon V}{L(z')k \cos kH} \sin k(H \mp z) \quad z \gtrless 0 \tag{3.13}$$

where it will be noted that for an infinitesimal radius a, $L(z')$ is a function both of the length of the dipole $2H$, and also of the point of observation z', which is equal to z. In order to avoid the complexity of this apparent constant $L(z')$ being in fact a function of position, it is customary to assume that an average value for it is known, so that the equation for the current can be written

$$I(z) = \frac{jV}{cL_{av}\cos kH}\sin k(H \mp z) \tag{3.14}$$

Since in addition the velocity of propagation along the dipole is that of light, c may be written as $1/[L_{av}C_{av}]^{\frac{1}{2}}$, and the expression for the current becomes

$$I(z) = j\frac{2V}{2Z_c\cos kH}\sin k(H \mp z) \quad z \gtrless 0 \tag{3.15}$$

where $2V$ is the potential difference across the generator terminals and $2Z_c$ is an average characteristic impedance for the dipole, defined through the equation for a monopole

$$Z_c = \left[\frac{L_{av}}{C_{av}}\right]^{\frac{1}{2}} \tag{3.16}$$

Similarly the potential distribution can be written in terms of the same constants

$$V(z) = \pm\frac{V}{\cos kH}\cos k(H \mp z) \quad z \gtrless 0 \tag{3.17}$$

The expressions for current and potential given by eqns (3.15) and (3.17) form an important base for deriving the most important properties of a filamentary dipole antenna, such as its radiation pattern, power gain and input impedance. Nevertheless it is clear from their algebraic form that the approximations introduced in their derivation have resulted in two obvious errors:

(a) that the input current is in phase quadrature with the input voltage, implying that no input power can be delivered to the antenna. Although this is a serious deficiency it may be shown that corrections can be made for the error. It is also true that the value for the calculated input reactance derived from these formulae is useful in practice. Thus

$$X_i = \frac{2V(0)}{I(0)} = -j2Z_c\cot kH; \tag{3.18}$$

(b) that when the dipole length is one wavelength, or a multiple thereof, the input reactance is infinite. In practice, lengths of dipoles immediately on both sides of these values must be avoided in the use of these approximate formulae.

3.3 Short-circuited reflecting dipole of length $2H$

A short-circuited reflecting dipole extending from $-H$ to $+H$, as shown in Fig. 3.2, is irradiated by an incoming uniform plane wave. It will be assumed initially that the direction of the incoming electric field is parallel to the axis of the dipole so that the excitation is everywhere cophasal. As a result of the excitation, current and charge distributions will be set up, which will in turn set up an electric field parallel to the dipole axis, such that the sum of this field on the surface together with the inducing field there is equal to zero. Thus

$$-j\omega A_z - \nabla_z V + E_z^i = 0 \tag{3.19}$$

Following the argument for the transmitting dipole this leads to the differential equation

$$\frac{\partial^2 A_z}{\partial z^2} + k^2 A_z = -j\frac{k^2 E_z^i}{\omega} \tag{3.20}$$

The solution of this differential equation consists of a complementary function and a particular integral. Instead of solving this equation, however, we shall consider more directly the corresponding differential equation for the scalar potential

$$\begin{aligned}\frac{\partial^2 V}{\partial z^2} &= -j\omega\frac{\partial A_z}{\partial z}\\ &= -k^2 V\end{aligned} \tag{3.21}$$

Fig. 3.2. Short-circuited dipole of length $2H$ irradiated by uniform plane wave.

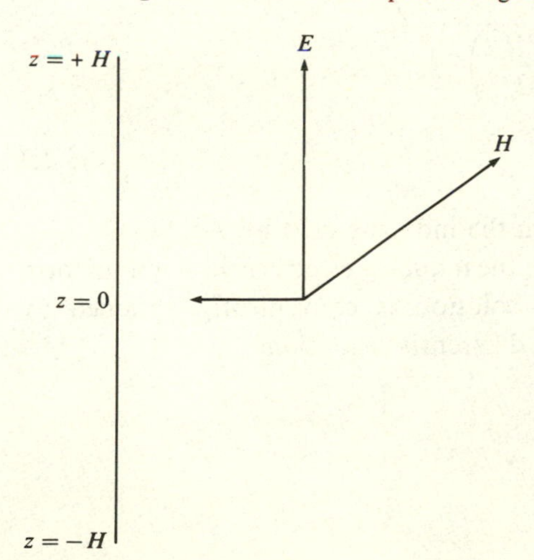

In this case no particular integral is required and the solution becomes

$$V = B_1 \cos kz + C_1 \sin kz \tag{3.22}$$

from which

$$-j\omega A_z = k(-B_1 \sin kz + C_1 \cos kz) - E_z^i$$

and hence since $A(z) \approx L_{av} I(z)$

$$I(z) = \frac{k}{-j\omega L_{av}}(-B_1 \sin kz + C_1 \cos kz) + \frac{E_z^i}{j\omega L_{av}} \tag{3.23}$$

Using the boundary conditions that the current is zero at the two ends $\pm H$ gives

$$B_1 = 0$$

and

$$C_1 = \frac{E_z^i}{k \cos kH}$$

Hence

$$I(z) = \frac{E_z^i}{j\omega L_{av}}\left(1 - \frac{\cos kz}{\cos kH}\right) \tag{3.24}$$

Since the reflecting dipole is not divided as it is in the transmitting case, a single equation describes the distribution of current over its full length. Again it will be noted that a pole occurs for a particular length of dipole, in this case when it is $\lambda/2$ in length. It is possible, however, to avoid this pole by taking account of the radiation resistance of the dipole. It will also be noted that when the dipole is small the current can be written in the form

$$I(z) = \frac{E}{j\omega L_{av}} \frac{\left(1 - \frac{k^2H^2}{2} - 1 + \frac{k^2z^2}{2}\right)}{\left(1 - \frac{k^2H^2}{2}\right)}$$

$$\approx \frac{jE}{\omega L_{av}} \frac{k^2}{2}(H^2 - z^2) \tag{3.25}$$

showing that the current leads on the inducing field by 90°.

For the more general case when the inducing electric field is not uniform along the reflecting dipole the solution is conveniently obtained by returning to the basic first order differential equations

$$\frac{\partial V(z)}{\partial z} = -j\omega A(z) + E_z^i$$

and

$$\frac{\partial A(z)}{\partial z} = -j\omega\mu\varepsilon V$$

Using now the approximation that the vector potential and current are related by

$$A(z) = L_{av} I(z)$$

gives

$$\frac{\partial V(z)}{\partial z} = -j\omega L_{av} I(z) + E_z^i \tag{3.26}$$

$$\frac{\partial I(z)}{\partial z} = -\frac{j\omega\mu\varepsilon}{L_{av}} V(z) \tag{3.27}$$

The solution to eqns (3.26) and (3.27) will be built up from the simpler problem shown in Fig. 3.3, where E_z^i is assumed to exist only over an increment of length dz' at $z = z'$. Consequently the two sections of the reflecting dipole on either side of dz' are sections on which no inducing field exists, and consequently approximate solution for the currents and potentials on these two sections are

$$\left.\begin{aligned} V_1 &= A \sin kz + B \cos kz \\ I_1 &= \frac{jk}{\omega L_{av}} (A \cos kz - B \sin kz) \end{aligned}\right\} \quad -H \leqslant z \leqslant z' \tag{3.28}$$

and

$$\left.\begin{aligned} V_2 &= C \sin kz + D \cos kz \\ I_2 &= \frac{jk}{\omega L_{av}} (C \cos kz - D \sin kz) \end{aligned}\right\} \quad z' \leqslant z \leqslant H \tag{3.29}$$

Fig. 3.3. Short-circuited dipole with electric field E_z applied over length dz'.

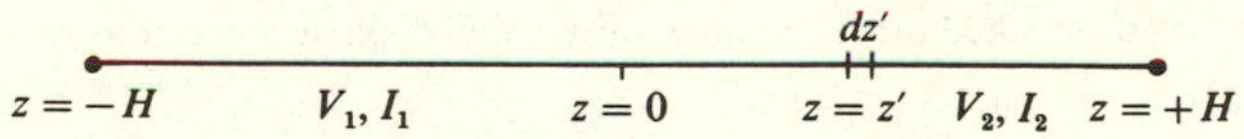

$z = -H$ $\quad V_1, I_1 \quad$ $z = 0$ $\quad$ dz' $\quad z = z' \quad V_2, I_2 \quad z = +H$

The boundary conditions are that there must be continuity of current at $z = z'$ and a discontinuity of potential there, given by

$$V_2 - V_1 = E_z^i \, dz' \tag{3.30}$$

Using these boundary conditions means that the constants C, D can be related to the constants A, B through the equations

$$C = A + E_z^i \sin(kz') \, dz' \tag{3.31}$$

$$D = B + E_z^i \cos(kz') \, dz' \tag{3.32}$$

and hence, over the whole conductor, only these constants A, B require to be found from the usual conditions that the current must fall to zero at both ends.

But to leave the solution for the original problem more general, we shall not apply these boundary conditions, and instead will generalise the above solution to solve the problem when E_z^i exists over the full length of the dipole. Writing the solution for the region $z' \leqslant z \leqslant H$ in terms of A, B gives

$$V_2 = A \sin kz + B \cos kz + E_z^i \mathrm{d}z' \cos k(z - z') \tag{3.33}$$

$$I_2 = \frac{jk}{\omega L_{av}} [A \cos kz - B \sin kz - E_z^i dz' \sin k(z - z')] \tag{3.34}$$

Now $E_z^i dz'$ is the integral of the electric field to the left of the region in which these equations apply. Consequently these solutions may be generalised to cover the case of excitation along the full dipole to give

$$V(z) = A \sin kz + B \cos kz + \int_{-H}^{z} E_z(z') \cos k(z - z')\, dz' \tag{3.35}$$

and

$$I(z) = \frac{jk}{\omega L_{av}} \left[A \cos kz - B \sin kz - \int_{-H}^{z} E_z(z') \sin k(z - z')\, dz' \right] \tag{3.36}$$

and these equations apply along the full length of the dipole. It is straightforward to confirm by differentiation that the solutions satisfy the original first order differential eqns (3.26) and (3.27).

As an example of the use of these solutions the problem of the reflecting dipole extending along the z-axis from $-H$ to $+H$ will now be generalised to deal with an incoming uniform phase wave arriving at an angle θ to the z-axis, with its electric field in the plane of the dipole, as shown in Fig. 3.4. The

Fig. 3.4. Short-circuited dipole irradiated by uniform plane wave at angle θ.

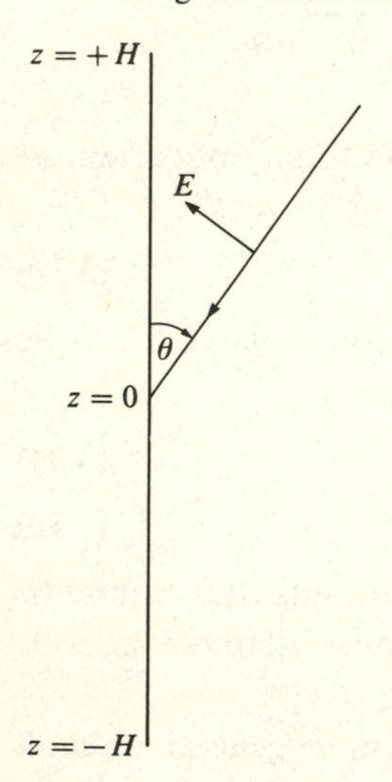

energising electric field as a function of position along the dipole is given by

$$E(z') = E \sin \theta e^{jkz' \cos \theta} \tag{3.37}$$

Consequently the equation for the current distribution is

$$I(z) = \frac{jk}{\omega L_{av}} \left[A \cos kz - B \sin kz - \int_{-H}^{z} E \sin \theta e^{jkz' \cos \theta} \sin k(z - z') \, dz' \right] \tag{3.38}$$

The integrand in z' is readily integrable by changing the trigonometric term to exponential form and the result is

$$I(z) = \frac{jk}{\omega L_{av}} \left\{ A \cos kz - B \sin kz - \frac{E}{k \sin \theta} \times [e^{jkz \cos \theta} - j \cos \theta \sin k(z + H) e^{-jkH \cos \theta} - \cos k(z + H) e^{-jkH \cos \theta}] \right\} \tag{3.39}$$

To find the constants A, B use is made as before of the boundary conditions that the current falls to zero at $\pm H$. Applying these conditions gives

$$A = \frac{E}{k \cos kH \sin \theta} [\cos(kH \cos\theta) - (\cos^2 kH - \sin kH \cos kH \cos \theta) e^{-jkH \cos \theta}]$$

and

$$B = -A \cot kH$$

Hence the current can be found at any point z along the axis. In particular the current at the centre of the reflecting dipole is

$$I(0) = \frac{jE}{\omega L_{av} \cos kH} \left[\frac{\cos(kH \cos \theta) - \cos kH}{\sin \theta} \right] \tag{3.40}$$

The expression in square brackets indicates the response of the reflecting dipole to a wave arriving at angle θ to the axis of the dipole. It is thus the receiving radiation pattern of the dipole.

3.4 Open-circuited linear dipole of length 2*H*

Let the dipole extend along the z-axis from $-H$ to $+H$, as shown in Fig. 3.5. The direction of the incoming electric field will be taken parallel to the axis, so as to produce the maximum output voltage from the open-circuited terminals at the origin. Then for this uniform, cophasal case the equations for the potentials and currents on the separate halves are

$$\left.\begin{aligned} V_1 &= B_1 \cos kz + C_1 \sin kz \quad &(3.41)\\ I_1 &= \frac{jk}{\omega L_{av}}(-B_1 \sin kz + C_1 \cos kz) + \frac{E_z^i}{j\omega L_{av}} \quad &(3.42)\end{aligned}\right\} 0 \leqslant z \leqslant H$$

and

$$\left.\begin{aligned} V_2 &= B_2 \cos kz + C_2 \sin kz \quad &(3.43)\\ I_2 &= \frac{jk}{\omega L_{av}}(-B_2 \sin kz + C_2 \cos kz) + \frac{E_z^i}{j\omega L_{av}} \quad &(3.44)\end{aligned}\right\} -H \leqslant z \leqslant 0$$

From the boundary conditions that the current is zero at the origin where the dipole is open-circuited it follows that

$$C_1 = C_2 = \frac{E_z^i}{k}$$

Since likewise at the top and bottom of the oc dipole the currents are again zero,

$$0 = kB_1 \sin kH - kC_1 \cos kH + E_z^i$$

and

$$0 = -kB_2 \sin kH - kC_2 \cos kH + E_z^i$$

so that

$$B_1 = -B_2 = -\frac{E_z^i(1 - \cos kH)}{k \sin kH} \tag{3.45}$$

Fig. 3.5. Opencircuited receiving dipole of length $2H$ with cophasal electric field parallel to dipole.

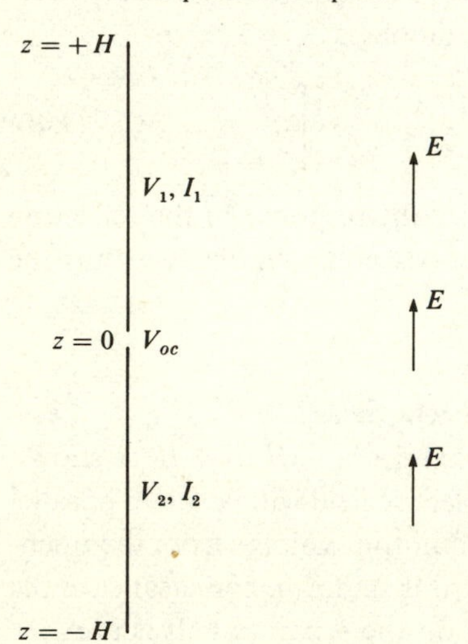

Hence the open-circuit voltage at the origin is

$$V_1(0) - V_2(0) = B_1 - B_2$$

$$= -\frac{2E_z^i(1 - \cos kH)}{k \sin kH} \quad (3.46)$$

When the dipole is short in terms of a wavelength the trigonometric terms may be replaced by their small argument approximations, giving

$$V_{oc_{l \ll \lambda}} = -E_z^i H \quad (3.47)$$

If the effective length of the dipole is defined as that length by which the electric field must be multiplied to give the open-circuit output voltage, then the effective length in this case is half the physical length of $2H$. For the common case of a half-wavelength dipole where H is $\lambda/4$,

$$V_{oc_{l=\lambda/2}} = -\frac{\lambda}{\pi} E_z^i \quad (3.48)$$

so that the effective length is

$$l_{eff} = \frac{\lambda}{\pi} \quad (3.49)$$

It will be noted that the negative sign associated with the output voltage arises because for an electric field in the positive z-direction, positive charge is driven away from the top terminal along the top half of the dipole, and towards it along the bottom half. It will also be observed from the above analysis that because a dipole is open-circuited at its terminals this does not mean that the current is zero elsewhere along the dipole, though when the dipole is short the current will be small.

3.5 Output impedance of linear dipole of length $2H$

If the dipole is regarded at its terminals as being a passive network, the ratio of an externally applied voltage to the input current produced by that voltage would be the impedance of the network. Similarly for a network including a generator, the source or output impedance is defined as

$$Z_0 = -\frac{V_{oc}}{I_{sc}} \quad (3.50)$$

where the current I_{sc} flows out of the network when its terminals are short circuited. Applying this result to the linear dipole for the case of cophasal excitation gives

$$Z_0 = \frac{2E_z^i(1 - \cos kH)}{k \sin kH} \frac{\omega L_{av} \cos kH}{jE_z^i(1 - \cos kH)}$$

$$= -j\frac{2\omega L_{av}}{k}\cot kH$$

$$= -j2Z_c \cot kH \tag{3.51}$$

where $2Z_c$ is the characteristic impedance of the dipole defined through the equation for a monopole

$$Z_c = \left[\frac{L_{av}}{C_{av}}\right]^{\frac{1}{2}} \tag{3.52}$$

This result for output impedance agrees with the source impedance previously calculated from the transmitting viewpoint.

3.6 Impedance loaded linear dipole of length $2H$

The two previous cases of short-circuited and open-circuited linear dipoles lead naturally to the more general problem of an impedance loaded dipole. Usually this impedance loading is imposed at the output terminals, but in some situations there are advantages in incorporating an impedance elsewhere in the dipole. This can have the effect, for example, of producing an antenna with a more uniform, as opposed to triangular, transmitting current distribution.

Consider first the case of a dipole with a load impedance Z_L inserted at its mid-point. If this were treated by the method adopted for the open-circuited dipole, there would be two unknown constants for each of the top and bottom sections, which would then be related to each other by the continuity of current and discontinuity of potential through and across the load impedance. The two remaining constants would then be found from the boundary conditions for current at the ends of the dipole.

It is, however, much simpler to make use of a Thévenin equivalent circuit for the dipole, consisting of an emf equal to its open-circuit output voltage, in series with its output and load impedances. The output voltage is then the voltage which exists across the load impedance, i.e.

$$V_0 = \frac{V_{oc}}{Z_0 + Z_L} Z_L \tag{3.53}$$

Consider now the case when two equal impedances Z are inserted, one at a distance h from the origin and the other, symmetrically, at $-h$. Assume also that a single impedance Z_L is connected across the output terminals at the origin. It would be possible to divide the dipole into four sections and associate with each of these sections two unknown constants, and then find all eight constants from the boundary conditions at the end of each section. It is better, however, to make use of the simplification afforded by considering impedances, and a suitable form of these will now be defined.

At every point along the filamentary dipole both a potential and a current exist. This potential is not electrostatic – it is time varying, but associated with charges in the usual way. Consequently at an observation point symmetrically placed with regard to a symmetrical distribution of charge the potential will be zero. Such a situation would exist, at the origin, for example, if the load impedance Z_L were replaced by a balanced generator, or in the receiving situation if the dipole were excited by an incoming electromagnetic wave with its electric vector parallel to the dipole axis. But it would not exist if the incoming wave arrived at an angle other than 90° to the dipole axis, because then the exciting electric field would not be in phase at geometrically symmetrical points along the dipole. Consequently the charge distribution would not be symmetrical and the potential would not then be zero at the origin.

It is in fact more direct to note that since scalar and vector potentials on the dipole itself are related by

$$j\omega\mu\varepsilon V = \nabla \cdot \mathbf{A} = \frac{\partial A_z}{\partial z}$$

the scalar potential is zero where the vector potential goes through a turning value. Or since the vector potential is being approximated to being directly proportional to the current, the scalar potential is zero at every current maximum, and at minima other than end points.

Associated with the value of potential at every point on the dipole an impedance will now be defined in a particular direction as the ratio of potential at that point to the current passing through the point. Thus,

$$Z(z) = \frac{V(z)}{I(z)} \tag{3.54}$$

Clearly therefore this impedance is infinite at the ends of the dipole where the current is zero. The advantage of introducing the concept of potential to dipoles is clearly illustrated by considering how this impedance changes between the two ends of an impedance Z. If at the end where the current flows out of the network impedance the calculated impedance is Z_A, then at the other end it is simply $(Z_A + Z)$ since the current is the same at both ends. Thus the current and potential boundary conditions are automatically satisfied without having to introduce two new unknown constants in moving to a new section of the dipole.

Returning now to the case of the dipole of length $2H$, with equal impedances Z inserted symmetrically at $\pm h$ from the origin, as shown in Fig. 3.6, the impedance at the top point H of the dipole is clearly infinite since the current there is zero. The observation point along the dipole axis is now brought into the region $h \leqslant z \leqslant H$ where the current distribution must be of the form

$$I(z) = A \sin k(H - z) \tag{3.55}$$

where A is a constant, and the potential distribution is obtained from

$$\begin{aligned} V(z) &= -\frac{L_{av}}{j\omega\mu\varepsilon}\frac{\partial}{\partial z}I(z) \\ &= -jZ_c A \cos k(H - z) \end{aligned} \tag{3.56}$$

Thus at the point A immediately above the impedance at $z = h$,

$$Z(z) = -jZ_c \cot k(H - h) \tag{3.57}$$

and at the point B immediately below it,

$$Z(z) = Z - jZ_c \cot k(H - h) \tag{3.58}$$

As the observation point is progressively lowered to the origin this impedance is then transformed in the same way as an impedance on a transmission line is transformed. It is therefore convenient in dealing with numerical problems to use a Smith impedance chart to transform the impedance $[Z - jZ_c \cot k(H - h)]$ by rotating it through twice the electrical angle kh to obtain the impedance at the origin. This represents the impedance of the top half of the dipole only, so that the total input impedance is twice this value.

It will be noted that, although this method provides a rapid means of estimating the input impedance and, more particularly, the input reactance

Fig. 3.6. Transmitting dipole of length $2H$ with equal impedances Z at distances $\pm h$ from feed.

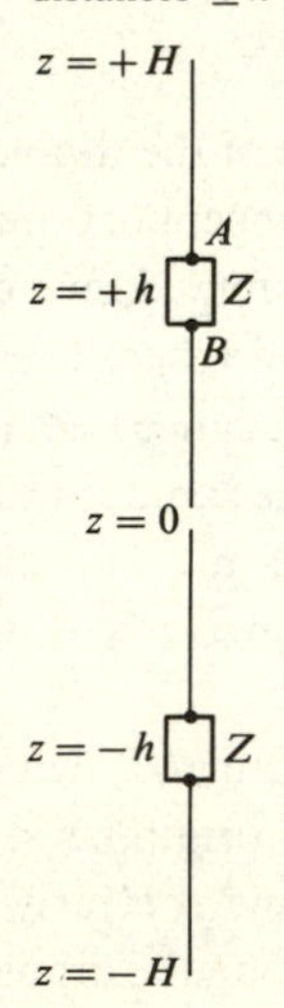

of a loaded dipole antenna, if the current at a point other than the origin were required it would be necessary to return to the two unknown constants per section type of solution. However, for short dipoles in particular the effect of impedance loading on the input impedance is much greater than on the radiation pattern.

3.7 Travelling wave current on filamentary dipole

The exact wave equation for the vector potential on the surface of a transmitting perfectly conducting linear dipole has been shown to be

$$\frac{\partial^2 A_z}{\partial z^2} + k^2 A_z = 0 \tag{3.59}$$

Using the approximation that the vector potential on the surface is related to the current by

$$A(z) \approx L_{av} I(z)$$

gives the solution for the current in its usual form as

$$I(z) = A \sin kz + B \cos kz \tag{3.60}$$

But this solution of the wave equation may also be expressed as

$$I(z) = A_1 e^{-jkz} + B_1 e^{+jkz} \tag{3.61}$$

where the physical meaning of the separate terms is that they represent travelling waves of current in the positive and negative z-directions respectively. If then the coefficients A_1 and B_1 are equal this new solution takes the same form as the preceding one with A equal to zero. Likewise, if A_1 is equal to $-B_1$ the new solution is again of the same form with B equal to zero.

For a dipole with a single generator at the origin it is to be expected that the constant A_1 will be finite since this represents a wave, characterised in this equation by a current travelling out from the generator in the positive z-direction. If a wave of amplitude B_1 exists it can do so only by virtue of reflection of the outgoing wave. Hence to produce a single outgoing wave there must be a mechanism for preventing such a wave from returning.

An approximate means of doing this is shown in Fig. 3.7, where it will be

Fig. 3.7. Transmitting dipole with travelling wave of current from $z = 0$ to $z = R$ and standing wave from R to H.

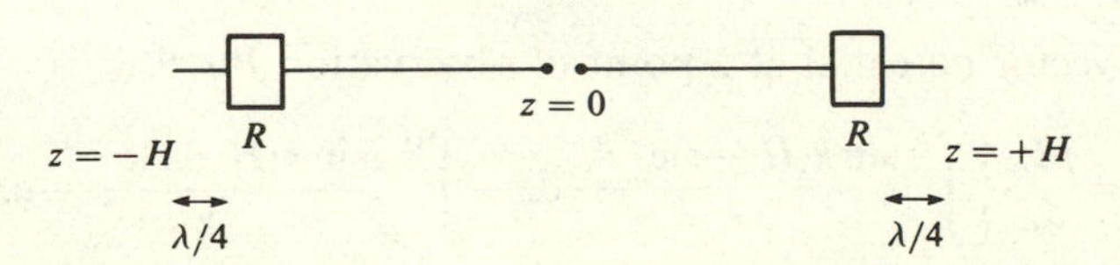

seen that a resistor R is placed at a distance of $\lambda/4$ from each end of the dipole. These would normally be expected to be positions of maximum current so that there would be maximum dissipation of reflected and incident energy there.

Consider now the impedance at the top end of the resistance R, looking towards the top of the dipole at $z = H$. This impedance is given by

$$\begin{aligned} Z &= -jZ_c \cot k\frac{\lambda}{4} \\ &= 0 \end{aligned}$$

Consequently the impedance at the lower end of the resistance R is

$$Z = R \tag{3.62}$$

If now this resistance R is selected to be equal to the characteristic impedance of the top half of the dipole, this means that the input impedance of the dipole will be seen to be matched, and consequently there will exist only an outgoing wave from the generator to the load resistor R. At the top end of R, however, it has been shown that the impedance looking to the open end is zero, and consequently there will be complete reflection of the wave going from R to the open end, setting up a standing wave of current on that quarter-wavelength of conductor. Thus there is a travelling wave of current from the generator up to the resistor R and a standing wave with peak amplitude equal to that of the travelling wave thereafter. The numerical value of R required is typically 300Ω–500Ω depending on the dimensions of the dipole.

3.8 Radiation pattern of linear dipole of length $2H$ with standing wave of current

The linear dipole is assumed to lie along the z-axis from $-H$ to $+H$, and is energised at its mid-point so that the current distribution along it is of the form developed in eqn (3.15), viz.

$$I(z) = j\frac{2V}{2Z_c \cos kH} \sin k(H \mp z) \quad z \gtrless 0 \tag{3.63}$$

where $2V$ is the potential difference applied across the antenna terminals. Writing the maximum amplitude of this current as I_m gives

$$I(z) = I_m \sin k(H \mp z) \quad z \gtrless 0 \tag{3.64}$$

so that the vector potential at a point of observation $P(r, \theta, \phi)$ is

$$A_z = \frac{\mu I_m}{4\pi}\left\{\int_0^H \frac{\sin k(H-z)e^{-jkr}}{r}dz + \int_{-H}^0 \frac{\sin k(H+z)e^{-jkr}}{r}dz\right\}$$

In the far field,

$$r \approx r_0 - z\cos\theta$$

so that substituting this in the exponential term and $r \approx r_0$ in the denominator gives

$$A_z \approx \frac{\mu I_m}{4\pi r_0} e^{-jkr_0}\left\{\int_0^H \sin k(H-z)e^{jkz\cos\theta}\,dz + \int_{-H}^0 \sin k(H+z)e^{jkz\cos\theta}\,dz\right\}$$

$$= \frac{\mu I_m}{4\pi r_0} e^{-jkr_0}[I_1 + I_2] \tag{3.65}$$

where

$$I_1 = \frac{\sin k(H-z)e^{jkz\cos\theta}}{jk\cos\theta} + k\int_0^H \frac{e^{jkz\cos\theta}}{jk\cos\theta}\cos k(H-z)\,dz$$

$$= \frac{\sin k(H-z)e^{jkz\cos\theta}}{jk\cos\theta} + \frac{1}{j\cos\theta}\left[\cos k(H-z)\frac{e^{jkz\cos\theta}}{jk\cos\theta} - \frac{I_1}{j\cos\theta}\right]$$

i.e.

$$I_1\left(1 - \frac{1}{\cos^2\theta}\right) = \frac{e^{jkz\cos\theta}}{jk\cos\theta}\left[\sin k(H-z) - \frac{j}{\cos\theta}\cos k(H-z)\right]$$

or

$$I_1 = \frac{1}{\sin^2\theta}\left[\frac{e^{jkz\cos\theta}}{k}\cos k(H-z) + j\frac{e^{jkz\cos\theta}}{k}\cos\theta\sin k(H-z)\right]_0^H$$

$$= \frac{1}{k\sin^2\theta}[e^{jkH\cos\theta} - \cos kH - j\cos\theta\sin kH] \tag{3.66}$$

Similarly

$$I_2 = \frac{1}{k\sin^2\theta}[e^{-jkH\cos\theta} - \cos kH + j\cos\theta\sin kH] \tag{3.67}$$

Hence

$$A_z = \frac{\mu I_m}{2\pi}\frac{e^{-jkr_0}}{r_0}\frac{1}{k\sin^2\theta}[\cos(kH\cos\theta) - \cos kH] \tag{3.68}$$

This value for A_z is now conveniently transformed to spherical coordinates using

$$A_r = A_z\cos\theta$$

and

$$A_\theta = -A_z\sin\theta$$

and the magnetic intensity **H** is derived from

$$H_\phi = \frac{1}{\mu}\left[\frac{A_\theta}{r} + \frac{\partial A_\theta}{\partial r} - \frac{1}{r}\frac{\partial A_r}{\partial \theta}\right]$$

$$\approx \frac{1}{\mu}\frac{\partial A_\theta}{\partial r}$$

in the far field. This gives

$$H_\phi = \frac{jI_m}{2\pi}\frac{e^{-jkr_0}}{r_0}\left[\frac{\cos(kH\cos\theta) - \cos kH}{\sin\theta}\right] \tag{3.69}$$

and

$$E_\theta = Z_0 H_\phi \tag{3.70}$$

From eqn (3.69) it will be seen that the field H_ϕ in the equatorial plane for a half-wave dipole has the same magnitude as that of an infinite filament carrying the same direct current.

3.9 General expression for fields of linear dipole of length $2H$ with standing wave of current

It has previously been shown that the vector potential set up at an observation point $P(r, \theta, \phi)$ by a linear dipole extending from $-H$ to $+H$, and carrying a current $I_m \sin k(H \mp z)$, is

$$A_z = \frac{\mu I_m}{4\pi}\left\{\int_0^H \frac{\sin k(H - z')e^{-jkr}}{r}dz' + \int_{-H}^0 \frac{\sin k(H + z')e^{-jkr}}{r}dz'\right\} \tag{3.71}$$

We now wish to evaluate the fields due to this current at any point in space, and not just in the radiation pattern zone. Since the fields are symmetrical in azimuth it is sufficient to consider a point of observation in the yz-plane where

$$r^2 = (z - z')^2 + y^2$$

Substituting exponential terms for the trigonometric gives

$$A_z = \frac{\mu I_m}{j8\pi}\left\{\int_0^H [e^{jkH}e^{-jk(r+z')} - e^{-jkH}e^{-jk(r-z')}]\frac{dz'}{r} + \int_{-H}^0 [e^{jkH}e^{-jk(r-z')} - e^{-jkH}e^{-jk(r+z')}]\frac{dz'}{r}\right\} \tag{3.72}$$

Although these integrands may be evaluated in series form it is possible to obtain expressions for H and E in closed form by taking appropriate derivatives.

Thus

$$H_\phi = -H_x = -\frac{1}{\mu}\frac{\partial A_z}{\partial y} \tag{3.73}$$

Since the limits in the integrals for A_z are constants, the derivatives with respect to y may be taken inside the integral sign to give

$$\begin{aligned} H_\phi = \frac{-\mu I_m}{j8\pi}\Bigg\{ & e^{jkH}\int_0^H \frac{\partial}{\partial y}(e^{-jk(r+z')})\frac{dz'}{r} \\ & - e^{-jkH}\int_0^H \frac{\partial}{\partial y}(e^{-jk(r-z')})\frac{dz'}{r} \\ & + e^{jkH}\int_{-H}^0 \frac{\partial}{\partial y}(e^{-jk(r-z')})\frac{dz'}{r} \\ & - e^{-jkH}\int_{-H}^0 \frac{\partial}{\partial y}(e^{-jk(r+z')})\frac{dz'}{r}\Bigg\} \\ = H_{\phi 1} + H_{\phi 2} + H_{\phi 3} + H_{\phi 4} & \end{aligned} \tag{3.74}$$

where

$$\begin{aligned} H_{\phi 1} &= -\frac{\mu I_m}{j8\pi}e^{jkH}\int_0^H \frac{re^{-jk(r+z')}(-jk)\frac{y}{r} - e^{-jk(r+z')}\frac{y}{r}}{r^2}dz' \\ &= \frac{\mu I_m y}{j8\pi}e^{jkH}\int_0^H e^{-jk(r+z')}\left(\frac{jk}{r^2}+\frac{1}{r^3}\right)dz' \qquad \text{etc.} \end{aligned}$$

It can readily be verified that this integrand is the derivative with respect to z' of the function

$$\frac{e^{-jk(r+z')}}{r(r+z'-z)}$$

Hence

$$H_{\phi 1} = \frac{\mu I_m y}{j8\pi}e^{jkH}\left[\frac{e^{-jk(r+z')}}{r(r+z'-z)}\right]_0^H \tag{3.75}$$

and similarly for $H_{\phi 2}$ to $H_{\phi 4}$. Adding the four terms gives the total magnetic field as

$$H_\phi = \frac{jI_m}{4\pi y}[e^{-jkr_1} + e^{-jkr_2} - 2\cos kH e^{-jkr_0}] \tag{3.76}$$

where

$$r_1^2 = y^2 + (z-H)^2 \tag{3.77}$$

$$r_2^2 = y^2 + (z+H)^2 \tag{3.78}$$

$$r_0^2 = y^2 + z^2 \tag{3.79}$$

Since H_ϕ is the total magnetic field all the components of electric field can be derived from it. Thus in cylindrical coordinates, replacing y by ρ,

$$E_\rho = -\frac{1}{j\omega\varepsilon}\frac{\partial H_\phi}{\partial z}$$

$$= j\frac{Z_0}{4\pi}\frac{I_m}{\rho}\left[(z-H)\frac{e^{-jkr_1}}{r_1} + (z+H)\frac{e^{-jkr_2}}{r_2} - 2z\cos kH\frac{e^{-jkr_0}}{r_0}\right] \tag{3.80}$$

and

$$E_z = \frac{1}{j\omega\varepsilon}\frac{1}{\rho}\frac{\partial}{\partial\rho}(\rho H_\phi)$$

$$= -j\frac{Z_0}{4\pi}I_m\left[\frac{e^{-jkr_1}}{r_1} + \frac{e^{-jkr_2}}{r_2} - 2\cos kH\frac{e^{-jkr_0}}{r_0}\right] \tag{3.81}$$

In general the E_ρ-component at a conducting surface is much more than the parallel component of field, except at the origin where the radial component is zero, due to the charge density on the surface changing polarity as the observation point goes through the origin.

Since these fields apply at all positions, the far field results previously derived can be deduced from them. Thus in the far field

$$r_1 \approx r_0 - H\cos\theta$$

$$r_2 \approx r_0 + H\cos\theta$$

so that the magnetic field component becomes

$$H_\phi \approx \frac{jI_m e^{-jkr_0}}{4\pi r_0 \sin\theta}[e^{jkH\cos\theta} + e^{-jkH\cos\theta} - 2\cos kH]$$

$$= \frac{jI_m}{2\pi}\frac{e^{-jkr_0}}{r_0}\left[\frac{\cos(kH\cos\theta) - \cos kH}{\sin\theta}\right] \tag{3.82}$$

in agreement with eqn (3.69).

3.10 Transmitting and receiving power gains of dipole and monopole antennas

In many transmitting situations it is required that the field strength of the radiated wave in a particular direction should be much larger than the average field strength over all directions. In terms of power density the maximum power density has to be much larger than the average, and the ratio of these two quantities is the transmitting power gain of the antenna. That is,

$$G_{TX} = \frac{\text{maximum radiation intensity}}{\text{average radiation intensity over spherical surface}} = \frac{\Phi_{max}}{\Phi_{av}} \tag{3.83}$$

The units of the radiation intensity may be watts per unit solid angle, or watts per square metre.

From eqn (3.83) an alternative definition of the transmitting power gain of a given antenna is that it represents the number of times the input power to a matched isotropic radiator would have to be increased in order to produce the same field strength, or radiation intensity, at a fixed observation point, as the given antenna. When lossless antennas are being considered an antenna with a narrow beam will have a larger power gain than an antenna with a broad beam. In the lower limit when the broad beam becomes isotropic, the power gain falls to unity.

Correspondingly, in a receiving situation, since it has been shown that the transmitting and receiving radiation patterns of a dipole antenna are identical, it might be expected than an antenna with a large transmitting power gain would deliver more signal power to a receiver than an antenna with a smaller transmitting power gain. This is in fact the case, and by suitable definition the receiving power gain of any passive antenna can be defined to be equal to its transmitting power gain. Thus we define

$$G_{RX} = \frac{\text{power delivered to the matched load of the test antenna}}{\text{power delivered to the matched load of an isotropic antenna}} \tag{3.84}$$

Although an isotropic antenna does not exist in practice, it must, from eqn (3.84), have a receiving power gain of unity. It will be shown in Chapter 11 that it is convenient to define an effective area A_e for any antenna through the equation

$$G = \frac{4\pi A_e}{\lambda^2} \tag{3.85}$$

where G is its transmitting or receiving power gain. Hence from eqn (3.85) for a power gain of unity the effective receiving area of an isotropic antenna is

$$A_e = \frac{\lambda^2}{4\pi} \tag{3.86}$$

This gives for the power delivered to the matched load of an isotropic antenna

$$\begin{aligned} P_{iso} &= \text{incident power density} \cdot A_e \\ &= \frac{E_i^2}{2Z_0}\frac{\lambda^2}{4\pi} \end{aligned} \tag{3.87}$$

Hence the receiving power gain for any antenna is

$$G_{RX}=\frac{V_{oc}^2}{8R_r}\cdot\frac{8\pi Z_0}{E_i^2\lambda^2} \tag{3.88}$$

3.10.1 *Transmitting power gain of half-wave dipole in free space and quarter-wave monopole above ground*

From eqn (3.69) the maximum magnetic field strength radiated by a half-wave dipole is given by

$$|H_{\phi max}|_d=\frac{I_{md}}{2\pi r} \tag{3.89}$$

where r is the radial distance in the equatorial plane to the point of observation, and I_{md} is the maximum current on the dipole. The maximum radiation intensity then becomes

$$\Phi_{max\cdot d}=\frac{I_{md}^2}{4\pi^2 r^2}\frac{Z_0}{2}$$

and the average radiation intensity is

$$\Phi_{av\cdot d}=\frac{P_{rad}}{4\pi r^2} \tag{3.90}$$

where P_{rad} is the total power radiated by the dipole. This power expressed in terms of the dipole radiation resistance R_{rd} is given by

$$P_{rad}=\tfrac{1}{2}I_{md}^2 R_{rd}$$

so that the transmitting power gain, from eqn (3.83), is

$$G_{TX\cdot d}=\frac{Z_0}{\pi R_{rd}}=1.64 \tag{3.91}$$

For the case of a quarter-wave monopole which is considered to be radiating the same total power P_{rad} as the above dipole, the maximum current in the monopole I_{mm} will be $\sqrt{2}$ times the maximum current on the dipole I_{md}. This is because for equal radiated powers

$$\tfrac{1}{2}I_{mm}^2 R_{rm}=\tfrac{1}{2}I_{md}^2 R_{rd}$$

where R_{rm} is the radiation resistance of the monopole, which is one-half that of the corresponding dipole. Consequently the maximum magnetic field strength will be

$$|H_{\phi max}|_m=\frac{I_{md}}{\sqrt{2}\pi r} \tag{3.92}$$

and the maximum radiation intensity is

$$\Phi_{max\cdot m}=\frac{I_{md}^2}{2\pi^2 r^2}\frac{Z_0}{2}=2\Phi_{max\cdot d}$$

This same power P_{rad} when distributed uniformly over a spherical surface of the same radius r would produce an average radiation intensity

$$\Phi_{av} = \frac{1}{2}\frac{I_{md}^2}{4\pi r^2} R_{rd}$$

Hence the power gain of the monopole according to the definition given in eqn (3.83) is

$$G_{TXm} = 3.28 \qquad (3.93)$$

It would, however, be possible to define the average radiation intensity for a monopole in terms of the total radiated power divided by the area of the hemisphere over which the radiation occurs. Using this definition would make the power gain of the monopole equal to that of the dipole, but the definition given by eqn (3.83) has now been established by custom.

3.10.2 *Receiving power gains of half-wave dipole in free space and quarter-wave monopole above ground*

It has been shown in eqn (3.48) that the open-circuit voltage obtainable from a half-wave dipole placed in a uniform cophasal field is given by

$$V_{oc \cdot d} = -\frac{\lambda}{\pi} E^i$$

Hence the receiving power gain from eqn (3.88) becomes

$$G_{RX \cdot d} = \frac{Z_0}{\pi R_{rd}} = 1.64 \qquad (3.94)$$

which establishes equality with the transmitting power gain given in eqn (3.91).

In the case of the quarter-wave monopole above an infinite ground plane the open-circuit voltage will be

$$V_{oc \cdot m} = -\frac{\lambda}{2\pi} E^i \qquad (3.95)$$

so that the signal power delivered to a matched load is

$$P_s = \frac{V_{oc \cdot m}^2}{8R_{rm}} \qquad (3.96)$$

where R_{rm} is the radiation resistance of the monopole. This power is then one-half of the signal power delivered to the corresponding dipole.

To show the equality of the power gain with that of the transmitting case, account has to be taken of the fact that the effective receiving area of an isotropic antenna placed at the location of the monopole is one-quarter that of an isotropic antenna in free space. This is because the incoming power exists over one hemisphere only and the matched load will be one-half that of the isotropic antenna in free space. Consequently,

$$A_{e\cdot m} = \frac{\lambda^2}{16\pi} \tag{3.97}$$

and

$$G_{RX\cdot m} = \frac{V_{o\cdot c\cdot m}^2}{8R_{r\cdot m}} \frac{2Z_0 16\pi}{E^{i2}\lambda^2} = \frac{Z_0}{\pi R_{r\cdot m}} = 3.28 \tag{3.98}$$

In practice it is more important to note that the received power is one-half that of the corresponding dipole, than that the power gain is twice as great as that of the dipole.

3.11 Filamentary half-wave dipole in corner reflector

In situations where the power gain of an isolated $\lambda/2$ dipole is insufficient, the forward gain may be increased by placing the dipole in front of a reflecting sheet at a distance of $\lambda/4$, $3\lambda/4$, etc., from it. In such cases the forward field is doubled for the case of an infinite reflector, and the increase in power gain is significant even when the input resistance may be adversely affected by the proximity of the sheet. When the effect on input resistance is neglected the improvement in power gain is 6dB.

This power gain improvement may be increased still further by bending the reflecting sheet forward, and the particular case of such a 90° corner reflector will now be considered. Again if the reflecting sheets are infinite in extent, the improvement in forward field strength can be calculated from image theory, as shown in Fig. 3.8, to be a factor of four times when the driven dipole is placed at a distance of 0.5λ, 1.5λ, etc., from the apex.

Fig. 3.8. 90° corner reflector with $\lambda/2$ dipole feed and three images.

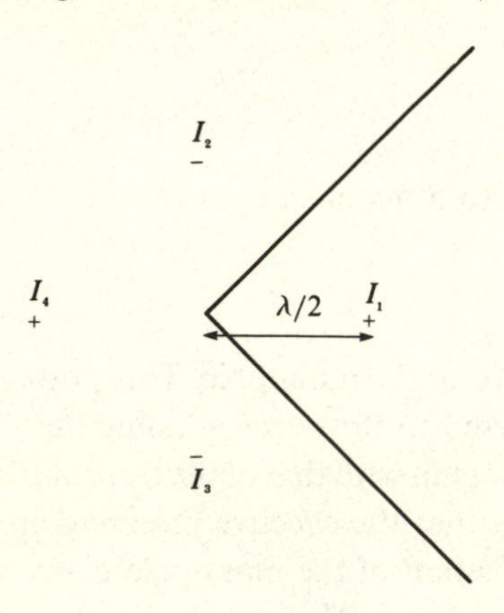

The input impedance of the driven dipole may be found from

$$V_1 = I_1 Z_{11} + I_2 Z_{12} + I_3 Z_{13} + I_4 Z_{14} \tag{3.99}$$

But

$$I_1 = -I_2 = -I_3 = I_4$$

so that

$$\frac{V_1}{I_1} = Z_{11} - Z_{12} - Z_{13} + Z_{14} \tag{3.100}$$

Since the input reactance can be tuned out at the terminals of the driven antenna, attention will be restricted to the input or radiation resistance component, given by

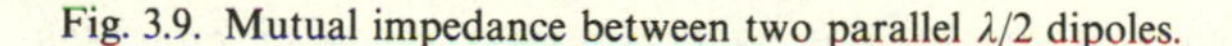

$$R_i = R_{11} - R_{12} - R_{13} + R_{14} \tag{3.101}$$

In this equation R_{11} is the input resistance of the driven dipole in isolation, R_{12} and R_{13} are the mutual resistances between $\lambda/2$ dipoles spaced 0.707 λ apart, and R_{14} the mutual resistance between similar dipoles spaced 1.0λ apart. Thus, from Fig. 3.9,

$$\begin{aligned} R_i &\approx 73 + 24 + 24 + 3.8 \\ &= 124.8\Omega \end{aligned} \tag{3.102}$$

The power gain of the antenna is given by

$$G = \frac{\text{maximum radiation intensity}}{\text{average radiation intensity}} = \frac{\Phi_{max}}{\Phi_{av}}$$

Fig. 3.9. Mutual impedance between two parallel $\lambda/2$ dipoles.

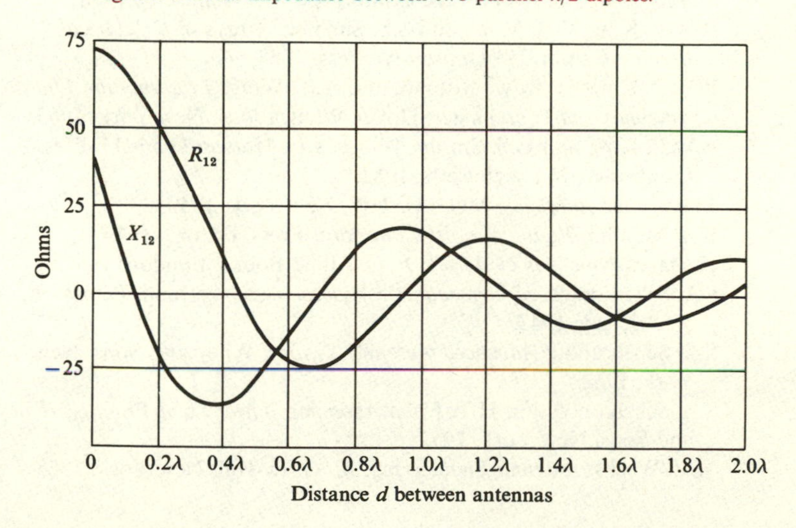

The maximum radiation intensity is

$$\Phi_{max} = \frac{|H_{\phi_{max}}|^2 Z_0}{2} \tag{3.103}$$

where $H_{\phi_{max}}$ is four times that of an isolated $\lambda/2$ dipole. Likewise the average intensity is

$$\Phi_{av} = \frac{1}{2} \frac{I^2 R_i}{4\pi r^2}$$

Hence the power gain becomes

$$G = \frac{1920}{124.8} = 15.4 \text{ or } 11.9\text{dB} \tag{3.104}$$

Had the fed dipole been placed at 1.5λ from the apex, instead of 0.5λ, the field reinforcement would have remained unchanged, but the input resistance would have been such as to increase the power gain to approximately 15dB.

Further reading

C.A. Balanis: *Antenna Theory: Analysis and Design*: Harper and Row, New York, 1982.

J. Galejs: *Antennas in Inhomogeneous Media*: Pergamon Press, Oxford, 1969.

R.F. Harrington: *Field Computation by Moment Methods*: Macmillan, New York, 1968.

R.W.P. King: *The Theory of Linear Antennas*: Harvard University Press, Cambridge, Massachusetts, 1956.

R.W.P. King and C.W. Harrison: *Antennas and Waves; A Modern Approach*: The MIT Press, Cambridge, Massachusetts, 1968.

R.W.P. King, R.B. Mack and S.S. Sandler: *Arrays of Cylindrical Dipoles*: Cambridge University Press, 1968.

R.W.P. King, H. Rowe Mimno, and A.H. Wing: *Transmission Lines, Antennas and Waveguides*: Dover Publications, New York, 1965.

R.W.P. King and G.S. Smith: *Antennas In Matter*: The MIT Press, Cambridge, Massachusetts, 1981.

J.D. Kraus: *Antennas*: McGraw-Hill, New York, 1950.

E.B. Moullin: *Radio Aerials*: Clarendon Press, Oxford, 1949.

H. Page: *Principles of Aerial Design:* Iliffe Books, London, 1966.

S.A. Schelkunoff: *Electromagnetic Waves*: Van Nostrand, Princeton, New Jersey, 1943.

S.A. Schelkunoff: *Advanced Antenna Theory*: Wiley and Sons, New York, 1952.

S.A. Schelkunoff and H.T. Friis: *Antennas, Theory and Practice*: Wiley and Sons, New York, 1952.

W.L. Weeks: *Antenna Engineering*: McGraw-Hill, New York, 1968.

4

+−+

Computer solutions of dipole and monopole antennas

4.1 Computer solutions of linear dipoles

The solutions derived for the filamentary dipoles in the preceding chapter have provided a basic knowledge of the results to be expected from such antennas within the conventional range of dipole lengths up to about one wavelength. Approximate distributions of current for both transmitting and receiving conditions, together with satisfactory radiation patterns and values of power gain for most applications, have enabled such antennas to be designed to a large extent theoretically. Nevertheless for particular values of dipole length when, for example, the antenna is nearly antiresonant, and a number of such dipoles may be connected in parallel to provide a good power match, it is worthwhile to obtain more accurate solutions. For this and other purposes computer solutions are highly cost effective, in comparison with more refined analytical techniques.

The increased accuracy of the computer solutions is, however, not the only advantage that accrues from this type of solution. For some purposes the breadth of information which is available once the dipole has been characterised for the computer is even greater. Thus once the data representing the dipole is available, it becomes a trivial change to alter the operation from transmitting to receiving, and because of the computer's high speed calculating ability, information about current distribution, impedance and radiation pattern for any position of the feed terminals is rapidly available. In this context it should be noted that the filamentary solutions require to be re-solved for every change in position of the feed terminals.

The characterisation of a dipole antenna for computer purposes is conveniently made on the basis of its impedance matrix. Thus the dipole is conceptually divided into a large number of equal length segments as

Fig. 4.1. Linear dipole divided into N equal segments.

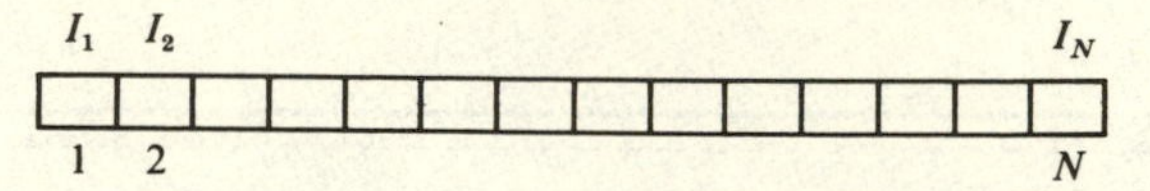

shown in Fig. 4.1. Each segment possesses both a self impedance and a mutual impedance with every other segment. If it is then assumed that a voltage source is applied to each segment, and that each segment carries a constant current over its elemental length, the electromagnetic problem of the dipole is simplified to that of a multiport network. For such a network it is possible to write down a set of simultaneous equations of the form

$$\left.\begin{aligned} V_1 &= I_1 Z_{11} + I_2 Z_{12} + \cdots + I_N Z_{1N} \\ V_2 &= I_1 Z_{21} + I_2 Z_{22} + \cdots + I_N Z_{2N} \\ &\vdots \\ V_N &= I_1 Z_{N1} + I_2 Z_{N2} + \cdots + I_N Z_{NN} \end{aligned}\right\} \tag{4.1}$$

This may be abbreviated to its matrix form

$$[V] = [Z][I] \tag{4.2}$$

where $[V]$, $[I]$ are column matrices of the applied voltage and the current in each segment, and $[Z]$ is a square matrix containing the self impedance of each segment and the mutual impedances between segments.

From either form of the above equations, the unknown currents $I_1, \ldots, I_N$ may be found, provided the impedance matrix and the applied voltage to each segment have been specified, so that the solutions can be written

$$\left.\begin{aligned} I_1 &= V_1 Y_{11} + V_2 Y_{12} + \cdots + V_N Y_{1N} \\ I_2 &= V_1 Y_{21} + V_2 Y_{22} + \cdots + V_N Y_{2N} \\ &\vdots \\ I_N &= V_1 Y_{N1} + V_2 Y_{N2} + \cdots + V_N Y_{NN} \end{aligned}\right\} \tag{4.3}$$

or, in abbreviated matrix form,

$$[I] = [Y][V] \tag{4.4}$$

The solution of the equations for current, or at least the inversion of the impedance matrix $[Z]$ to the admittance matrix $[Y]$, is carried out by the computer. When this inversion has been completed eqn (4.3) shows that the current for either transmitting or receiving operation can readily be found, as outlined in the following sections.

4.2 Transmitting operation

When the antenna is used for transmitting there will normally be a single applied voltage connected across its centre terminals r, as shown in

Fig. 4.2, so that the voltage column matrix becomes

$$[V] = \begin{bmatrix} 0 \\ 0 \\ \cdot \\ V_r \\ \cdot \\ 0 \\ 0 \end{bmatrix} \tag{4.5}$$

Fig. 4.2. Transmitting dipole with voltage V_r across its centre terminals.

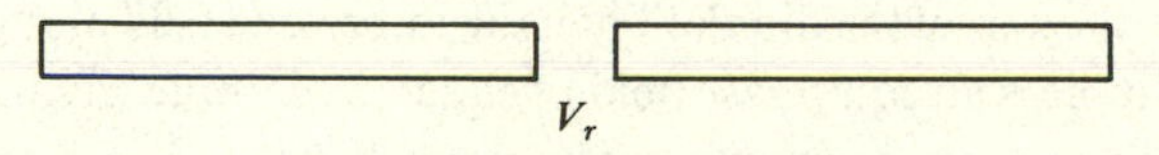

Hence the currents in each segment become

$$\left.\begin{aligned} I_1 &= V_r Y_{1r} \\ &\vdots \\ I_r &= V_r Y_{rr} \\ &\vdots \\ I_N &= V_r Y_{Nr} \end{aligned}\right\} \tag{4.6}$$

It will be noted that because Y is a complex admittance, both the in-phase and quadrature components of the currents are given, and the input admittance of the antenna is simply the Y_{rr} component of the admittance matrix. Not only so, but, for any feeding position j of this antenna, the input admittance is simply Y_{jj}. Hence the elements of the main diagonal of the admittance matrix are the input admittance of the antenna when fed at each segment in turn.

Provided N is sufficiently large the numerical solutions for the currents given by eqn (4.6) will allow a graph of the transmitting current along the antenna to be drawn when it is fed at its centre element r. Likewise when the antenna is fed at any other position j, the current in element one will be $V_j Y_{1j}$, etc., so that column j of the admittance matrix, with each term multiplied by the applied voltage V_j, enables the transmitting current along the antenna to be found when it is fed at element j. Consequently each column of the admittance matrix has the physical significance that it represents the transmitting current distribution along the antenna when it is fed with unit voltage at the element corresponding to that column number. Each term of the admittance matrix has, therefore, at least one physical meaning, with the main diagonal elements having two.

It would be straightforward, using this technique, to analyse antennas with more than one feeding point, and such antennas have been used.

Clearly also, increased accuracy can be obtained by increasing the number of segments on the dipole, but it is found in practice that ten segments per wavelength is sufficient for most purposes.

4.3 Receiving operation

Under normal receiving conditions a load impedance Z_L is connected to the terminals of the antenna, and the output voltage or load current is required when the antenna is placed in an electric field which has a component along the axis of the dipole. The simplest case to consider is when the direction of propagation of the incoming radio wave is perpendicular to the axis of the dipole. Then the impressed field along every segment of the dipole is an equiphase field of equal magnitude, giving equal voltage excitation of every segment, as shown in Fig. 4.3.

Fig. 4.3. Receiving dipole with equal voltages across each segment.

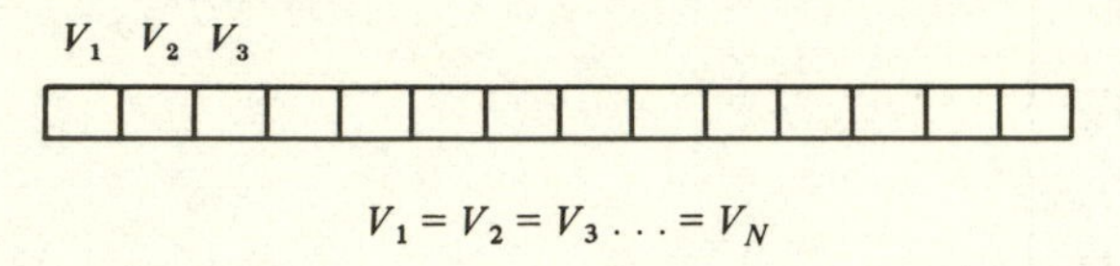

It would be possible to solve this problem by taking the impedance matrix for the dipole, adding the load impedance Z_L to the self impedance of the element at which it is inserted, and then inverting this matrix to obtain a new admittance matrix for the problem. If the voltage excitation of each segment is then El, where E is the incoming field strength and l is the length of each segment, and the load impedance is inserted at segment r, then the current through the load would be

$$I_L = I_r = El[Y^1_{r1} + Y^1_{r2} + \cdots + Y^1_{rN}] \tag{4.7}$$

where $Y^1_{r1}, \ldots, Y^1_{rN}$ are components of the inverted impedance matrix which includes Z_L.

It is, however, possible to solve this problem without going through a matrix inversion process if the original admittance matrix, corresponding to zero impedance load Z_L, has already been obtained. In such a case the current through the short-circuited element r will be

$$I_{sc_r} = I_r = El[Y_{r1} + Y_{r2} + \cdots + Y_{rN}] \tag{4.8}$$

But this short-circuited current is related to the input admittance of the antenna at element r through the equation

$$Y_{rr} = -\frac{I_{sc_r}}{V_{oc_r}} \tag{4.9}$$

where V_{oc_r} is the open-circuited voltage of the antenna when it is placed in the same field that produces I_{sc_r}.

Hence this open-circuited voltage is given by

$$V_{oc_r} = -\frac{El}{Y_{rr}}[Y_{r1} + Y_{r2} + \cdots + Y_{rN}] \tag{4.10}$$

Then by the application of Thévenin's theorem the output voltage for a load impedance Z_L inserted at element r is

$$V_L = \frac{V_{oc_r}}{\left(Z_L + \dfrac{1}{Y_{rr}}\right)} Z_L \tag{4.11}$$

In the more general receiving situation when the direction of an incoming uniform plane wave makes an angle θ with the axis of the dipole, the applied voltage to segment r is, from Fig. 4.4,

$$El \sin\theta e^{jkz_r\cos\theta} \tag{4.12}$$

Hence from eqn (4.10) the open-circuit voltage is given by

$$V_{oc_r} = -\frac{El\sin\theta}{Y_{rr}}[Y_{r1}e^{jkz_1\cos\theta} + Y_{r2}e^{jkz_2\cos\theta} + \cdots + Y_{rN}e^{jkzN\cos\theta}]$$

and the output voltage follows from Thévenin's theorem in the usual way.

4.4 Monopole antenna

The use of monopole antennas with finite sized ground places of circular or polygonal shape is common practice when an antenna is to be

Fig. 4.4. Receiving dipole with incoming wave at angle θ to dipole axis.

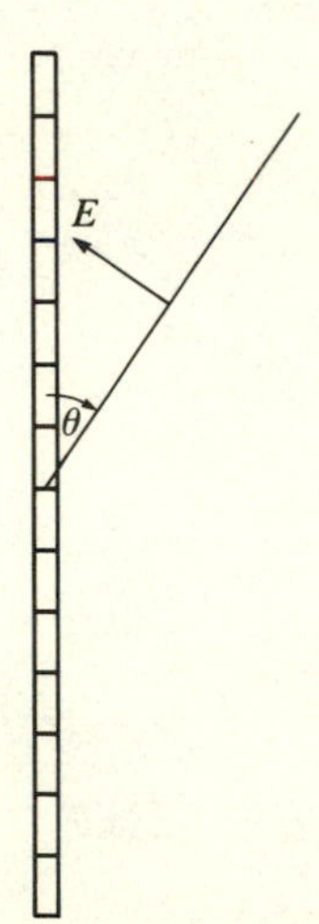

used close to the ground or to a conducting surface. The effect of the different configuration affects both the input impedance and the radiation pattern and is a function of both the shape and size of the ground plane. The problem will be considered in two stages – first the effect of an infinite ground plane will be discussed, and this will be followed by an approximate analysis of the effect of the boundary edge.

4.4.1 *Monopole on infinite ground plane*

Let the monopole in Fig. 4.5 be divided into N elements. Each element has associated with it a constant current I which produces an in-phase image of equal magnitude. Thus the number of unknown currents remains at N, and the equations to be solved are

$$\left.\begin{aligned} V_1 &= I_1(Z_{11}+Z_{11'})+I_2(Z_{12}+Z_{12'})+\cdots+I_N(Z_{1N}+Z_{1N'}) \\ V_2 &= I_1(Z_{21}+Z_{21'})+I_2(Z_{22}+Z_{22'})+\cdots+I_N(Z_{2N}+Z_{2N'}) \\ &\vdots \\ V_N &+ I_1(Z_{N1}+Z_{N1'})+I_2(Z_{N2}+Z_{N2'})+\cdots+I_N(Z_{NN}+Z_{NN'}) \end{aligned}\right\} \tag{4.13}$$

where each of the impedances denoted by a prime is associated with an image current. This system of equations can be expressed in matrix form as

$$[V]=[Z'][I] \tag{4.14}$$

where $[Z']$ is an impedance matrix with elements $(Z_{ij}+Z_{ij'})$. Inverting the square matrix $[Z']$ gives, in the usual way,

$$[I]=[Y'][V] \tag{4.15}$$

Fig. 4.5. Monopole with N elements above ground plane.

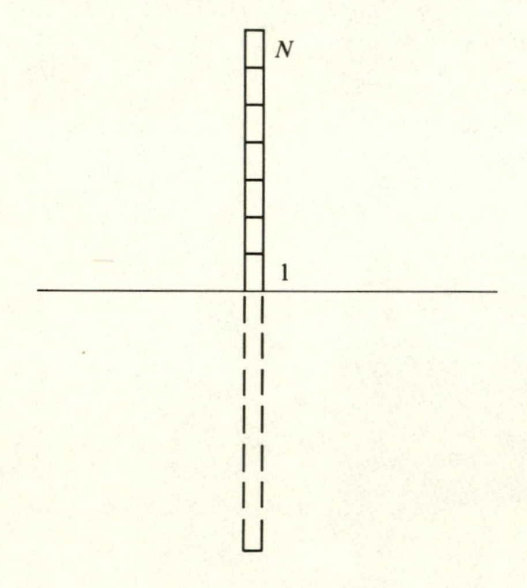

and hence the current distribution along the monopole, and its input admittance can be found. The current distribution will be the same as that along the top half of the corresponding dipole of twice the length, and the input impedance one-half of that of the same dipole.

With regard to the radiation pattern this will be the top half of the corresponding pattern for the full dipole, as can be understood from image theory. However, looking at it from the point of view of radiation from physical currents, it is clear that the pattern is the superposition of the radiation from each element on the monopole and on the infinite sized ground plane. The type of integration required for these latter currents is discussed in the following section, but the evaluation of the integral for the infinite ground plane case has not been accomplished analytically.

4.4.2 *Monopole on finite circular ground plane – input impedance*

The effect of truncating the infinite ground plane is more marked on the radiation pattern of the antenna than it is on the input impedance. The general effect on the pattern is that the direction of maximum radiation is tilted up from the direction of the ground plane where the new field value is approximately one-half of the maximum value, and a number of auxiliary lobes are introduced both above and below it. The larger the ground plane the greater the number of these lobes.

For the infinite ground plane case the matrix equation was written as

$$[V]=[Z'][I]$$

where $[Z']$ was composed of elements Z_{ij} defined as

$$Z_{ij}=-\frac{(E_{ij}+E_{i'j})}{I(i)}\,dz \tag{4.16}$$

In eqn (4.16) E_{ij} is the electric field tangential to the surface of the jth element due to the current $I(i)$ in the ith element, and $E_{i'j}$ is the corresponding electric field due to the image of $I(i)$ in an infinitely large ground plane.

When the ground plane is finite the component of electric field tangential to the jth element must have two additional components superimposed on it: firstly, a component $E_{a\infty}$ due to the ground plane current from its edge at radius a to infinity, since this component no longer flows when the ground plane is truncated; and secondly a component E_{0a}, associated with the reflected current on the top surface of the ground plane, together with the corresponding transmitted current flowing on its underside. Thus the modified form of eqn (4.16) becomes

$$Z_{ij}=-\frac{(E_{ij}+E_{i'j}-E_{a\infty}+E_{0a})}{I(i)}\,dz \tag{4.17}$$

To find $E_{a\infty}$ consider first the surface current density J on an infinite ground plane due to the elemental length of current $I(i)\,dz$ at height z_i on the monopole, as shown in Fig. 4.6. This density at radius ρ on the ground plane is given by

$$\begin{aligned}\mathbf{J}(\rho) &= 2\mathbf{n} \times \mathbf{H}(\rho)\\ &= -\frac{\rho I(i)dz}{2\pi r_i^2}\left(jk + \frac{1}{r_i}\right)e^{-jkr_i}\hat{\boldsymbol{\rho}}\end{aligned} \tag{4.18}$$

where $\hat{\boldsymbol{\rho}}$ is a unit vector on the ground plane in the radial direction, and

$$r_i = (\rho^2 + z_i^2)^{\frac{1}{2}} \tag{4.19}$$

It is required to find the surface electric field E_z at the element j of the monopole which has to be subtracted in eqn (4.17) because of the absence of ground plane current from radius a to infinity.

To do this consider an outward flowing element of current on the ground plane $J\rho'd\rho'd\phi'$ which sets up a vector potential $\mathbf{dA}$ at any point (x, y, z) in space given by

$$\mathbf{dA} = \frac{\mu J\rho' d\rho' d\phi'}{4\pi r'}e^{-jkr'}\hat{\boldsymbol{\rho}} \tag{4.20}$$

where

$$r' = [(x - x')^2 + (y - y')^2 + z^2]^{\frac{1}{2}}$$

The magnetic field components are related to $\mathbf{dA}$ through the equation

$$\mathbf{dH} = \frac{1}{\mu}\nabla \times \mathbf{dA}$$

giving

$$dH_{\substack{x\\y}} = \pm\frac{J\rho' d\rho' d\phi'}{4\pi}\frac{z}{r'}\left(\frac{1}{r'^2} + \frac{jk}{r'}\right)\begin{matrix}\sin\phi'\\ \cos\phi'\end{matrix}\,e^{-jkr'} \tag{4.21}$$

Fig. 4.6. Elemental length of current $I(i)dz$ at height Z_i on monopole.

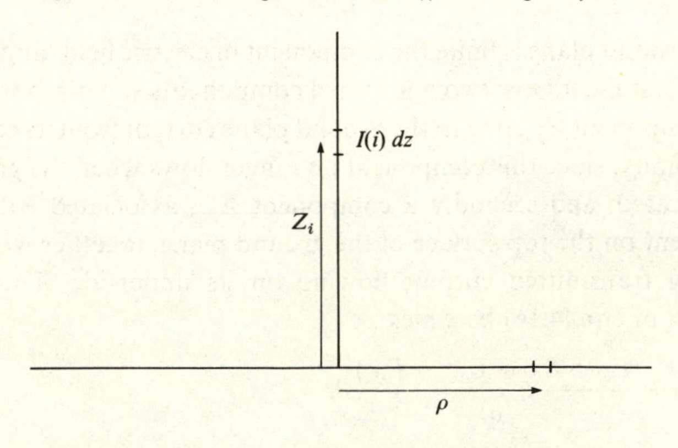

where ϕ' is the angular location of the current element measured from the x-axis.

The electric field E_z can now be found anywhere in space, using Maxwell's equations, to give

$$dE_z = \frac{j}{\omega\varepsilon}\frac{J\rho' d\rho' d\phi'}{4\pi}\left(-j\frac{3k}{r'} - \frac{3}{r'^2} + k^2\right)z$$
$$\times [(x - x')\cos\phi' + (y - y')\sin\phi']\frac{e^{-jkr'}}{r'^3} \tag{4.22}$$

In particular, on the monopole axis where $x = y = 0$, this gives

$$dE_z = -\frac{JZ_0\rho'^2 d\rho' d\phi'}{4\pi}z\left(\frac{3}{r'} - \frac{j3}{kr'^2} + jk\right)\frac{e^{-jkr'}}{r'^3} \tag{4.23}$$

Integrating with respect to ϕ' gives the axial electric field E_z due to a ring of current at radius ρ' as

$$E_z = -\frac{JZ_0\rho'^2 z}{2}\left(\frac{3}{r'} - \frac{j3}{kr'^2} + jk\right)\frac{e^{-jkr'}}{r'^3}d\rho' \tag{4.24}$$

Hence the total axial effect of the ground plane from radius a to infinity at element z_j is

$$E_z = \int_a^\infty \frac{J(\rho)Z_0\rho'^2 z_j}{2}\left[\frac{3}{r_j} + j\left(k - \frac{3}{kr_j^2}\right)\right]\frac{e^{-jkr_j}}{r_j^3}d\rho' \tag{4.25}$$

where

$$r_j = (\rho'^2 + r_j^2)^{\frac{1}{2}}$$

The upper limit of integration in eqn (4.25) can in practice be truncated at six wavelengths for ground planes of up to three wavelengths radius.

The second component of axial electric field to be superimposed, E_{0a}, arises because of the reflected current on the top of the ground plane, and the transmitted current underneath, both flowing in the direction opposite to the main ground plane current. The magnitudes of these two additional currents depend on both the reflection coefficient at the edge and the way that these currents fall to zero at the centre of the ground plane. This boundary condition is necessary because on the top surface the current on the monopole has initially been postulated, and hence the current at zero radius on the ground plane cannot be affected by reflection. On the underneath side the current flows radially to this centre point, where it must therefore have zero amplitude.

As far as the reflection coefficient is concerned, an exact theoretical solution has been worked out for particular cases, and has given an answer of 0.5. This is identical with the reflection coefficient for the current on a half-plane when it is irradiated at any angle of incidence by a uniform plane wave with its magnetic field parallel to the edge. For the circular ground this condition of parallel incidence is always satisfied. Hence it appears reasonable to adopt this figure of 0.5 as a universal reflection coefficient for all sizes of ground plane.

With regard to the radial variation in amplitude and phase of the reflected and transmitted currents at the edge, clearly a variety of propagation constants could be assumed which would tend to zero amplitude at the centre of the ground plane. For simplicity, however, a linear law will be assumed so that both reflected and transmitted current amplitude will be written

$$J_{r/t} = -\frac{\rho}{2a} J_{\infty}(a) e^{-jk(a-\rho)} \tag{4.26}$$

Substituting from eqn (4.18) gives

$$J_{r/t} = \frac{\rho I(i)}{4\pi(a^2 + z_i^2)} \left(jk + \frac{1}{(a^2 + z_i^2)^{\frac{1}{2}}} \right) e^{-jk(a^2 + z_i^2)^{\frac{1}{2}}} e^{-jk(a-\rho)} dz \tag{4.27}$$

Both currents travel in the same physical direction, and although they travel on opposite sides of the conducting ground plane this is no different from the case of currents flowing on the top and bottom surfaces of a sphere or spheroid. The radiation from currents over the whole surface radiates into free space so that the two current densities may be added to give the additional component of electric field tangential to the surface of element z_j on the monopole, due to the current element $I(i)$ at position z_i, as

$$E_{0a} = -\int_0^a \frac{[J_r(\rho) + J_t(\rho)] Z_0 \rho'^2 z_j}{2} \left[\frac{3}{r_j} + j\left(k - \frac{3}{k r_j^2} \right) \right] \frac{e^{-jkr_j}}{r_j^3} d\rho' \tag{4.28}$$

Hence from eqn (4.17) all the components of [Z] can be calculated, and the matrix may be inverted to give the admittance matrix in the usual way. As an example, the input impedance of a 0.224λ monopole of radius 0.003λ has been calculated as a function of ground plane radius, and is shown in Fig. 4.7. The results are compared with published experimental values and agree well in shape and reasonably well in magnitude. It will be noted that, as the ground plane radius increases calculated and measured values tend to a constant value which is one-half that of the corresponding dipole of appropriate length and diameter. The discrepancy that exists between theory and experimental results is certainly partly due to the very simple law which has been assumed for the propagation constants for the reflected and transmitted currents.

4.4.3 *Monopole on finite circular ground plane – radiation pattern*

The radiation pattern of the antenna is formed from the sum of the pattern of the monopole and the ground plane. Since the current distribution along the monopole has been found in the process of finding the input impedance of the antenna, this enables the current density on both sides of the ground plane to be determined from eqns (4.18) and (4.27)

Fig. 4.7. Input impedance of 0.224λ monopole as function of ground plane radius. (After Awadalla and Maclean, 1979.)

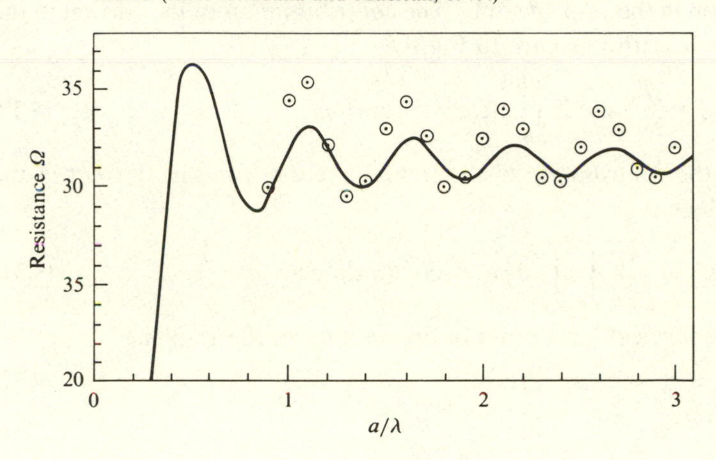

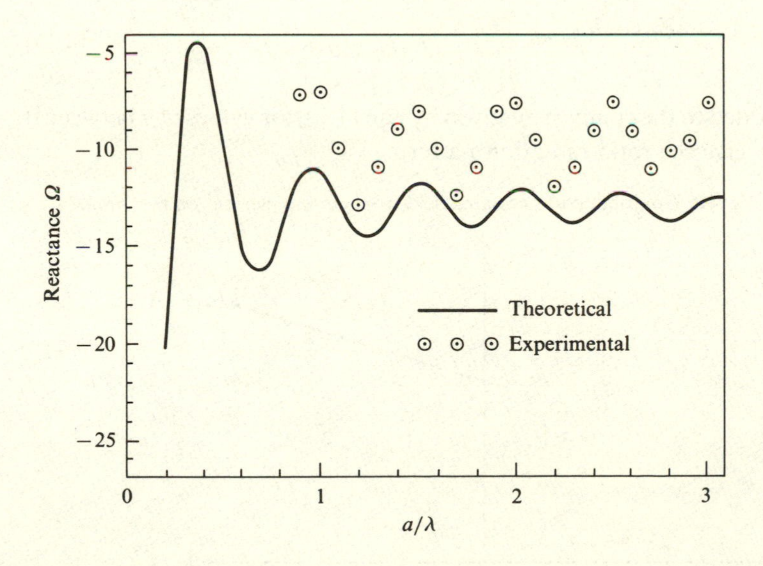

Combining these into a single sheet of current gives

$$J(\rho') = -\frac{1}{2\pi}\int_0^H \rho' I(z)\left\{\left(jk + \frac{1}{(\rho'^2 + z_i^2)^{\frac{1}{2}}}\right)\frac{e^{-jk(\rho'^2 + z_i^2)^{\frac{1}{2}}}}{(\rho'^2 + z_i^2)}\right.$$
$$\left. - \left(jk + \frac{1}{(a^2 + z_i^2)^{\frac{1}{2}}}\right)\frac{e^{-jk(a^2 + z_i^2)^{\frac{1}{2}}}}{(a^2 + z_i^2)} e^{-jk(a-\rho')}\right\} dz_i \tag{4.29}$$

since the currents on the two sides due to the reflection and transmitted components can be added directly.

To find the radiation pattern it is sufficient to consider a point of observation in the yz-plane only. The contribution from the current in the monopole is, with reference to Fig. 4.8,

$$E_{\theta 1} = \frac{jZ_0}{2\lambda r_0} e^{-jkr_0} \int_0^H I(z) e^{jkz\cos\theta} \sin\theta \, dz \tag{4.30}$$

Likewise the contribution in the same direction from the current in the ground plane is

$$E_{\theta 2} = \frac{jZ_0}{2\lambda r_0} \int_{\rho'} \int_{\phi'} J_{y'} e^{-jkr} \cos\theta \rho' \, d\rho' \, d\phi' \tag{4.31}$$

But in the far field for a point of observation in the yz-plane:

$$r \approx r_0 - \rho' \sin\theta \sin\phi' \tag{4.32}$$

Hence

$$E_{\theta 2} \approx \frac{jZ_0 \cos\theta}{2\lambda r_0} e^{-jkr_0} \int_{0 \, \rho'}^{a} \int_{0 \, \phi'}^{2\pi} J(\rho') \sin\phi' e^{jk\rho' \sin\theta \sin\phi'} \rho' \, d\rho' \, d\phi'$$
$$= \frac{kZ_0 \cos\theta}{2r_0} e^{-jkr_0} \int_0^a J(\rho') J_1(k\rho' \sin\theta) \rho' \, d\rho' \tag{4.33}$$

This is added to the component given by eqn (4.30) for values of θ between 0 and π to give the total radiation pattern.

Fig. 4.8. Contribution to radiated field from current element on monopole.

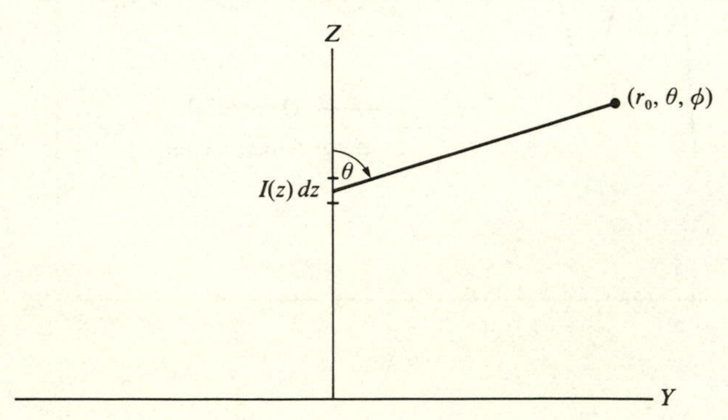

This radiation pattern for a monopole of length 0.224λ and 0.003λ radius located at the centre of a circular ground plane of 6λ diameter has been calculated and is compared with published experimental data in Fig. 4.9. Similarly the corresponding pattern for a $\lambda/4$ monopole of radius $\frac{1}{32}''$ placed on a circular ground plane of 1.2λ diameter is shown in Fig. 4.10. In both

Fig. 4.9. Radiation pattern of 0.224λ monopole on circular ground plane of diameter 6λ. (After Awadalla and Maclean, 1979.)

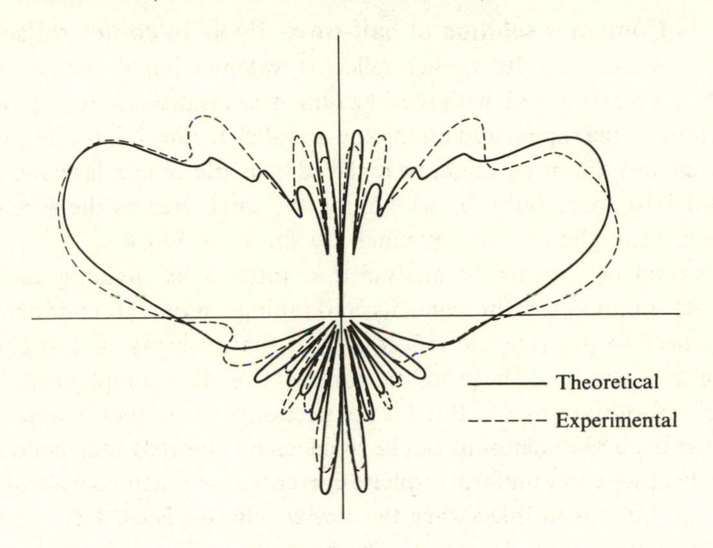

Fig. 4.10. Radiation pattern of 0.25λ monopole on circular ground plane of diameter 6λ. (After Awadalla and Maclean, 1979.)

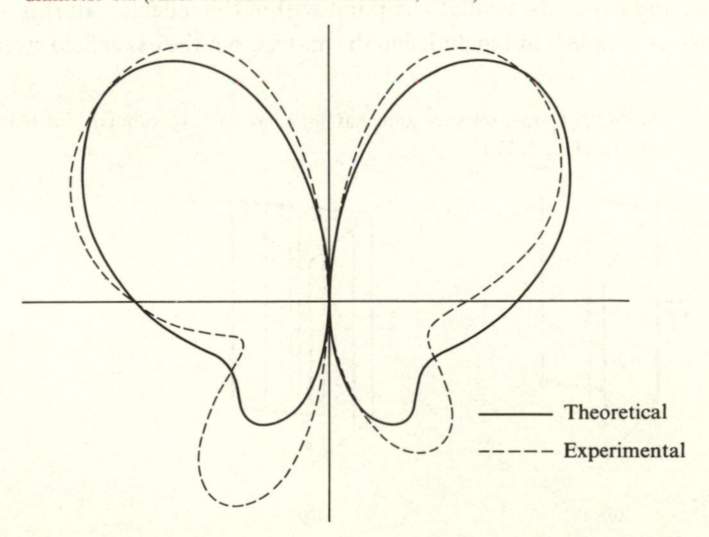

cases the agreement with experiment is excellent. It will be noted that as the ground plane diameter increases the number of auxiliary lobes in the radiation pattern likewise increases. It may be shown that the angles at which nulls appear in the pattern occur at values of θ, given approximately by

$$\theta = \sin^{-1}\frac{n\lambda}{2a} n = 1, 2 \cdots \tag{4.34}$$

4.5 Computer solution of half-wave dipole in corner reflector

A practical 90° corner reflector antenna might consist of two sheets, each $1.0\lambda \times 0.8\lambda$, with the 0.8λ dimension transverse to the forward direction of maximum radiation, and parallel to the $\lambda/2$ feeding dipole. Alternatively the sheets might be replaced by a grid of tubular conductors, spaced $\lambda/10$ apart, but otherwise filling the same area as the sheets they replace. Examples of these antennas are shown in Fig. 4.11.

It would be possible to analyse this antenna by dividing each 0.8λ conductor into, say, eight segments, and taking ten tubular conductors for each sheet to describe the reflector by a matrix array of 160 complex elements. There would be in addition at least five other complex elements to describe the driven dipole. But it is not necessary to use such a large matrix because the 0.8λ dimensions can be replaced by infinitely long conductors, provided they carry uniform in-phase currents along their complete length. This condition is satisfied when the corner reflector is used as a receiving antenna with normal incidence. The problem is thus reduced to a two-dimensional one which allows the currents in the infinite conductors to be found, and hence the field at any point within the reflector. In this analysis the driven dipole is not included in the matrix, but the exact field strength in

Fig. 4.11. Corner reflector antennas: (*a*) solid sheet; (*b*) tubular grid. (After Morris *et al.*, 1977.)

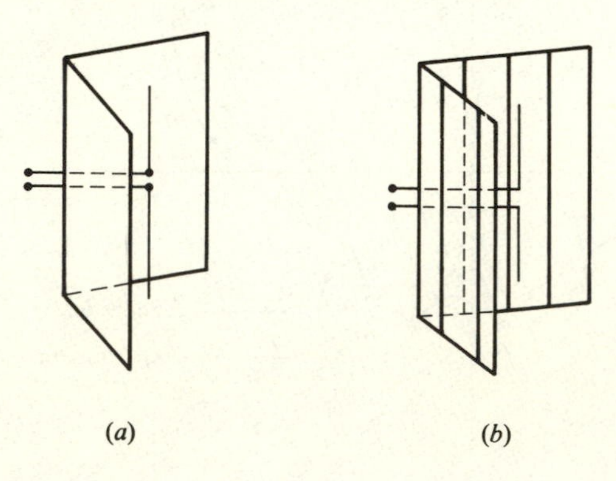

which the dipole is placed can be calculated for any length of plates corresponding to the 1.0λ dimension described above. The power delivered to a matched load connected to this dipole can then be calculated provided that its input resistance is known.

To illustrate this attack on the problem, consider that the corner reflector is constructed from parallel wires of radius b and infinite length, as in Fig. 4.12. The self impedance per unit length of each such conductor is

$$Z_{11} = \frac{\omega\mu}{4}[J_0(kb) - jY_0(kb)] \tag{4.35}$$

If adjacent wires touch each other the mutual impedance per length between wire 1 and wire n on the same plane is

$$Z_{1n} = \frac{\omega\mu}{4}\{J_0[2(n-1)kb] - jY_0[2(n-1)kb]\} \tag{4.36}$$

Likewise between wire 1 on the first plane and wire n' on the other, the mutual impedance per unit length is

$$Z_{1n'} = \frac{\omega\mu}{4}\{J_0\sqrt{[(2n-1)^2 + 1^2]}kb - jY_0\sqrt{[(2n-1)^2 + 1^2]}kb\} \tag{4.37}$$

Similarly the mutual impedance per unit length between any other two conductors on the same sheet or on different sheets is readily found, and thus the impedance matrix for the reflecting sheets is known.

To find the current on each of the conductors let the reference phase for the incoming wave be taken at the vertex, so that the impressed field for

Fig. 4.12. Cross section of 90° corner reflector divided into parallel strips.

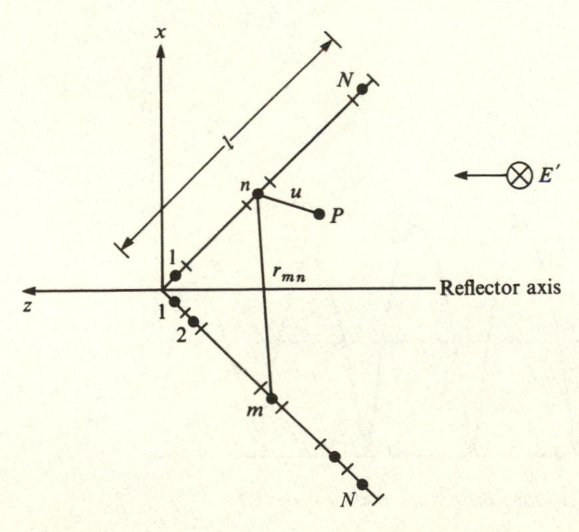

conductor 1 is $Ee^{j(kb/\surd 2)}$, and likewise for the remaining conductors. Then the system of simultaneous equations to be solved is

$$\left.\begin{aligned} Ee^{j(kb/\surd 2)} &= I_1Z_{11} + I_2Z_{12} + \cdots + I_NZ_{1N} \\ Ee^{j(3kb/\surd 2)} &= I_1Z_{21} + I_2Z_{22} + \cdots + I_NZ_{2N} \\ &\vdots \\ Ee^{j(2(N-1)b/\surd 2)} &= I_1Z_{N1} + I_2Z_{N2} + \cdots + I_NZ_{NN} \end{aligned}\right\} \tag{4.38}$$

Inverting the impedance matrix gives the admittance matrix, from which the current in each wire can be found. Thus

$$I_1 = Y_{11}Ee^{j(kb/\surd 2)} + Y_{12}Ee^{j(3kb/\surd 2)} + \cdots + Y_{1N}Ee^{j((2N-1)kb/\surd 2)} \tag{4.39}$$

and similarly for the other conductors.

As a numerical example the case of a 90° corner reflector with sides of length λ has been taken. Each sheet is considered to be divided into ten segments, each of which was represented by a cylindrical conductor of radius b. When the reflector was excited by a uniform plane wave travelling along the positive z-axis in Fig. 4.12 the currents were calculated from equations of the form of eqn (4.39). The results are shown in Fig. 4.13 where the essential feature is seen to be the standing wave nature of the current along the sheet. This can be understood to arise from the interference

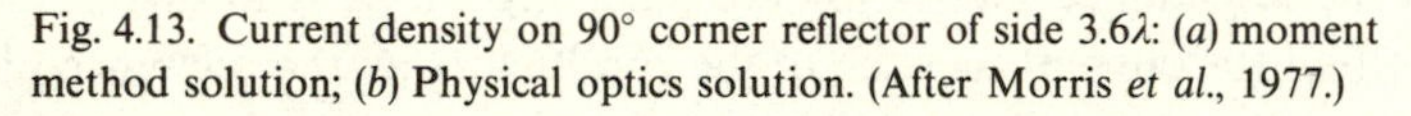

Fig. 4.13. Current density on 90° corner reflector of side 3.6λ: (a) moment method solution; (b) Physical optics solution. (After Morris *et al.*, 1977.)

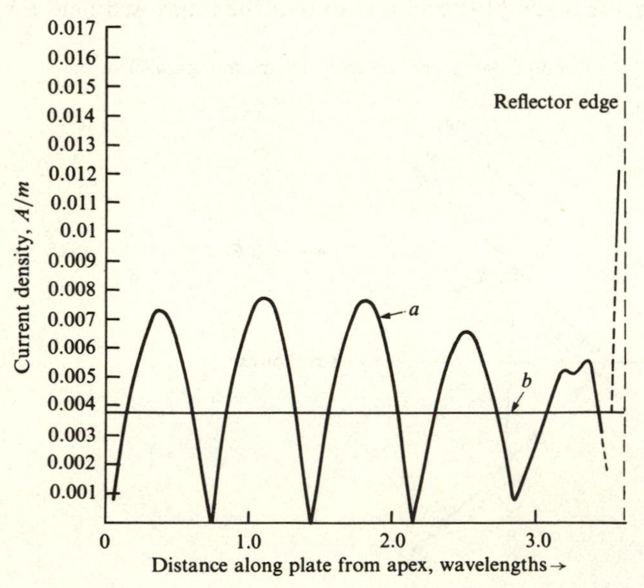

pattern set up by the incoming wave at one sheet and the wave reflected at 45° by the second sheet. The diagram also shows clearly the form of the edge currents at the end of the sheet, though these are not significant in altering the field pattern set up well inside the reflector where the load dipole is likely to be placed.

To calculate this field pattern, the resultant field is obtained from the sum of the contributions of the separate conductors plus the incident field. For example, at a point p distance u from conductor n, the field due to this conductor is

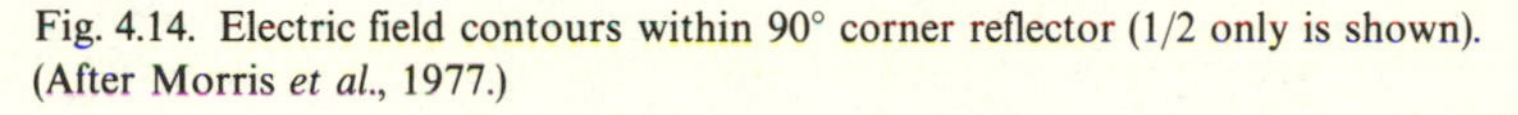

$$-\frac{\omega\mu}{4}I_n[J_0(ku)-jY_0(ku)].$$

The total fields obtained in this way are shown in Fig. 4.14 where the regular pattern of the islands of fields displays graphically the resonator type action of the reflector. Close to the aperture of the antenna the fields reduce in magnitude, and this result can be interpreted in terms of antiphase edge currents. It will also be seen that the positions of maximum field values on axis occur at distances from the apex of approximately 0.5λ, 1.5λ, etc.

The power gain of a $\lambda/2$ dipole placed 0.5λ from the apex can now be calculated approximately as follows. For a receiving antenna this gain will be defined as

Fig. 4.14. Electric field contours within 90° corner reflector (1/2 only is shown). (After Morris *et al.*, 1977.)

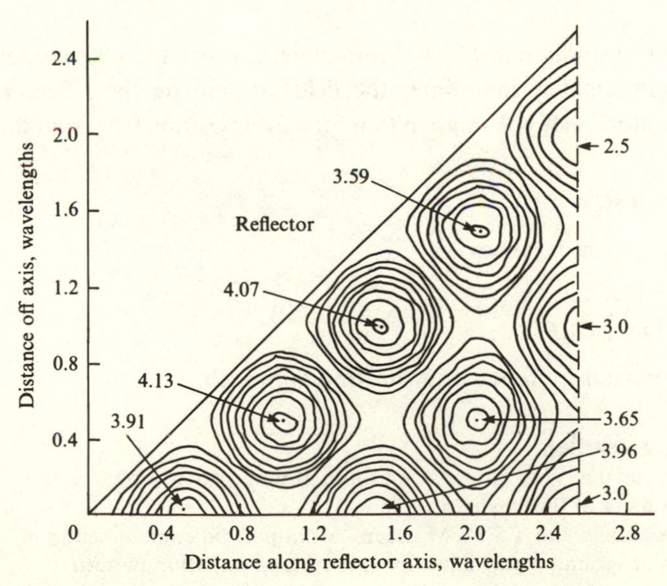

$$G = \frac{\text{power delivered to matched load connected to antenna}}{\text{power delivered to matched load connected to isotropic antenna}}$$

$$= \frac{W}{W_i} \tag{4.40}$$

Since for an isotropic antenna the power gain, by definition, is unity, and since for all antennas

$$G = \frac{4\pi}{\lambda^2} A_{eff}$$

where A_{eff} is the effective aperture of the antenna, it follows that for an isotropic antenna

$$W_i = \frac{E^2}{2Z_0} A_{eff}$$

$$= \frac{E^2}{2Z_0} \frac{\lambda}{4\pi}$$

The power to the matched load connected to the $\lambda/2$ dipole is

$$W = \frac{V_{oc}^2}{8R_r} \tag{4.41}$$

where V_{oc} is the open-circuited voltage developed by the dipole when it is placed in the corner reflector, and R_r is its radiation resistance in this position. But, for a $\lambda/2$ dipole,

$$V_{oc} = \frac{\lambda}{\pi} E$$

where E is the field strength in which it is immersed, which for a 90° corner reflector is approximately four times the field incident on the reflector. Likewise it has previously been shown that for a feed position 0.5λ from the apex,

$$R_r \approx 124.8\Omega$$

Hence the power gain becomes

$$G = \frac{\lambda^2}{\pi^2} \frac{16E^2 2Z_0 4\pi}{8(124.8)E^2 \lambda^2} = 15.4 \quad \text{or} \quad 11.9\text{dB} \tag{4.42}$$

as calculated previously from a transmitting approach.

Further reading

K.H. Awadalla: Wire antennas on finite ground planes: PhD thesis, University of Birmingham, 1978.

K.H. Awadalla and T.S.M. Maclean: 'Monopole antenna at centre of circular ground plane: input impedance and radiation pattern':

Trans. IEEE (*APG*), **AP-27**, 1979. (Copyright C, 1979, *IEEE*.)
C.A. Balanis: *Antenna Theory: Analysis and Design*: Harper and Row, New York, 1982.
H.V. Cottony and A.C. Wilson: 'Radiation patterns of finite-size corner reflector antennas': *IRE Trans. Antennas and Propagation*, **AP-8**, 1960, 144–57.
R.F. Harrington: *Field Computations by Moment Methods*: Macmillan, New York, 1968.
G. Morris, T.S.M. Maclean and K.R.G. Bailey: 'Corner reflector with improved power gain': Microwaves, Optics and Acoustics; **1**, 1977. (Copyright Controller, HMSO, London, 1977.)
A.W. Rudge, K. Milne, A.D. Olver and P. Knight (Ed.): *The Handbook of Antenna Design*: Peter Peregrinus, 1982.
W.L. Stutzman, G.A. Thiele: *Antenna Theory and Design*: Wiley and Sons, New York, 1981.

5

+−+

Loop antennas

The loop antenna, shown in Fig. 5.1 is perhaps the most commonly used type of wire antenna after the dipole. It is most frequently used, for reasons which will be shown later in the chapter, as a receiving antenna. Indeed its structure is such that it does not require any exciting radio wave for it to produce an output voltage. It will produce an output voltage if a time-varying magnetic flux, with no significant electric field associated with it, threads the loop. For this reason it is sometimes said that a small loop antenna responds only to the magnetic intensity **H** and not to the electric field **E**. This, however, is misleading and it will be shown later that the operation of the loop antenna can be explained fully in terms of the incident

Fig. 5.1. Loop antennas: (*a*) rectangular; (*b*) circular.

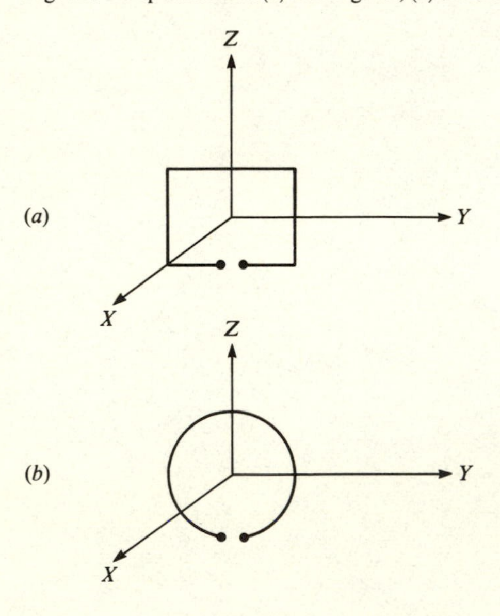

electric field **E**, in exactly the same way that the operation of the electric dipole has been explained.

5.1 Flux linking approach to rectangular receiving loop antenna

Referring to Fig. 5.2 let an electromagnetic wave be considered to travel along the y-axis with its magnetic intensity **H** parallel to the z-axis, so that **H** threads the loop. If the loop size were so small that the approximation could be made that **H** was uniform over the area of the loop, then from Faraday's law the emf available from the loop terminals would be given by

$$V = -\frac{\partial \Phi}{\partial t} = -j\omega\mu H l d$$

$$= -j\omega\mu H \times \text{loop area} \tag{5.1}$$

If the dimension d of the loop in the direction of wave travel were such that the value of H_z could no longer be described as being uniform over the length d, then the flux linking the loop is given by

$$\Phi = \mu H l \int_{-d/2}^{+d/2} e^{-jky}\, dy$$

$$= \mu H l d \frac{\sin\left(k\frac{d}{2}\right)}{k\frac{d}{2}}$$

Hence the output voltage becomes

$$V = -j\omega\mu H(\text{loop area}) \frac{\sin\left(k\frac{d}{2}\right)}{k\frac{d}{2}} \tag{5.2}$$

Fig. 5.2. Induced emf in loop by flux threading.

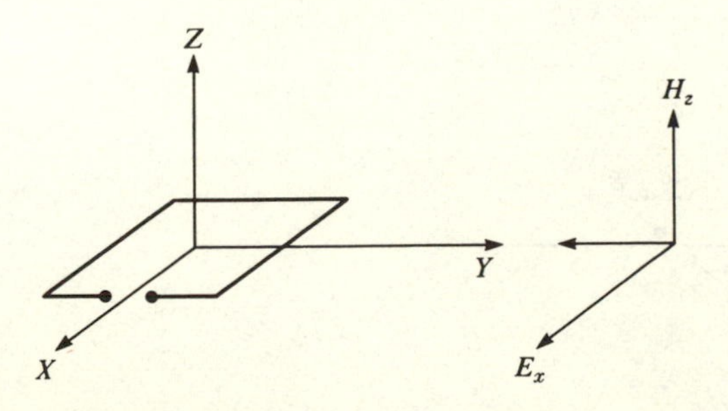

This may then be generalised to the case where the incoming electromagnetic wave makes an angle θ with the z-axis to give

$$V = -j\omega\mu H(\text{loop area})\frac{\sin\left(k\frac{d}{2}\right)}{k\frac{d}{2}}\sin\theta \tag{5.3}$$

5.2 Flux cutting approach to rectangular receiving loop antenna

When the incoming plane wave is travelling along the z-axis towards the origin with the magnetic intensity **H** directed along the x-axis, so that no flux threads the loop, it is clear that the sides AB and CD each of length d are going to be cut by the magnetic intensity H_x. Consequently an emf will be induced in these two sides of the loop, but in the same direction in both sides, so that the integral of the total emf round the loop will be zero, in accordance with the result derived from the flux linking approach. Nevertheless because there are these two oppositely directed emf's in opposite sides of the loop, as shown in Fig. 5.3, the question must be asked whether any current flows in the loop.

If Fig. 5.3 represented an electrical network, rather than a loop antenna, it would be correct to say that no current would flow in any part of the circuit, since the current is uniform in a series network. But the diagram in Fig. 5.3 is more correctly thought of as a transmission line rather than a network, since the loop antenna occupies physical length in the x-direction, along which opposing currents might be expected to flow from the opposing generators parallel to the y-direction. From these opposing emf's it is intuitively to be expected that the current is zero at the mid-points of sides AD and BC and maximum at the mid-points of the orthogonal pair of sides AB and CD. But to find out what that maximum is and the value of current elsewhere in the loop, it is necessary to take into account

Fig. 5.3. Induced emf in loop by flux cutting.

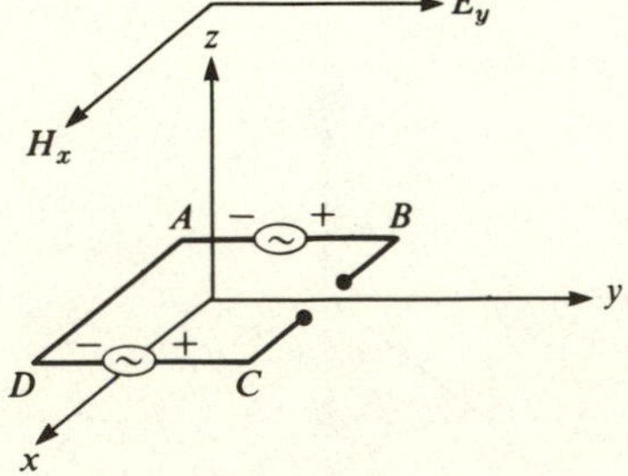

analytically the effect of the inducing electric field in the incoming radio wave.

5.3 Electric field approach to rectangular receiving loop antenna

5.3.1 *Broadside incidence with loop terminals short-circuited*

The rectangular loop shown in Fig. 5.4(*a*) is irradiated by a uniform plane wave travelling along the z-axis in the negative z-direction, with its electric vector $\mathbf{E}^i$ along the x-axis. To find the current distribution round the short-circuited loop the problem is first simplified by developing the loop into a rectilinear conductor of length $(2l + 2d)$ equal to that of the loop perimeter, as shown in Fig. 5.4(*b*). This removes the problem of taking into account the effect of the corners of the loop on the current distribution.

In the developed form of the loop shown in Fig. 5.4(*b*), where an origin is taken for convenience at the mid-point of CD, it is clear that an impressed electric field of uniform, in-phase distribution will exist only over the sides AD and BC, i.e. over the length $2l$ only. Consequently for this developed diagram the potential and current distributions are obtained from eqns (3.35) and (3.36) for $z \geqslant [2l + (3d/2)]$ as

$$V(z) = A \sin kz + B \cos kz + \int_{d/2}^{(l+d/2)} E(z') \cos k(z - z')\, dz'$$
$$- \int_{(l+3d/2)}^{(2l+3d/2)} E(z') \cos k(z - z')\, dz' \qquad (5.4)$$

Fig. 5.4. Induced emf in loop by inducing electric field.

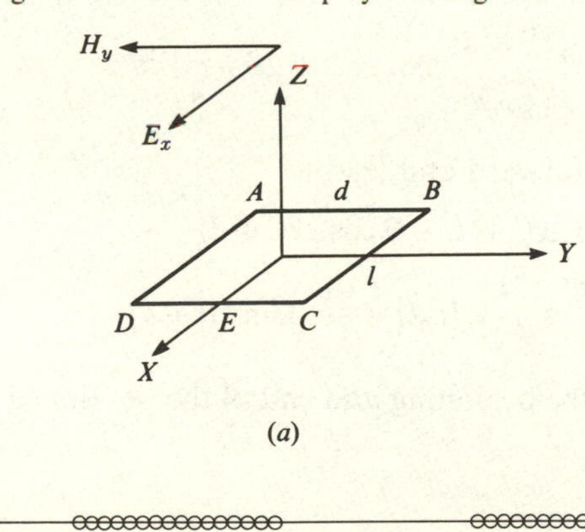

and

$$I(z) = \frac{j}{Z_c}\left[A\cos kz - B\sin kz - \int_{d/2}^{(l+d/2)} E(z')\sin k(z-z')\,dz' \right.$$

$$\left. + \int_{(l+3d/2)}^{(2l+3d/2)} E(z')\sin k(z-z')\,dz' \right] \tag{5.5}$$

where in eqns (5.4) and (5.5) the signs associated with the consecutive integrals in each of the equations are reversed because the electric fields in the two active sides oppose each other as far as the flow of current is concerned.

In a short-circuited loop the two ends of the developed diagram must be considered to be a single point, which is the mid-point of side *CD*. The corresponding electrical boundary conditions are that the potentials and currents at the two ends which coalesce into this single point are identical. Thus

$$V(0) = V(2l+2d) \tag{5.6(a)}$$

and

$$I(0) = V(2l+2d) \tag{5.6(b)}$$

This gives from eqn (5.6(a))

$$V(0) = B \tag{5.7}$$

and

$$V(2l+2d) = A\sin 2k(l+d) + B\cos 2k(l+d)$$

$$+ E^i \int_{d/2}^{(l+d/2)} \cos k(2l+2d-z')\,dz'$$

$$- E^i \int_{(l+3d/2)}^{(2l+3d/2)} \cos k(2l+2d-z')\,dz' \tag{5.8}$$

The integrals are straightforward and lead to

$$V(2l+2d) = A\sin 2k(l+d) + B\cos 2k(l+d)$$

$$- \frac{4E^i}{k}\sin\frac{kl}{2}\sin k\left(\frac{l+d}{2}\right)\sin k(l+d) \tag{5.9}$$

Similarly the currents at the beginning and end of the developed loop are

$$I(0) = \frac{j}{Z_c}A \tag{5.10}$$

and

$$I(2l+2d) = \frac{j}{Z_c}\left[A\cos 2k(l+d) - B\sin 2k(l+d) \right.$$

$$-E^i\int_{d/2}^{(l+d/2)}\sin k(2l+2d-z')\,dz'$$

$$\left. +E^i\int_{(1+3d/2)}^{(2l+3d/2)}\sin k(2l+2d-z')\,dz'\right] \qquad (5.11)$$

Eqn (5.11) leads to

$$I(2l+2d)=\frac{j}{Z_c}\left[A\cos 2k(l+d)-B\sin 2k(l+d)\right.$$

$$\left.-\frac{4E^i}{k}\sin\frac{kl}{2}\sin\frac{k(l+d)}{2}\cos k(l+d)\right] \qquad (5.12)$$

Applying the boundary conditions of eqns (5.6) gives for the two unknowns

$$A=0 \qquad (5.13)$$

$$B=-\frac{E^i}{k}\frac{\sin\left(\dfrac{kl}{2}\right)}{\cos k\left(\dfrac{l+d}{2}\right)} \qquad (5.14)$$

Hence the current is zero at the mid-point of side *CD* where the origin has been taken. Since the constants *A*, *B* are known it is now straightforward to find the current at any other point on the loop. For example, half-way round the loop from the origin the current is given by

$$I(l+d)=\frac{j}{Z_c}\left[A\cos k(l+d)-B\sin k(l+d)\right.$$

$$\left.-E\int_{d/2}^{(l+d/2)}\sin k(l+d-z')\,dz'\right]$$

which again leads to zero current.

It is of interest also to evaluate the current at the mid-point of one of the sides on which the electric field is impressed. This current is given by

$$I\left(\frac{l+d}{2}\right)=\frac{j}{Z_c}\left[A\cos k\left(\frac{l+d}{2}\right)-B\sin\frac{k(l+d)}{2}\right.$$

$$\left.-E\int_{d/2}^{(l+d/2)}\sin k\left(\frac{l}{2}+\frac{d}{2}-z'\right)dz'\right] \qquad (5.15)$$

Evaluation of this equation gives

$$I\left(\frac{l+d}{2}\right)=\frac{j}{Z_c}\frac{2E^i}{k}\frac{\sin\dfrac{kl}{4}\sin k\left(\dfrac{l+2d}{4}\right)}{\cos k\left(\dfrac{l+d}{2}\right)} \qquad (5.16)$$

When the loop is small so that the trigonometric functions can be replaced by the first terms in their power series, this gives

$$I\left(\frac{l+d}{2}\right)=\frac{j}{Z_c}\frac{E^i}{k}\frac{k^2ld}{4}=\frac{j}{Z_c}\frac{kE^i(\text{area})}{4}$$

It will be noted that the current round the loop is in phase quadrature with the impressed field $\mathbf{E}^i$ for this broadside excitation. The current magnitude is maximum at the mid-points of the sides on which the impressed electric field operates and then falls to zero at the mid-points of the orthogonal sides. At these points alone the loop can therefore be open-circuited without affecting the distribution of current on it.

5.3.2 *Broadside incidence with loop terminals open-circuited*

Let the loop terminals now be open-circuited at the mid-point of CD, so that the boundary condition at z equal to zero becomes

$$I(0)=0$$

But this again makes the constant A equal to zero, so that the open-circuit voltage from the loop is

$$V_{oc}=V(0)-V(2l+2d)$$

From eqn (5.9) this gives

$$V_{oc}=B-B\cos 2k(l+d)+\frac{4E^i}{k}\sin\frac{kl}{2}\sin\frac{k(l+d)}{2}\sin k(l+d) \quad (5.17)$$

where the constant B is found from the boundary condition that

$$I(2l+2d)=0$$

This gives, as in eqn (5.14),

$$B=-\frac{E^i}{k}\frac{\sin\dfrac{kl}{2}}{\cos k\left(\dfrac{l+d}{2}\right)}$$

Hence the open-circuit voltage becomes

$$V_{oc}=-\frac{E^i}{k}\frac{\sin\dfrac{kl}{2}}{\cos k\left(\dfrac{l+d}{2}\right)}\cdot 2\sin^2 k(l+d)$$

$$+\frac{4E^i}{k}\sin\frac{kl}{2}\sin k\frac{(l+d)}{2}\sin k(l+d)=0 \quad (5.18)$$

in agreement with the flux linking approach.

Let the incoming electric vector now be rotated through $\pi/2$ so that it is directed along the y-axis. Then since the current is zero at z equal to zero, the constant A is again zero and the open-circuit voltage becomes

$$V_{oc} = V(0) - V(2l + 2d)$$

where $V(0) = B$ and

$$\begin{aligned} V(2l + 2d) = B\cos 2k(l + d) &+ E^i \int_0^{d/2} \cos k(2l + 2d - z')\,dz' \\ &- E^i \int_{l+d/2}^{l+3d/2} \cos k(2l + 2d - z')\,dz' \\ &+ E^i \int_{2l+3d/2}^{2l+2d} \cos k(2l + 2d - z')\,dz' \end{aligned} \tag{5.19}$$

The integrations are straightforward and lead to

$$\begin{aligned} V(2l + 2d) &= B\cos 2k(l + d) \\ &\quad - \frac{8E^i}{k}\sin\frac{kd}{4}\sin k\left(\frac{l + d}{2}\right)\sin k\left(\frac{2l + d}{2}\right)\cos k(l + d) \end{aligned} \tag{5.20}$$

The constant B is obtained from the boundary condition that

$$I(2l + 2d) = 0$$

This gives

$$\begin{aligned} &- B\sin 2k(l + d) - E^i \int_0^{d/2} \sin k(2l + 2d - z')\,dz' \\ &+ E^i \int_{l+d/2}^{l+3d/2} \sin k(2l + 2d - z')\,dz' \\ &- E^i \int_{2l+3d/2}^{2l+2d} \sin k(2l + 2d - z')\,dz' = 0 \end{aligned}$$

and hence

$$B = \frac{4E^i}{k}\,\frac{\sin\frac{kd}{4}\sin k\left(\frac{l + d}{2}\right)\sin k\left(\frac{2l + d}{4}\right)}{\cos k(l + d)} \tag{5.21}$$

Substituting for B in eqn (5.20) gives

$$V_{oc} = \frac{8E^i}{k}\,\frac{\sin\frac{kd}{4}\sin k\left(\frac{l + d}{2}\right)\sin k\left(\frac{2l + d}{4}\right)}{\cos k(l + d)} \tag{5.22}$$

a result which can also be obtained from the use of eqn (5.16) with l and d interchanged, together with the knowledge of the input impedance of the loop to be given later, and the equation

$$V_{oc} = -I_{sc} Z_i \tag{5.23}$$

It is thus clear from a comparison of eqns (5.18) and (5.22) that the open-circuited output voltage from the loop differs according to the position of the output terminals. In both cases no magnetic flux links the loop, but nevertheless an output voltage is obtained when the terminals are positioned away from the middle of the sides which are not being excited.

For the case of a small square loop eqn (5.22) reduces to

$$V_{oc} = \frac{3E^i k^2 d^3}{2} = kE^i d^2 \times (\tfrac{3}{2}kd) \tag{5.24}$$

This is $\frac{3}{2}kd$ times the voltage due to flux linking, so that for a small loop its magnitude is very small.

5.3.3 *Endfire incidence with rectangular loop terminals short-circuited*

Referring again to Fig. 5.2 the rectangular loop in the xy-plane is irradiated by a uniform plane wave travelling along the y-axis in the negative y-direction with its electric vector $\mathbf{E}^i$ directed along the positive x-axis. Considering now the developed form of the loop as shown in Fig. 5.4, the potential and current equations for points $z \geqslant [2l + (3d/2)]$ remain the same as eqns (5.4) and (5.5), with $E(z')$ replaced by $E(z')e^{jkd/2}$ in each of the first integrals, and by $E(z')e^{-jkd/2}$ in each of the second integrals. Thus the equations for potential and current at the two ends of the loop become

$$V(0) = B \tag{5.25}$$

$$V(2l+2d) = A \sin 2k(l+d) + B \cos 2k(l+d) + \frac{2E^i}{k} \sin\frac{kl}{2}\left[e^{jkd/2} \cos k\frac{3(l+d)}{2} - e^{-jkd/2} \cos k\left(\frac{l+d}{2}\right)\right] \tag{5.26}$$

and

$$I(0) = \frac{jA}{Z_c} \tag{5.27}$$

$$I(2l+2d) = \frac{j}{Z_c}\left[A \cos 2k(l+d) - B \sin 2k(l+d) - \frac{2E^i}{k} \sin\frac{kl}{2}\left[e^{jkd/2} \sin k\frac{3(l+d)}{2} - e^{-jkd/2} \sin k\left(\frac{l+d}{2}\right)\right]\right] \tag{5.28}$$

The boundary conditions remain as in eqns (5.6(a) and (b)) so that these equations become

$$A \sin 2k(l+d) + B[\cos 2k(l+d) - 1]$$
$$= -\frac{2E^i}{k}\sin\frac{kl}{2}\left[e^{jkd/2}\cos\frac{3k(l+d)}{2} - e^{-jkd/2}\cos\frac{k(l+d)}{2}\right]$$

and

$$A[\cos 2k(l+d) - 1] - B\sin 2k(l+d)$$
$$= \frac{2E^i}{k}\sin\frac{kl}{2}\left[e^{jkd/2}\sin\frac{3k(l+d)}{2} - e^{-jkd/2}\sin\frac{k(l+d)}{2}\right]$$

Hence the constants A, B may be simplified to

$$A = -jE^i\frac{\sin\frac{kl}{2}\sin\frac{kd}{2}}{k\sin k\left(\frac{l+d}{2}\right)} \tag{5.29}$$

and

$$B = -\frac{E^i}{k}\sin\frac{kl}{2}\frac{\cos\frac{kd}{2}}{\cos k\left(\frac{l+d}{2}\right)} \tag{5.30}$$

In the case of small loops these reduce to

$$A = -j\frac{E^i ld}{2(l+d)} \tag{5.31}$$

and

$$B = -\frac{E^i l}{2} \tag{5.32}$$

Hence from eqn (5.29) the current at the mid-point of CD is given under short-circuited conditions by

$$I(0) = \frac{E^i}{kZ_c}\frac{\sin\frac{kl}{2}\sin\frac{kd}{2}}{\sin k\left(\frac{l+d}{2}\right)} \tag{5.33}$$

in place of the zero value for broadside incidence. It will be noted that the current at this point is in phase with the impressed electric field $\mathbf{E}^i$.

5.3.4 *Endfire incidence with rectangular loop terminals open-circuited*

The general equations for current and potential remain as in Section 5.3.3 but the boundary condition for an open circuit at the mid-point of CD implies that

$$A = 0 \tag{5.34}$$

Hence the difference in potential across the open-circuited terminals is

$$V_{oc} = V(0) - V(2l + 2d)$$

This gives

$$V_{oc} = B - B\cos 2k(l+d) - \frac{2E^i}{k}\sin\frac{kl}{2}\left[e^{jkd/2}\cos\frac{3k(l+d)}{2} - e^{-jkd/2}\cos\frac{k(l+d)}{2}\right] \tag{5.35}$$

where B can be found from the condition that

$$I(2l + 2d) = 0$$

This gives

$$B = -\frac{2E^i \sin\frac{kl}{2}}{k\sin 2k(l+d)}\left[e^{jkd/2}\sin\frac{3k(l+d)}{2} - e^{-jkd/2}\sin\frac{k(l+d)}{2}\right] \tag{5.36}$$

from which the open-circuit voltage simplifies to

$$V_{oc} = -j\frac{4E^i}{k}\frac{\sin\frac{kl}{2}\sin\frac{kd}{2}\cos k\left(\frac{l+d}{2}\right)}{\cos k(l+d)} \tag{5.37}$$

When the loop is very small this again reduces to

$$V_{oc} = -j\omega\mu H l d \tag{5.38}$$

as in eqn (5.1).

If the incoming wave now travels along the positive x-axis with a y-directed electric field, then again

$$A = 0$$

and

$$V(0) = B$$

The new value of potential at the end of the loop wire is now

$$V(2l+2d) = B\cos 2k(l+d) + E^i e^{jkl/2}\int_0^{d/2}\cos k(2l+2d-z')\,dz' - E^i e^{-jkl/2}\int_{l+d/2}^{l+3d/2}\cos k(2l+2d-z')\,dz' + E^i e^{jkl/2}\int_{2l+3d/2}^{2l+2d}\cos k(2l+2d-z')\,dz' \tag{5.39}$$

This gives

$$V(2l+2d) = B\cos 2k(l+d) + \frac{4E^i}{k}\sin\frac{kd}{4}\cos k(l+d) \times \left\{ e^{jkl/2}\cos k\left(l+\frac{3d}{4}\right) - e^{-jkl/2}\cos\frac{kd}{4} \right\} \quad (5.40)$$

where the constant B is again found from the boundary condition that

$$I(2l+2d) = 0 \quad (5.41)$$

This gives

$$-B\sin 2k(l+d) - E^i e^{jkl/2}\int_0^d \sin k(2l+2d-z')\,dz' + E^i e^{-jkl/2}\int_{l+d/2}^{l+3d/2} \sin k(2l+2d-z')\,dz' - E^i e^{jkl/2}\int_{2l+3d/2}^{2l+2d} \sin k(2l+2d-z')\,dz' = 0 \quad (5.42)$$

Performing the integrations leads to

$$B = -\frac{2E^i}{k}\,\frac{\sin\frac{kd}{4}\left[e^{jkl/2}\cos k\left(l+\frac{3d}{4}\right) - e^{-jkl/2}\cos\frac{kd}{4}\right]}{\cos k(l+d)} \quad (5.43)$$

Hence the open-circuit voltage due to this orthogonally polarised wave becomes

$$V_{oc} = -\frac{4E^i}{k}\,\frac{\sin\frac{kd}{4}\left[e^{jkl/2}\cos k\left(l+\frac{3d}{4}\right) - e^{-jkl/2}\cos\frac{kd}{4}\right]}{\cos k(l+d)} \quad (5.44)$$

When the loop is small in terms of wavelengths this reduces for a square loop to

$$V_{oc} = -jE^i kd^2[1 + j\tfrac{3}{2}kd] \quad (5.45)$$

Hence by comparison with eqn (5.38) an additional term exists in the output voltage due to the orthogonal polarisation or, what amounts to the same thing, due to the output terminals being placed orthogonal to the position where the output voltage is due only to the magnetic flux linking the loop. It is sometimes said that the second term is due to the electric field operating on the loop but this explanation is to be deprecated since both terms have been derived by considering only the electric field impressed on the loop.

5.4 Input impedance of rectangular loop

Using the expression

$$Z_i = -\frac{V_{oc}}{I_{sc}} \tag{5.46}$$

we obtain, from eqns (5.33) and (5.37),

$$Z_i = \frac{j\dfrac{4E^i}{k}\dfrac{\sin\dfrac{kl}{2}\sin\dfrac{kd}{2}\cos k\left(\dfrac{l+d}{2}\right)}{\cos k(l+d)}}{\dfrac{E^i}{kZ_c}\dfrac{\sin\dfrac{kl}{2}\sin\dfrac{kd}{2}}{\sin k\left(\dfrac{l+d}{2}\right)}}$$

i.e.

$$Z_i = j2Z_c \tan k(l+d) \tag{5.47}$$

As with the linear dipole this approach ignores the real component of input impedance associated with the radiation resistance of the loop.

5.4.1 *Radiation resistance of small rectangular loop*

Consider a single turn of the rectangular loop $OPQR$, as shown in Fig. 5.5. It will be assumed that this turn is sufficiently small in terms of a wavelength so that the current has the same amplitude and phase round its periphery. Because of this there are no charges anywhere on the loop, and hence the electric field everywhere in space can be derived from vector potentials in the x- and y-directions only.

The vector potential at a point A on the surface of the wire, due to x-

Fig. 5.5. Single turn rectangular loop carrying uniform current.

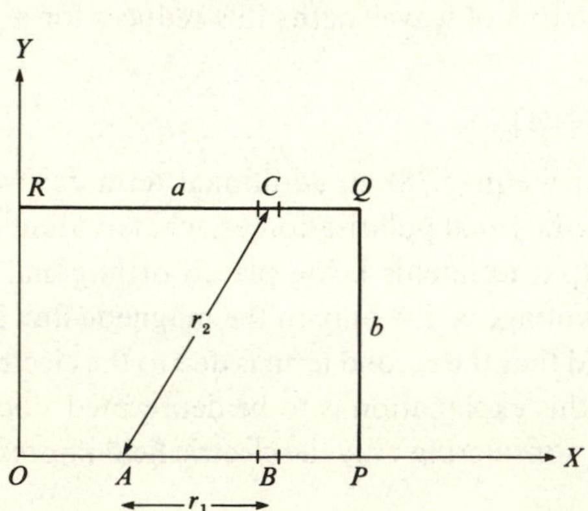

directed current elements at B and C, is also x-directed and is given in magnitude by

$$dA_x = \frac{\mu I}{4\pi}\left[\frac{e^{-jkr_1}}{r_1} - \frac{e^{-jkr_2}}{r_2}\right]dx \tag{5.48}$$

where r_1, r_2 are equal to the lengths AB and AC respectively. If the exponential terms are expanded in their power series this gives

$$dA_x \approx \frac{\mu I}{4\pi}\left\{k\left(\frac{1}{kr_1} - \frac{kr_1}{2}\right) - jk\left(1 - \frac{k^2r_1^2}{6}\right)\right.$$
$$\left. - k\left(\frac{1}{kr_2} - \frac{kr_2}{2}\right) + jk\left(1 - \frac{k^2r_2^2}{6}\right)\right\}dx$$

The quadrature component only of this expression will contribute to the electric field working in opposition to the current flow I, and this component is

$$dA_{xq} = j\frac{\mu I k dx}{4\pi}\frac{k^2}{6}(r_1^2 - r_2^2)$$
$$= -j\frac{\mu I k^3 b^2}{24\pi} \tag{5.49}$$

where b is the loop dimension orthogonal to the current elements. The integration of eqn (5.49) along the side OP of the loop gives

$$A_{xq} = -\frac{j\mu I k^3 ab^2}{24\pi}$$

so that the antiphase electric field at the point of observation A is

$$E_x = -\frac{\omega\mu I k^3 ab^2}{24\pi} \tag{5.50}$$

The radiation resistance associated with the sides OP and QR is then

$$R = -\frac{2\int_0^a E_x dx}{I} = \frac{\omega\mu k^3 a^2 b^2}{12\pi}$$

and, since the value of radiation resistance associated with the sides PQ and OR is identical, the total radiation resistance of the loop antenna becomes

$$R_r = \frac{\omega\mu k^3 a^2 b^2}{6\pi} = 20k^4(\text{loop area})^2 \tag{5.51}$$

For n turns this result is multiplied by n^2.

5.4.2 *Loss resistance of small rectangular loop*

If the loop is constructed of conducting tape with a thickness very much greater than the skin depth $1/\sqrt{(\pi f \mu\sigma)}$ where σ is the conductivity of the metal used for the tape, then the loss resistance, as shown in standard electromagnetic texts, is given by

$$R_1 = \frac{R_s \times \text{length of tape}}{2 \times \text{tape width}} \tag{5.52}$$

where R_s is the surface resistivity $\sqrt{[(\pi f \mu)/\sigma]}$, and the factor of 2 in the denominator is accounted for by the fact that current may be assumed to flow equally on both sides of the tape. For copper with a conductivity of 4×10^7 mhos/m, the skin depth at a frequency of 150 MHz is 6.5×10^{-6} m and the surface resistivity R_s is then $3.85 \times 10^{-3}\Omega$. Consequently for a loop with dimensions, for example 6.5×2.7 cm, and of tape width 0.6 cm, the loss resistance is found to be 0.059Ω. By comparison the radiation resistance from eqn (5.51) is 0.006Ω, and hence the efficiency is

$$\eta = \frac{R_r}{R_r + R_e} = 0.092$$

This represents a relatively high efficiency for a loop aerial and is a consequence of the very wide tape chosen for its construction.

5.4.3 *Average characteristic impedance of small rectangular loop*

In Section 5.4 the input reactance of a rectangular loop was given in terms of its average characteristic impedance by the equation

$$X_i = j2Z_c \tan k(l + d) \tag{5.53}$$

Since this input reactance is also equal to $j\omega L$, where L is the inductance of the loop, an expression for the average characteristic Z_c can readily be found for small rectangular loops. Thus for small loops, for which the current round the periphery is constant in magnitude and phase, the inductance of a loop with sides l and d, wound with wire of radius a, is shown in standard electromagnetic texts to be given by

$$L = \frac{\mu}{\pi}\left\{ l \cdot \ln \frac{2ld}{a(l + \sqrt{(d^2 + l^2)})} + d \cdot \ln \frac{2ld}{a(d + \sqrt{(d^2 + l^2)})} + 2(a + \sqrt{(d^2 + l^2)} - d - l) \right\} \tag{5.54}$$

Hence the average characteristic impedance is given by

$$Z_c = \frac{\omega L}{2 \tan k(l + d)} \tag{5.55}$$

5.5 Electric field approach to circular receiving loop antenna

5.5.1 *Broadside incidence with loop terminals short-circuited*

With reference to Fig. 5.6, using s as the parameter of length along the circular arc,

$$V(s) = A\sin ks + B\cos ks + \int_0^s E(s')\cos k(s-s')\,ds' \tag{5.56}$$

$$I(s) = \frac{j}{Z_c}\left[A\cos ks - B\sin ks - \int_0^s E(s')\sin k(s-s')\,ds'\right] \tag{5.57}$$

where for broadside incidence with the electric field being x-directed $E(s') = E^i \sin\phi'$, so that

$$\begin{aligned}
V(a\phi) &= A\sin ka\phi + B\cos ka\phi + aE^i\int_0^\phi \sin\phi'\cos ka(\phi-\phi')\,d\phi' \\
&= A\sin ka\phi + B\cos ka\phi + \frac{aE^i}{2}\int_0^\phi [\sin(\phi'+ka\phi-ka\phi') \\
&\quad + \sin(\phi'-ka\phi+ka\phi')]\,d\phi' \\
&= A\sin ka\phi + B\cos ka\phi + \frac{aE^i}{2}\left[-\frac{\cos(\phi'+ka\phi-ka\phi')}{(1-ka)}\right. \\
&\quad \left. - \frac{\cos(\phi'-ka\phi+ka\phi')}{(1+ka)}\right]_0^\phi \\
&= A\sin ka\phi + B\cos ka\phi - \frac{aE^i}{2}\left[\frac{\cos\phi-\cos ka\phi}{1-ka}\right. \\
&\quad \left. + \frac{\cos\phi-\cos ka\phi}{1+ka}\right]
\end{aligned}$$

Fig. 5.6. Short-circuited circular loop: broadside incidence.

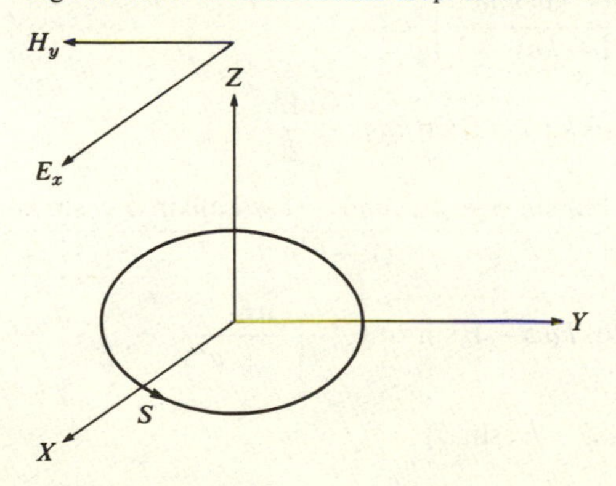

$$= A \sin ka\phi + B \cos ka\phi - aE^i \frac{\cos \phi - \cos ka\phi}{(1 - k^2a^2)} \tag{5.58}$$

Hence

$$V(0) = B$$

$$V(2\pi) = A \sin 2\pi ka + B \cos 2\pi ka - \frac{aE^i}{(1 - k^2a^2)}(1 - \cos 2\pi ka)$$

Similarly from

$$I(s) = \frac{j}{Z_c}\left[A \cos ks - B \sin ks - \int_0^s E(s') \sin k(s - s')\, ds' \right]$$

we obtain for broadside incidence

$$\begin{aligned}
I(a\phi) &= \frac{j}{Z_c}\Bigg[A \cos ka\phi - B \sin ka\phi \\
&\quad - aE^i \int_0^\phi \sin \phi' \sin ka(\phi - \phi')\, d\phi' \Bigg] \\
&= \frac{j}{Z_c}\Bigg\{ A \cos ka\phi - B \sin ka\phi - \frac{aE^i}{2} \int_0^\phi [\cos(\phi' - ka\phi \\
&\quad + ka\phi') - \cos(\phi' + ka\phi - ka\phi')]\, d\phi' \Bigg\} \\
&= \frac{j}{Z_c}\Bigg\{ A \cos ka\phi - B \sin ka\phi - \frac{aE^i}{2}\Bigg[\frac{\sin(\phi' - ka\phi + ka\phi')}{(1 + ka)} \\
&\quad - \frac{\sin(\phi' + ka\phi - ka\phi')}{(1 - ka)} \Bigg]_0^\phi \Bigg\} \\
&= \frac{j}{Z_c}\Bigg\{ A \cos ka\phi - B \sin ka\phi - \frac{aE^i}{2}\Bigg[\frac{(\sin \phi + \sin ka\phi)}{(1 + ka)} \\
&\quad - \frac{(\sin \phi - \sin ka\phi)}{(1 - ka)} \Bigg] \Bigg\} \\
&= \frac{j}{Z_c}\Bigg\{ A \cos ka\phi - B \sin ka\phi - \frac{aE^i}{2} \\
&\quad \times \frac{[(1 - ka)(\sin \phi + \sin ka\phi) - (1 + ka)(\sin \phi - \sin ka\phi)]}{(1 - k^2a^2)} \Bigg\} \\
&= \frac{j}{Z_c}\Bigg\{ A \cos ka\phi - B \sin ka\phi - \frac{aE^i}{(1 - k^2a^2)} \\
&\quad \times (\sin ka\phi - ka \sin \phi) \Bigg\}
\end{aligned} \tag{5.59}$$

Hence

$$I(0) = \frac{j}{Z_c} A$$

$$I(2\pi) = \frac{j}{Z_c}\left[A \cos 2\pi ka - B \sin 2\pi ka - \frac{aE^i}{(1-k^2a^2)} \sin 2\pi ka \right]$$

Applying the boundary conditions for the short-circuited loop

$$I(0) = I(2\pi)$$
$$V(0) = V(2\pi)$$

gives

$$A[\cos 2\pi ka - 1] - B \sin 2\pi ka = \frac{aE^i}{(1-k^2a^2)} \sin 2\pi ka$$

and

$$A \sin 2\pi ka + B[\cos 2\pi ka - 1] = \frac{aE^i}{(1-k^2a^2)}(1 - \cos 2\pi ka)$$

Hence

$$A[\cos 2\pi ka - 1]^2 - B \sin 2\pi ka(\cos 2\pi ka - 1)$$
$$= \frac{aE^i}{(1-k^2a^2)} \sin 2\pi ka(\cos 2\pi ka - 1)$$

and

$$A \sin^2 2\pi ka + B \sin 2\pi ka(\cos 2\pi ka - 1)$$
$$= \rightarrow \frac{aE^i}{(1-k^2a^2)} \sin 2\pi ka(1 - \cos 2\pi ka)$$

Adding gives

$$A(1 - 2\cos 2\pi ka) = \frac{aE^i}{(1-k^2a^2)} \sin 2\pi ka \cdot 0$$

i.e.

$$A = 0 \tag{5.60}$$

and

$$B = -\frac{aE^i}{(1-k^2a^2)} \tag{5.61}$$

Half-way round the loop at ϕ equals π,

$$I(a\pi) = \frac{j}{Z_c}\left[A \cos \pi ka - B \sin \pi ka - \frac{aE^i}{(1-k^2a^2)} \sin \pi ka \right]$$
$$= \frac{j}{Z_c} \frac{aE^i}{(1-k^2a^2)}(\sin \pi ka - \sin \pi ka)$$
$$= 0 \tag{5.62}$$

so that the loop may be broken at $\phi=0°$ or $180°$ without affecting its operation.

To evaluate the current at $\pi/2$ radians from the zero current positions we have

$$I\left(\frac{a\pi}{2}\right)=\frac{j}{Z_c}\left\{A\cos\frac{\pi}{2}ka-B\sin\frac{\pi}{2}ka\right.$$
$$\left.-\frac{aE^i}{(1-k^2a^2)}\left[\sin\frac{\pi}{2}ka-ka\right]\right\}$$
$$=\frac{j}{Z_c}\frac{ka^2E^i}{(1-k^2a^2)} \tag{5.63}$$

When the loop is small this gives

$$I\left(a\frac{\pi}{2}\right)\approx\frac{j}{Z_c}\frac{E^i}{\lambda}2(\pi a^2)=\frac{jk}{Z_c}\frac{E^i}{\pi}(\text{area}) \tag{5.64}$$

5.5.2 *Endfire incidence with loop terminals short-circuited*

For end fire incidence let $E(s')=E^i\sin\phi' e^{jky'}=E^i\sin\phi' e^{jka\sin\phi'}$. Hence

$$V(a\phi)=A\sin ka\phi+B\cos ka\phi$$
$$+aE^i\int_0^{\phi}\sin\phi' e^{jka\sin\phi'}\cos ka(\phi-\phi')\,d\phi'$$
$$=A\sin ka\phi+B\cos ka\phi$$
$$+\frac{aE^i}{2}\int_0^{\phi}e^{jka\sin\phi'}\{\sin[(1-ka)\phi'+ka\phi]$$
$$+\sin[(1+ka)\phi'-ka\phi]\}\,d\phi' \tag{5.65}$$

Restricting attention to small loops for which $ka\ll 1$, in order to obtain a solution in closed form gives

$$V(a\phi)\approx A\sin ka\phi+B\cos ka\phi+aE^i$$
$$\times\left\{\int_0^{\phi}e^{jka\sin\phi'}\sin\phi'\cos ka\phi\,d\phi'\right\} \tag{5.66}$$

When ϕ is equal to 2π this gives

$$V(2\pi a)=A\sin 2\pi ka+B\cos 2\pi ka+j2\pi aE^i\cos 2\pi ka J_1(ka) \tag{5.67}$$

and

$$V(0)=B$$

Similarly

$$I(2\pi a)=\frac{j}{Z_c}\left[A\cos 2\pi ka-B\sin 2\pi ka\right.$$

$$- aE^i \int_0^{2\pi} \sin\phi' e^{jka\sin\phi'} \sin ka(\phi - \phi')\, d\phi' \Big]$$

$$= \frac{j}{Z_c}\Big[A\cos 2\pi ka - B\sin 2\pi ka - \frac{aE^i}{2}\int_0^{2\pi} e^{jka\sin\phi'}$$

$$\times \Big\{ \cos[(1+ka)\phi' - ka\phi] - \cos[(1-ka)\phi'$$

$$+ ka\phi] \Big\} d\phi' \Big]$$

$$\approx \frac{j}{Z_c}\Big[A\cos 2\pi ka - B\sin 2\pi ka$$

$$- aE^i \int_0^{2\pi} e^{jka\sin\phi'} \sin\phi' \sin ka\phi\, d\phi' \Big]$$

$$= \frac{j}{Z_c}[A\cos 2\pi ka - B\sin 2\pi ka - j2\pi aE^i \sin 2\pi ka J_1(ka)] \tag{5.68}$$

and

$$I(0) = \frac{jA}{Z_c}$$

Using the boundary conditions for the short-circuited loop

$$I(0) = I(2\pi a)$$

and

$$V(0) = V(2\pi a)$$

gives

$$A[\cos(2\pi ka) - 1] - B\sin 2\pi ka = j2\pi aE^i \sin 2\pi ka J_1(ka) \tag{5.69}$$

and

$$A\sin 2\pi ka + B[\cos(2\pi ka) - 1] = -j2\pi aE^i \cos 2\pi ka J_1(ka) \tag{5.70}$$

i.e.

$$A[\cos 2\pi ka - 1]^2 - B\sin 2\pi ka[\cos(2\pi ka) - 1]$$
$$= j2\pi aE^i \sin 2\pi ka[\cos(2\pi ka) - 1]J_1(ka) \tag{5.71}$$

and

$$A\sin^2 2\pi ka + B\sin 2\pi ka[\cos(2\pi ka) - 1]$$
$$= -j2\pi aE^i \sin 2\pi ka \cos 2\pi ka J_1(ka) \tag{5.72}$$

Adding eqns (5.71) and (5.72) gives

$$A[2 - 2\cos 2\pi ka] = -j2\pi aE^i J_1(ka)\sin(2\pi ka)$$

i.e.

$$A = -j\pi aE^i J_1(ka)\cot \pi ka \tag{5.73}$$

Hence from eqn (5.69)

$$B = -j3\pi aE^iJ_1(ka) \tag{5.74}$$

The current at ϕ equal to zero is then

$$\begin{aligned} I(0) &= \frac{\pi aE^i}{Z_c} J_1(ka) \cot \pi ka \\ &\approx \frac{aE^i}{2Z_c} \end{aligned} \tag{5.75}$$

which is in phase with the impressed electric field $\mathbf{E}^i$, as with eqn (5.33) for the rectangular loop.

5.5.3 *Endfire incidence with loop open-circuited – small loop case*

In this case using the boundary condition that the current is zero at the open-circuit position z equals zero gives

$$A = 0 \tag{5.76}$$

Hence the difference in potential across the open-circuited terminals is

$$V_{oc} = V(0) - V(2l + 2d)$$

This gives

$$\begin{aligned} V_{oc} &= B - B\cos 2\pi ka - aE^i \int_0^{2\pi} e^{jka\sin\phi'} \sin\phi' \cos 2\pi ka \, d\phi' \\ &= B - B\cos 2\pi ka - j2\pi aE^i \cos 2\pi kaJ_1(ka) \end{aligned} \tag{5.77}$$

where B can be found from the condition that

$$I(2\pi a) = 0$$

This gives

$$\begin{aligned} B\sin 2\pi ka &= -aE^i \int_0^{2\pi} e^{jka\sin\phi'} \sin\phi' \sin 2\pi ka \, d\phi' \\ &= -j2\pi aE^i \sin 2\pi kaJ_1(ka) \end{aligned}$$

or

$$B = -j2\pi aE^iJ_1(ka) \approx -j2\pi^2 \frac{a^2}{\lambda} E^i = -j\frac{k^2a^2}{2\lambda} E^i \tag{5.78}$$

Then

$$\begin{aligned} V_{oc} &= -j2\pi aE^iJ_1(ka)[1 - \cos 2\pi ka] - j2\pi aE^i \cos 2\pi kaJ_1(ka) \\ &= -j2\pi aE^iJ_1(ka) \\ &\approx -j\frac{k^2a^2}{2\lambda} E^i = -j\omega\mu H\pi a^2 \end{aligned} \tag{5.79}$$

5.5.4 *Input impedance of small circular loop*

From the equation

$$Z_i = -\frac{V_{oc}}{I_{sc}}$$

where V_{oc}, I_{sc} are given by the full forms of eqns (5.75) and (5.79) we obtain

$$Z_i = j2Z_c \tan(\pi ka) \tag{5.80}$$

Since the inductance of a small circular loop of radius a, constructed of wire of radius b, is

$$L = \mu a \left[\ln\frac{8a}{b} - 2 \right]$$

this provides an expression for the characteristic impedance of a small circular loop, namely

$$Z_c = \frac{\omega\mu a \left[\ln\frac{a}{b} - 2 \right]}{2 \tan \pi ka} \tag{5.81}$$

$$\approx \frac{Z_0}{2} \left[\ln\frac{a}{b} - 2 \right] \tag{5.82}$$

5.6 Rectangular loop antenna as magnetic field probe

In the preceding sections two approaches were used to find the open-circuit voltage developed by a loop antenna when it was irradiated by a uniform electromagnetic field having a planar wavefront. The first approach used the concept of the induced emf being caused by a time-varying flux linking the loop. The second approach used the idea of an impressed electric field being responsible for the production of an electric current in the loop, and the variation of this current round the loop setting up a potential which is proportional to the space derivation of the current. The output voltage is then the difference of two such potentials spaced an infinitesimal distance apart.

The two answers may be different, depending on the position of the terminals in the loop. The difference arises because the magnetic flux approach provides a line integral of electric field round the complete loop, and thus gives no information about electric fields which exist at points round the loop but whose line integral is zero. Such electric fields are associated with opposing currents round the loop periphery, of the type illustrated for broadside incidence by eqns (5.16) and (5.63), where this current is in phase quadrature with the impressed electric field E^i.

These quadrature currents are distinct from the component of current in phase with the impressed electric field $\mathbf{E}^i$, which is provided by both the magnetic flux and the impressed electric field methods. This current is often referred to as the loop current whereas the quadrature current is called the dipole current.

Because the loop current component is provided as the total current by the magnetic flux approach, it may be used as a measure of the magnetic flux density at an observation point in a magnetic field. Although in principle this loop current could be separated from the dipole current because of their quadrature relationship, it is simpler in practice to ensure that the dipole component of current is much less than the loop component. Or, in terms of open-circuit voltage, the component of open-circuit voltage exclusively associated with electric field, i.e. the real component in eqn (5.45), is much less than the quadrature component in the same equation. This is clearly achieved by making the electrical length kd of the square loop as small as possible, consistent with obtaining sufficient output voltage from the quadrature component.

5.7 Loop antenna as direction finder

Consider a small rectangular loop antenna positioned in the xz-plane, as shown in Fig. 5.7. Let a uniform wave travel in the negative x-direction with its electric field oriented along the z-axis. The *oc* output voltage from terminals located at the mid-point of side BC is then

Fig. 5.7. Open-circuited rectangular loop in xz-plane: end-fire incidence.

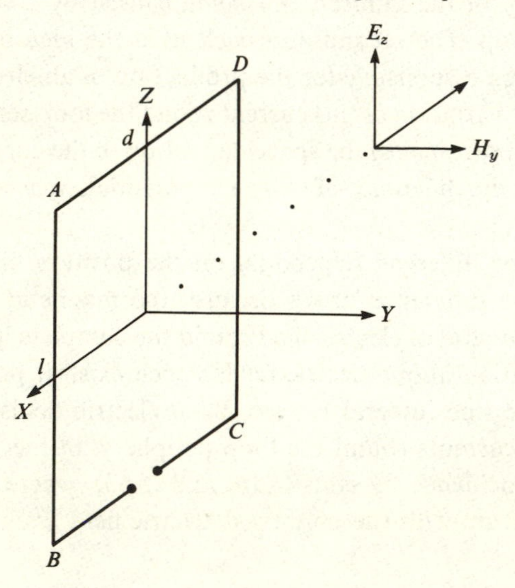

maximum and equal to

$$V_{oc} = -j\omega\mu Hld$$

Now let the angle of incidence of the incoming plane wave make an angle ϕ with the xz-plane, while the electric field remains oriented along the z-axis. Then the open-circuited output voltage becomes

$$V_{oc} = -j\omega\mu Hld\cos\phi \tag{5.83}$$

Hence, provided it is known that the electric field is oriented in this particular direction, then the direction of the angle ϕ that the incoming wave makes with the xz-plane may be found by rotating the loop so as to obtain zero output voltage. But the knowledge that the direction of the electric field be vertical in order to obtain the angle ϕ is a severe restriction on the usefulness of the loop antenna as a three-dimensional direction finder, and the error resulting from not knowing this information in practice leads to what is referred to as polarisation error.

As a first example of polarisation error let the incoming uniform plane wave initially be considered to travel in the xy-plane along the negative y-direction so that ϕ is equal to $\pi/2$, with its magnetic vector being x-directed. Then the output voltage is now zero. Now let the angle of elevation of the incoming wave measured from the z-axis be reduced from $\pi/2$ to θ so that, although the direction of the electric field necessarily changes, the associated magnetic intensity H is still in the x-direction and therefore still no magnetic flux links the loop and the output voltage is again zero. Hence if the electric field had been wrongly assumed to have remained z-directed the direction of the incoming wave (θ, ϕ) would be calculated as $(\pi/2, \pi/2)$ instead of the correct direction where θ can have any value between 0° and π with ϕ remaining at $\pi/2$.

As a further example of polarisation error let the wave travelling along the negative y-axis have its electric field in the xz-plane, making an angle of 45° with the z-axis. The magnetic field intensity is likewise in the xz-plane, orthogonal to the electric field and again making an angle of 45° with the z-axis. Since no flux links the loop there is zero output voltage from it. Now let the angle of elevation of the incoming wave be decreased from θ equals $\pi/2$ to zero so that both the electric and magnetic fields now lie in the xy-plane making angles of $-45°$ and $-135°$ with the x-axis. If the plane of the loop is now rotated from ϕ equals 0° to ϕ equals $+45°$ no flux links the loop and the output voltage is zero. But the incoming wave is at an angle (θ, ϕ) of $(0, \pi/2)$, whereas if vertical polarisation had been arbitrarily assumed in a direction finding situation the wave would have been interpreted as coming from the direction $(\pi/2, -45°)$. This error has resulted from the existence of a horizontal component of electric field in the *xy*-plane acting on the loop, and to avoid this error, direction finding antennas are frequently of dipole rather than loop types.

5.8 Screened loop antenna

5.8.1 *Balanced screen loop*

Referring to Fig. 5.1(*a*) the vertical sides in the diagram are considered to be excited by an incident electric field. These two sides, along with the orthogonal sides, function also as a transmission line in delivering a signal to the load connected across the output terminals. If any asymmetry to ground exists between the two halves of the loop, then with broadside operation there will no longer be equal and opposite currents flowing towards the output terminals so that the direction finding properties of the loop are impaired.

It has been found experimentally that by using a screened form of construction for the loop, as shown in Fig. 5.8, this impairment is reduced. To see why this should be the case consider the situation when the load terminals in Fig. 5.1(*a*) and the gap terminals in Fig. 5.8 are short-circuited. Then the two loops of Fig. 5.1(*a*) and the outer of Fig. 5.8 are identical, with the sides *CD* and *BA* acting as transmission lines which load the generating sides *BC* and *AD*. These transmission line loads are in parallel and hence may be treated independently in network terms. Consequently any unbalance to ground in the two halves of side *BA* will affect the current in that side only but not in side *CD*. Since, for a vertical loop, side *BA* is closer to the ground than side *CD*, unbalance because of the presence of feeder

Fig. 5.8. Balanced screened loop.

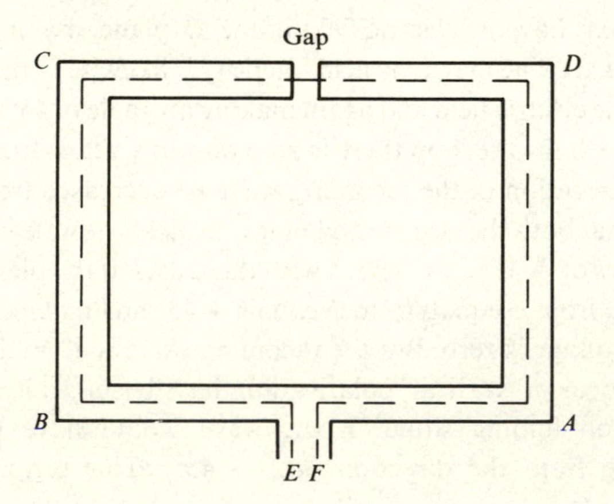

cables on one side rather than the other of the loop, for example, will be less when the output is taken from side *CD* rather than *BA*. But this output voltage taken across the gap in side *CD* is readily transferred through the inside of the coaxial cable to the new terminals *EF* in Fig. 5.8, leading to its more accurate direction finding properties.

5.8.2 *Unbalanced screen loop*

Since the output voltage from an antenna is normally taken by way of a coaxial cable to the receiver it is convenient to have an unbalanced form of screened loop which allows a direct connection to such a cable. This is achieved in the form of the unbalanced screened loop shown in Fig. 5.9 which is derived from Fig. 5.8 through the application of image plane techniques.

5.8.3 *Operation of balanced screened rectangular loop*

The output voltage from the terminals *EF* of the balanced screened loop antenna is obtained entirely from the voltage which exists across the gap at the top of the outer screen. Denoting this voltage by V_g gives, for the voltage across *EF*,

$$V_{EF} = V_g \cos \beta(l + d) - j2I_g Z_{0i} \sin \beta(l + d) \qquad (5.84)$$

where $(l + d)$ is the length of the coaxial transmission line between the gap and the output terminals at *EF*, Z_{0i} is the characteristic impedance of the coaxial line forming the screened loop, and I_g is the input current to the transmission line at the gap. The factor of two in the second term on the right-hand side of eqn (5.84) arises since in applying transmission line equations to this problem the total characteristic impedance is taken

Fig. 5.9. Unbalanced screened loop.

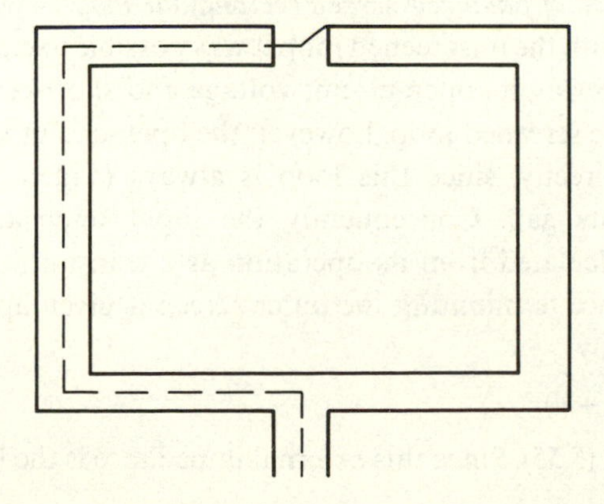

between the inner conductors of two coaxial lines, which is twice that of the single coaxial characteristic impedance Z_{0i}, as shown in Fig. 5.10.

The current I_g at the gap is given by

$$I_g = \frac{V_g}{Z_g} \tag{5.85}$$

Fig. 5.10. Transmission line developed form of balanced screened loop.

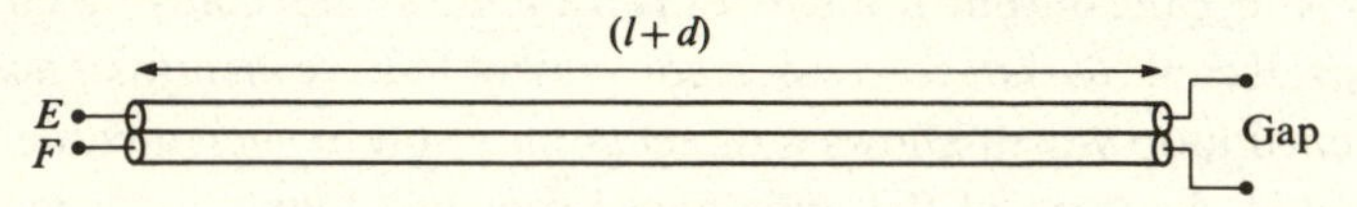

where Z_g is the input impedance at these terminals, which is related to the load impedance Z_L connected across EF by the transmission line equation

$$Z_g = 2Z_{0i}\frac{Z_L \cos \beta(l+d) + j2Z_{0i} \sin \beta(l+d)}{2Z_{0i} \cos \beta(l+d) + jZ_L \sin \beta(l+d)} \tag{5.86}$$

5.8.4 *Calculation of voltage V_g across gap in outer screen*

The outer surface of the screen is excited by the incoming wave, and behaves as an unscreened loop with a load equal to Z_g at the middle of the top side. Hence the voltage across the gap for a vertically polarised wave exciting the sides orthogonal to the gap is given by

$$V_g = \frac{V_{oc}}{Z_i + Z_g} Z_g$$

where V_{oc}, Z_i and Z_g are defined by eqns (5.37), (5.47) and (5.86) respectively.

When the direction of the incoming wave is perpendicular to that previously considered, so that the horizontal sides only are excited, then the same equation for the gap voltage applies, with V_g now given by eqn (5.44).

5.8.5 *Source impedance of balanced screen rectangular loop*

When dealing with the unscreened loop it was possible to calculate its input impedance knowing its open-circuit voltage and short-circuited current. In the case of the screened loop, however, the open-circuit voltage cannot be calculated directly, since this loop is always loaded by the impedance Z_g across its gap. Consequently the input impedance at terminals EF will be calculated from its operation as a transmitter.

The external impedance terminating the outer screen is given approximately from eqn (5.47) by

$$Z = j2Z_c \tan k(l+d)$$

where Z_c is given by eqn (5.55). Since this external impedance is the load at the end of the transmission line fed at the terminals EF this gives, for the input impedance at EF,

$$Z_i = j4Z_{0i}\frac{Z_c \tan k(l+d) + Z_{0i} \tan \beta(l+d)}{Z_{0i} - Z_c \tan k(l+d) \tan \beta(l+d)} \tag{5.87}$$

5.9 Transmitting loop antenna

Although most loop antennas are used for receiving purposes they can in principle be used also for transmitting. Properties such as radiation pattern, power gain and input impedance are identical in both cases, though the current distributions will in general be different.

5.9.1 *Radiation pattern of circular loop antenna carrying uniform in-phase current*

Let the circular loop antenna of radius a be positioned in the xy-plane with its centre at the origin, as shown in Fig. 5.11. At an observation point $P(r, \theta, \phi)$ the magnetic vector potential due to a current flowing in the element $ad\phi'$ has components

$$\left.\begin{aligned} dA_x &= -\frac{\mu a I}{4\pi}\frac{e^{-jkr}}{r}\sin\phi'\,d\phi' \\ &\text{and} \\ dA_y &= \frac{\mu a I}{4\pi}\frac{e^{-jkr}}{r}\cos\phi'\,d\phi' \end{aligned}\right\} \tag{5.88}$$

where r is the distance from the source element to the point of observation. When this observation point is in the far field

$$r \approx r_0 - a\sin\theta\cos(\phi - \phi')$$

so that the total vector potential components become, for a constant, in-

Fig. 5.11. Transmitting circular loop with uniform current.

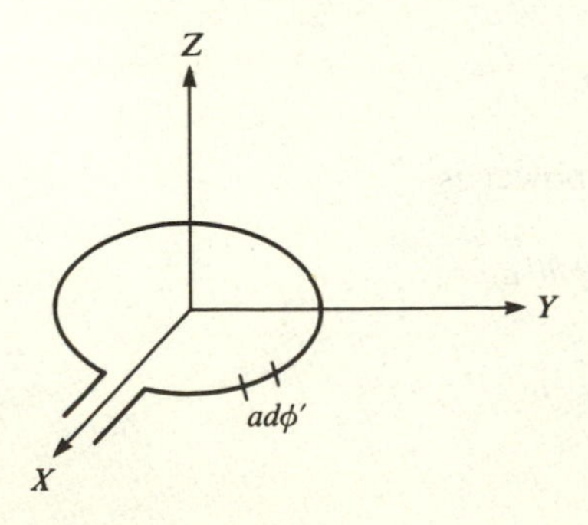

phase current in the loop

$$\left.\begin{aligned} A_x &= -\frac{\mu a I}{4\pi r_0} e^{-jkr_0} \int_0^{2\pi} e^{jka\sin\theta\cos(\phi-\phi')} \sin\phi'\, d\phi' \\ A_y &= \frac{\mu a I}{4\pi r_0} e^{-jkr_0} \int_0^{2\pi} e^{jka\sin\theta\cos(\phi-\phi')} \cos\phi'\, d\phi' \end{aligned}\right\} \tag{5.89}$$

Using the standard results

$$\int_0^{2\pi} e^{ju\cos(\phi-\phi')} \begin{matrix}\cos n\phi' \\ \sin n\phi'\end{matrix}\, d\phi' = 2\pi j^n \begin{matrix}\cos n\phi \\ \sin n\phi\end{matrix}\, J_n(u)$$

gives

$$A_x = -j\frac{\mu a I}{2}\frac{e^{-jkr_0}}{r_0} J_1(ka\sin\theta)\sin\phi$$

and

$$A_y = j\frac{\mu a I}{2}\frac{e^{-jkr_0}}{r_0} J_1(ka\sin\theta)\cos\phi \tag{5.90}$$

Converting to cylindrical coordinates using

$$\mathbf{a}_\phi = -\mathbf{a}_x \sin\phi + \mathbf{a}_y \cos\phi$$

gives for the total vector potential

$$A_\phi = j\frac{\mu a I}{2}\frac{e^{-jkr_0}}{r_0} J_1(ka\sin\theta) \tag{5.91}$$

Since for a constant current in the wire no charges exist on it, this means that the total electric field is given by

$$E_\phi = \frac{\omega\mu a I}{2}\frac{e^{-jkr_0}}{r_0} J_1(ka\sin\theta) \tag{5.92}$$

In order that the condition that the current is constant in magnitude and phase can be satisfied round the periphery of the loop, it is necessary that

$$ka \ll 1$$

For this situation

$$E_\phi \approx \frac{Z_0 k^2 a^2 I}{4}\frac{e^{-jkr_0}}{r_0}\sin\theta \tag{5.93}$$

Consequently the total radiated power is

$$\begin{aligned} W &= \int_0^{\pi}\int_0^{2\pi} \frac{|E_\phi|^2}{2Z_0} r_0^2 \sin\theta\, d\theta\, d\phi \\ &= \int_0^{\pi} \frac{\pi Z_0 k^4 a^4 I^2}{16} \sin^3\theta\, d\theta \\ &= 10\pi^2 (ka)^4 I^2 \end{aligned} \tag{5.94}$$

This gives the radiation resistance for the small loop as

$$R_r = 20k^4(\text{loop area})^2 \tag{5.95}$$

in agreement with eqn (5.51) for a small rectangular loop.

5.10 Radiation pattern of large rectangular loop

The large rectangular loop with sides of length l, d in the x- and y-directions, respectively, is positioned in the xy-plane symmetrically about the origins, as shown in Fig. 5.12. If the variable s is used to denote distance along the loop wire from the feed terminals, then an approximate general equation for the current in the loop is

$$I = I_0 \cos k(l + d - s) \tag{5.96}$$

where I_0 is the current at the short-circuited termination of the loop wire at the point furthest from these terminals. In particular, for each side of the loop the current is given by

$$\left.\begin{aligned}
I_1 &= I_0 \cos k\left(\frac{l+d}{2} - x'\right) && -\frac{l}{2} \leqslant x' \leqslant \frac{l}{2} \\
I_2 &= I_0 \cos k(l + d \pm y') && -\frac{d}{2} \leqslant y' \leqslant 0 \\
& && 0 \leqslant y' \leqslant \frac{d}{2} \\
I_3 &= -I_0 \cos k\left(\frac{l+d}{2} - x'\right) && -\frac{l}{2} \leqslant x' \leqslant \frac{l}{2} \\
I_4 &= I_0 \cos ky' && -\frac{d}{2} \leqslant y' \leqslant \frac{d}{2}
\end{aligned}\right\} \tag{5.97}$$

Due to these currents, vector potentials in the x- and y-directions are set up at every point in space so that, at a general observation point $P(r, \theta, \phi)$,

$$\begin{aligned}
A_x = \frac{\mu I_0}{4\pi}\left\{\int_{-l/2}^{l/2} \cos k\left(\frac{l+d}{2} - x'\right)\frac{e^{-jkr_1}}{r_1}\,dx' \right. \\
\left. - \int_{-l/2}^{l/2} \cos k\left(\frac{l+d}{2} - x'\right)\frac{e^{-jkr_2}}{r_2}\,dx'\right\}
\end{aligned} \tag{5.98}$$

Fig. 5.12. Transmitting rectangular loop with non-uniform current.

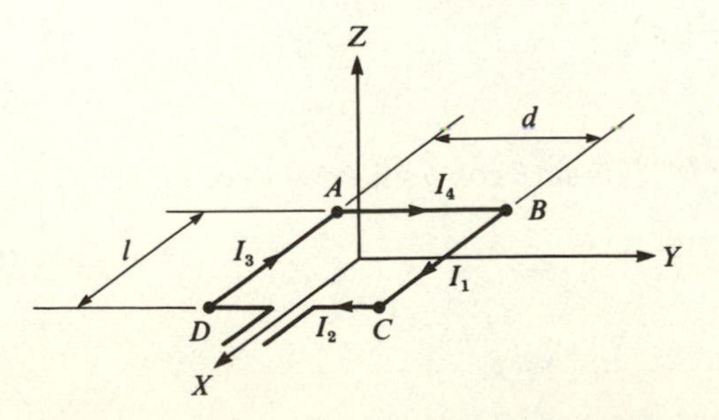

where $r_{\frac{1}{2}}$ are distances from the observation point to current elements $dx_{\frac{1}{2}}$ on wires AD and BC respectively. These distances are related in the far field to the distance r to the origin by the equations

$$\left.\begin{aligned} r_1 &= r - \frac{d}{2}\sin\theta\sin\phi - x'\sin\theta\cos\phi \\ r_2 &= r + \frac{d}{2}\sin\theta\sin\phi - x'\sin\theta\cos\phi \end{aligned}\right\} \tag{5.99}$$

Hence in the radiation field

$$\begin{aligned} A_x = \frac{\mu I_0}{4\pi r} e^{-jkr} \Bigg\{ & e^{jk(kd/2)\sin\theta\sin\phi} \int_{-l/2}^{l/2} \\ & \times \cos k\left(\frac{l+d}{2} - x'\right) e^{jkx'\sin\theta\cos\phi}\, dx' \\ & - e^{-j(kd/2)\sin\theta\sin\phi} \int_{-l/2}^{l/2} \cos k\left(\frac{l+d}{2} - x'\right) e^{jkx'\sin\theta\cos\phi}\, dx' \Bigg\} \\ = \frac{j\mu I_0}{2\pi r} & e^{-jkr} \sin\left(\frac{kd}{2}\sin\theta\sin\phi\right) \int_{-l/2}^{l/2} \\ & \times \cos k\left(\frac{l+d}{2} - x'\right) e^{jkx'\sin\theta\cos\phi}\, dx' \end{aligned} \tag{5.100}$$

Using the standard integrals

$$\int e^{ax}\cos kx\, dx = \frac{e^{ax}}{k^2 + a^2}(a\cos kx + k\sin kx)$$

and

$$\int e^{ax}\sin kx\, dx = \frac{e^{ax}}{k^2 + a^2}(a\sin kx - k\cos kx)$$

gives for the vector potential in the x-direction,

$$\begin{aligned} A_x = \frac{j\mu I_0}{2\pi} \frac{e^{-jkr}}{r} \frac{\sin\left(\dfrac{kd}{2}\sin\theta\sin\phi\right)}{k^2(1 - \sin^2\theta\cos^2\phi)} \Bigg\{ & \cos k\left(\frac{l+d}{2}\right) \\ & \times [e^{jkx'\sin\theta\cos\phi}(jk\sin\theta\cos\phi\cos kx' + k\sin kx')]_{-l/2}^{l/2} \\ & + \sin k\left(\frac{l+d}{2}\right) \\ & \times [e^{jkx'\sin\theta\cos\phi}(jk\sin\theta\cos\phi\sin kx' - k\cos kx')]_{-l/2}^{l/2} \Bigg\} \end{aligned} \tag{5.101}$$

This simplifies to

$$A_x = \frac{j\mu I_0}{\pi}\frac{e^{-jkr}}{r}\frac{\sin\left(\frac{kd}{2}\sin\theta\sin\phi\right)}{k(1-\sin^2\theta\cos^2\phi)}\left\{\cos k\left(\frac{l+d}{2}\right)\right.$$
$$\times\left[\sin\frac{kl}{2}\cos\left(\frac{kl}{2}\sin\theta\cos\phi\right)\right.$$
$$\left.-\sin\theta\cos\phi\cos\frac{kl}{2}\sin\left(\frac{kl}{2}\sin\theta\cos\phi\right)\right]$$
$$+j\sin k\left(\frac{l+d}{2}\right)\left[\sin\theta\cos\phi\sin\frac{kl}{2}\cos\left(\frac{kl}{2}\sin\theta\cos\phi\right)\right.$$
$$\left.\left.-\cos\frac{kl}{2}\sin\left(\frac{kl}{2}\sin\theta\cos\phi\right)\right]\right\} \tag{5.102}$$

Similarly the vector potential in the y-direction is

$$A_y = \frac{\mu I_0}{4\pi}\left\{\int_{-d/2}^{0} -\cos k(l+d+y')\frac{e^{-jkr_1}}{r_1}dy'\right.$$
$$-\int_{0}^{d/2}\cos k(l+d-y')\frac{e^{-jkr_1}}{r_1}dy'$$
$$\left.+\int_{-d/2}^{d/2}\cos ky'\frac{e^{-jkr_2}}{r_2}dy'\right\} \tag{5.103}$$

where $r_{\frac{1}{2}}$ are now distances from the observation point to current elements $dy_{\frac{1}{2}}$ on wires CD and BA respectively. In the far field the following relationships hold

$$r_1 = r - \frac{l}{2}\sin\theta\cos\phi - y'\sin\theta\sin\phi$$

$$r_2 = r + \frac{l}{2}\sin\theta\cos\phi - y'\sin\theta\sin\phi$$

Hence in this radiation zone A_y becomes

$$A_y = \frac{\mu I_0}{2\pi}\frac{e^{-jkr}}{r}\frac{1}{k(1-\sin^2\theta\sin^2\phi)}$$
$$\times\left\{e^{-j(kl/2)\sin\theta\cos\phi}\left[\sin\frac{kd}{2}\cos\left(\frac{kd}{2}\sin\theta\sin\phi\right)\right.\right.$$
$$\left.-\sin\theta\sin\phi\cos\frac{kd}{2}\sin\frac{kd}{2}\sin\theta\sin\phi\right]$$

$$- e^{(jkl/2)\sin\theta\cos\phi}\left[\cos k(l+d)\left[\sin\frac{kd}{2}\cos\left(\frac{kd}{2}\sin\theta\sin\phi\right)\right.\right.$$

$$\left. - \sin\theta\sin\phi\cos\frac{kd}{2}\sin\left(\frac{kd}{2}\sin\theta\sin\phi\right)\right]$$

$$+ \sin k(l+d)\left[1 - \sin\theta\sin\phi\sin\frac{kd}{2}\sin\left(\frac{kd}{2}\sin\theta\sin\phi\right)\right.$$

$$\left.\left.\left. - \cos\frac{kd}{2}\cos\left(\frac{kd}{2}\sin\theta\sin\phi\right)\right]\right]\right\} \tag{5.104}$$

Transforming to spherical coordinates using

$$A_\theta = A_x\cos\theta\cos\phi + A_y\cos\theta\sin\phi$$

$$A_\phi = -A_x\sin\phi + A_y\cos\phi$$

together with the far field approximations

$$H_{\theta_{FF}} \approx -\frac{1}{\mu}\frac{\partial A_\phi}{\partial r} = \frac{jk}{\mu}A_\phi$$

$$H_{\phi_{FF}} \approx \frac{1}{\mu}\frac{\partial A_\theta}{\partial r} = -\frac{jk}{\mu}A_\theta$$

leads to, for $\theta = \pi/2$, $\phi = 0°$, for the component of H_θ due to A_y,

$$H_{\theta_{1_{FF}}} = \frac{jI_0}{2\pi}\frac{e^{-jkr}}{r}\left\{e^{-j(kl/2)}\sin\frac{kd}{2} - e^{+j(kl/2)}\right.$$

$$\left.\times\left[\cos k(l+d)\sin\frac{kd}{2} + \sin k(l+d)\left(1 - \cos\frac{kd}{2}\right)\right]\right\} \tag{5.105}$$

and

$$H_{\phi_{1_{FF}}} = 0$$

For a small square loop with $kd \ll 1$, this reduces to

$$H_{\theta_{1_{FF}}} = \frac{jI_0 e^{-jkr}}{2\pi r}\frac{k^2d^2}{2}$$

A similar result is obtained in the direction $\theta = \pi/2$, $\phi = \pi/2$ from the vector potential associated with the x-directed current in the loop, with

$$H_{\theta_{2_{FF}}} = \frac{I_0}{\pi}\frac{e^{-jkr}}{r}\sin\frac{kd}{2}\cos k\left(\frac{l+d}{2}\right)\sin\frac{kl}{2} \tag{5.106}$$

and

$$H_{\phi_{2_{FF}}} = 0$$

For the case of a small square loop with $kd \ll 1$ this reduces to

$$H_{\theta_{2_{FF}}} = \frac{I_0 e^{-jkr}}{2\pi r}\frac{k^2d^2}{2}$$

The combined far field magnetic intensity is then given by the sum of eqns (5.105) and (5.106).

Further reading

C.A. Balanis: *Antenna Theory: Analysis and Design*: Harper and Row, New York, 1982.

R.E. Collin and F.J. Zucker: *Antenna Theory*: McGraw-Hill, New York, 1969.

R.W.P. King, H. Rowe Mimno and A.H. Wing: *Transmission Lines, Antennas and Waveguides*: Dover Publications, New York, 1965.

R.W.P. King and C.W. Harrison: *Antennas and Waves: A modern Approach*: The MIT Press, Cambridge, Massachusetts, 1968.

J.D. Kraus: *Antennas*: McGraw-Hill, New York, 1950.

L.G. Stass: Directional properties of wideband loop aerials: PhD thesis, University of Birmingham, 1969.

W.L. Stutzman and G.A. Thiele: *Antenna Theory and Design*: Wiley and Sons, 1981.

6

+ – + – + – + – + – + – + – + – + – + – + – + – + – + – + – + – + – + – + – +

Helical antennas

Although the helix is a regular geometric shape it was not until 1947 that any form of helical antenna was proposed. In that year designs for two basically different types of helical antenna were outlined as shown in Fig. 6.1. In one design radiation takes place normal to the axis of a small diameter helix, and in the other case along the axis of the helix which has a diameter of approximately one wavelength. Each of these two antennas will now be considered in turn.

6.1 Normal mode helical antennas

In these antennas which are used at a frequency where the length of one turn multiplied by the number of turns is much less than one wavelength, radiation takes place perpendicular to the helix axis. This is in keeping with the result to be expected from resolving each turn of the helix

Fig. 6.1. (*a*) normal mode helix; (*b*) axial mode helix.

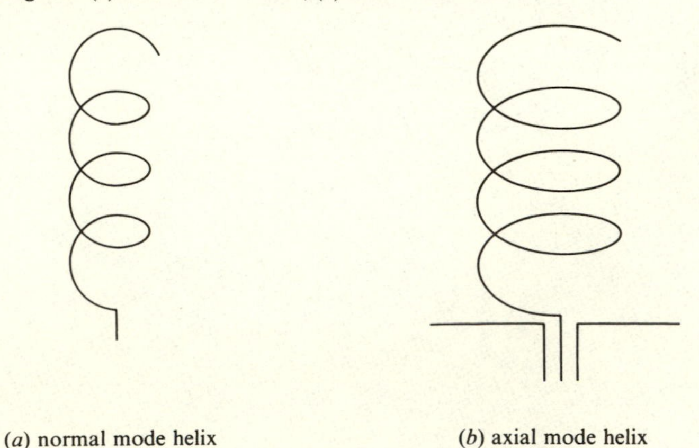

into the summation of a single turn loop plus a straight wire of length equal to the pitch of the helix, as shown in Fig. 6.2. Both these component radiators have their direction of maximum radiation perpendicular to the helix axis.

In Fig. 6.1(*a*) the input impedance of the helix shown in the diagram is clearly capacitive since the overall length of the antenna is much less than one wavelength. But an alternative design is possible in which the open helix of Fig. 6.1(*a*) is modified to a closed form of helix, as shown in Fig. 6.3. Again since the total length of the helix is much less than one wavelength, radiation takes place normal to the axis, but the input impedance between the terminals shown is now inductive and its resistive component differs from that of the open helix, as will be shown in Section 6.4.

Fig. 6.2. Helix resolved into loops plus straight wires equal to pitch length.

Fig. 6.3. Closed normal mode helix.

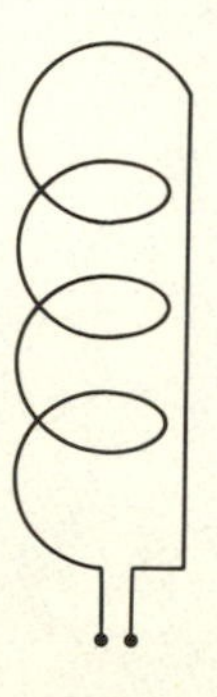

6.2 Analysis of small circular helical antenna – open helix

Referring to a single turn of the circular open helix of radius a and pitch S, as shown in Fig. 6.4, the coordinates of an element of wire ds are

$$\left.\begin{aligned} x &= a\cos u \\ y &= a\sin u \qquad -\pi \leqslant u \leqslant \pi \\ z &= \frac{S}{2\pi}u \end{aligned}\right\} \tag{6.1}$$

A constant current I flowing in this helix sets up a vector potential at an observation point $P(r_0, \theta, \phi)$ given by

$$\mathbf{A} = \frac{\mu I}{4\pi}\int \frac{e^{-jkr}}{r}\mathbf{ds} \tag{6.2}$$

Fig. 6.4. Single turn helical antenna: elevation and plan projections: open helix: AB; closed helix: ABA^1.

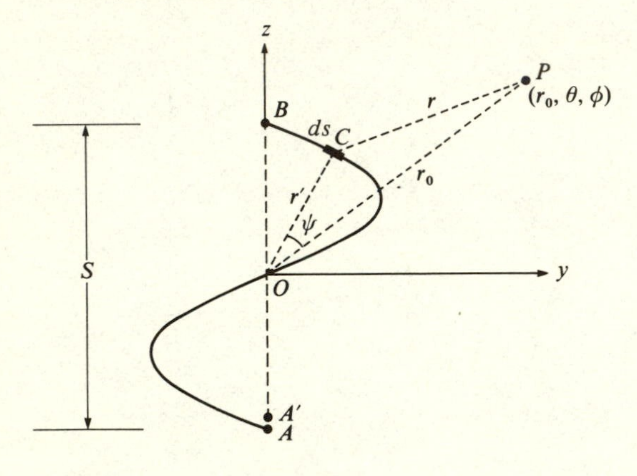

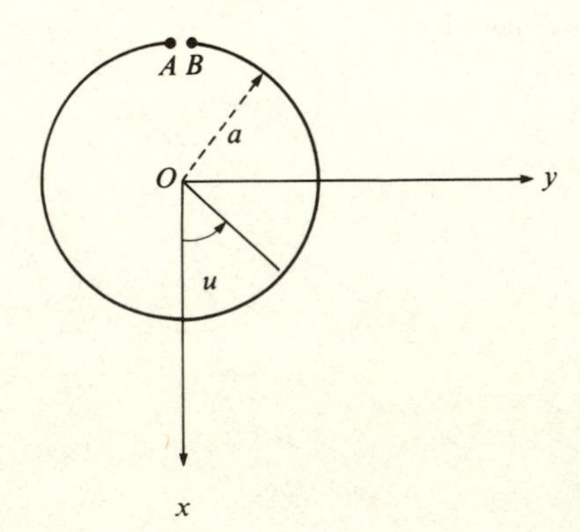

where r is the distance from element ds to P. In the far field,

$$r \approx r_0 - r' \cos\psi$$

where r' is the distance from the origin to ds, and ψ is $\widehat{COP}$ so that

$$r' \cos\psi = a \cos u \sin\theta \cos\phi + a \sin u \sin\theta \sin\phi + \frac{S}{2\pi} u \cos\theta$$

Then the components $A_{x,y,z}$ of this vector potential may be written

$$A_x = \frac{\mu I}{4\pi r_0} \cdot e^{-jkr_0} \cdot \int_{-\pi}^{+\pi} e^{jk(a\cos u \sin\theta\cos\phi + a\sin u \sin\theta \sin\phi + (S/2\pi)u\cos\theta)} \times (-) a \sin u \, du$$

$$A_y = \frac{\mu I}{4\pi r_0} \cdot e^{-jkr_0} \cdot \int_{-\pi}^{+\pi} e^{jk(a\cos u \sin\theta\cos\phi + a\sin u \sin\theta \sin\phi + (S/2\pi)u\cos\theta)} \cdot a \cos u \, du$$

$$A_z = \frac{\mu I}{4\pi r_0} \cdot e^{-jkr_0} \cdot \int_{-\pi}^{+\pi} e^{jk(a\cos u \sin\theta\cos\phi + a\sin u \sin\theta \sin\phi + (S/2\pi)u\cos\theta)} \cdot \frac{S}{2\pi} du \quad (6.3)$$

The integrals may readily be evaluated, subject to the approximation that $ka \ll 1$, by expanding the exponential in the first two terms of its power series to give

$$A_x = -\frac{j\mu\pi I}{2\pi r_0} \cdot e^{-jkr_0} \cdot \left(\frac{Sa}{\lambda} \cos\theta + \frac{ka^2}{2} \sin\theta \sin\phi \right) \quad (6.4)$$

$$A_y = \frac{j\mu\pi I}{4\pi r_0} \cdot e^{-jkr_0} \cdot ka^2 \sin\theta \cos\phi \quad (6.5)$$

$$A_z = \frac{\mu I S}{4\pi r_0} \cdot e^{-jkr_0} \quad (6.6)$$

Hence using $\mathbf{E} = -\nabla V - j\omega \mathbf{A}$ and noting that in spherical coordinates the scalar potential contribution affects the radial component only of the radiation electric field, the following results are obtained:

$$E_r = 0 \quad (6.7)$$

$$E_\theta = -\frac{k^2 Z_0 I}{4\pi r_0} \cdot e^{-jkr_0} \cdot \left(Sa \cos^2\theta \cos\phi - j\frac{S}{k} \sin\theta \right) \quad (6.8)$$

$$E_\phi = \frac{k^2 Z_0 I}{4\pi r_0} \cdot e^{-jkr_0} \cdot (Sa \cos\theta \sin\phi + \pi a^2 \sin\theta) \quad (6.9)$$

Eqns (6.8) and (6.9) show that when S is equal to πka^2, and for all points of observation in the xy-plane, where θ equals $\pi/2$, E_θ is equal to jE_ϕ, which is the condition for circular polarisation. In addition, since it has been specified that $ka \ll 1$, the second term in the final pair of brackets in both eqns (6.8) and (6.9) is much greater than the first term except for small values of θ, i.e. values close to the forward axis of the open helix, so that when S is equal to πka^2 the radiated field is approximately circularly polarised everywhere except where θ is small. But in the particular direction $\theta = \tan^{-1} ka$, $\phi = 3\pi/2$ it will be found that E_ϕ is zero, giving linear polarisation in the direction of this small angle. For a helix of n turns the above expressions for the radiated fields are multiplied by n.

6.3 Analysis of small circular helical antenna – closed helix

The closed helix is formed from the open helix by returning the conductor from its end point B in Fig. 6.4, in a straight line parallel to the z-axis, to the point A' beside its starting point A. Consequently the only vector term in the preceding analysis which requires to be modified is A_z, and this modification consists of adding a component given by

$$A_z' = -\frac{\mu I}{4\pi} \cdot e^{-jkr_0} \int_{-S/2}^{S/2} \frac{1}{r} e^{[jk(z\cos\theta - a\sin\theta\cos\phi)]}\, dz \tag{6.10}$$

The integration is straightforward and leads, in this case without consideration of charge, to

$$E_\theta = -\frac{k^2 Z_0 I}{4\pi r_0} \cdot e^{-jkr_0} \cdot Sa\cos\phi \tag{6.11}$$

Since E_ϕ necessarily remains unchanged it follows from the absence of the quadrature sign in eqn (6.11) that there is no possibility of obtaining circular polarisation from such a closed helix, for which the polarisation is always linear. In the yz-plane for which $\phi = \pm\pi/2$ it follows from eqn (6.11) that E_θ is identically zero, and if, in addition, $\theta = S/\pi a$ with the angle ϕ equal to $-\pi/2$, then it follows from eqn (6.9) that E_ϕ is zero also. Consequently there is a null in the radiation pattern in this direction, which can also be expressed in terms of the pitch angle α of the helix through the equation

$$\tan\theta_{(null)} = 2\tan\alpha$$

6.4 Radiation resistances of open and closed small helical antennas

The total power radiated by the open helix is

$$W = \int_{\phi=0}^{2\pi} \int_{\theta=0}^{\pi} \frac{|E_\theta|^2 + |E_\phi|^2}{2Z_0} r^2 \sin\theta\, d\theta\, d\phi \tag{6.12}$$

where E_θ, E_ϕ are given by eqns (6.8) and (6.9). This gives

$$W = \frac{k^4 Z_0 I^2}{32\pi^2} \int_0^{2\pi} \int_0^{\pi} \bigg(S^2 a^2 \cos^4\theta \cos^2\phi + \frac{S^2}{k^2} \sin^2\theta$$
$$+ S^2 a^2 \cos^2\theta \sin^2\phi + \pi^2 a^4 \sin^2\theta$$

$$+ 2\pi S a^3 \sin\theta \cos\theta \sin\phi \Big) r^2 \sin\theta \, d\theta \, d\phi$$

$$= \frac{k^4 Z_0 I^2}{32\pi^2} \int_0^{\pi} \left(\pi S^2 a^2 \sin\theta \cos^4\theta + \frac{2S^2\pi}{k^2} \sin^3\theta \right.$$

$$\left. + \pi S^2 a^2 \cos^2\theta \sin\theta + 2\pi^3 a^4 \sin^3\theta \right) d\theta \tag{6.13}$$

Using the standard integrals

$$\left.\begin{aligned} &\int_0^{\pi} \cos^4\theta \sin\theta \, d\theta = [-\tfrac{1}{5}\cos^5\theta]_0^{\pi} = \tfrac{2}{5} \\ &\int_0^{\pi} \sin^3\theta \, d\theta = -\tfrac{4}{3} \\ &\int_0^{\pi} \cos^2\theta \sin\theta \, d\theta = [-\tfrac{1}{3}\cos^3\theta]_0^{\pi} = \tfrac{2}{3} \end{aligned}\right\} \tag{6.14}$$

gives

$$W = 10k^2 S^2 I^2 \left[1 + k^2 \left(\frac{4a^2}{10} + \frac{\pi^2 a^4}{S^2} \right) \right] \tag{6.15}$$

Since this power is also equal to $\frac{1}{2}I^2 R_r$, this gives for the radiation resistance of the small open helical antenna

$$R_{ro} = 20k^2 S^2 \left[1 + k^2 a^2 \left(0.4 + \frac{\pi^2 a^2}{S^2} \right) \right] \tag{6.16}$$

Similarly for the closed helix where E_θ, E_ϕ are given by eqns (6.11) and (6.9) the radiated power becomes

$$W = 10\pi^2 k^4 a^4 I^2 \left[1 + \frac{S^2}{\pi^2 a^2} \right] \tag{6.17}$$

Hence the radiation resistance of the small closed helical antenna becomes

$$R_{rc} = 20k^4 \pi^2 a^4 [1 + 4\tan^2\alpha] \tag{6.18}$$

where α is the pitch angle of the helix.

6.5 Axial mode circular helical antenna

The axial mode helical antenna was proposed by Kraus from consideration of the helix used as a slow wave device in travelling wave tubes. Because antennas whose dimensions are comparable with, or greater than, a wavelength, radiate power more readily than antennas which are much smaller than a wavelength, he examined the radiating characteristics of a helix with a diameter equal to one wavelength. Because it was convenient to feed the helix from a coaxial cable, a ground screen was connected to the outer conductor of the cable and the inner conductor was connected to the beginning of the helix. The results obtained were remarkable with regard to the absence of anything which was critical in regard to the construction of the antenna. It is an antenna which, within broad limits, would be difficult to construct in such a way that it would not work.

Referring to Fig. 6.5 the main beam of radiation is directed along the forward axis, $\theta = 0°$, of the helix. This is to be expected since the helix consists approximately of an array of coaxial loops. Considering one of these loops with a travelling wave of current flowing circumferentially in it gives radiation along the axis of the loop since the current in two diametrically opposite elements is codirected when the circumference is one wavelength. But because of progressive phase delay between consecutive loops in travelling from the ground screen to the open end, the total radiation in the background direction, $\theta = \pi$, is cancelled, leaving forward radiation only.

It has been found experimentally that when the number of turns on the helix is greater than $3\frac{1}{2}$, the radiation from the antenna is circularly polarised along the forward axis. Elsewhere the polarisation is elliptical. As the number of turns is increased, the beamwidth for a 13° pitch angle helix decreases from approximately 60° between half-power points with $3\frac{1}{2}$ turns, to 42° between half-power points with ten turns. The cross-section of the helix may be circular, square or even triangular. As with similar travelling wave end-fire antennas such as the Yagi or dielectric rod, the first sidelobe

Fig. 6.5. Axial mode helix.

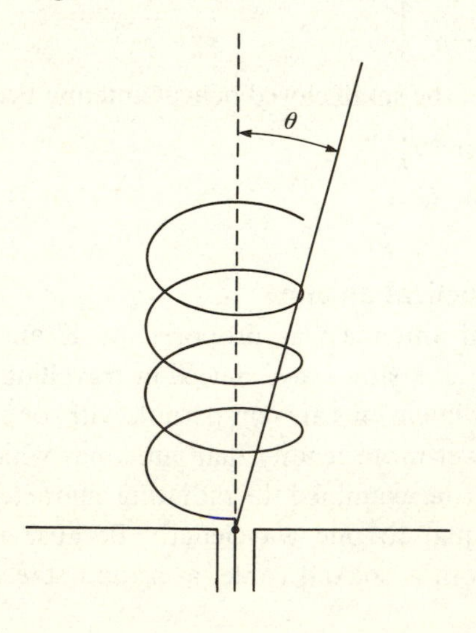

level is high by comparison with arrays of broadside construction, a typical value being 30% of the main beam in the forward direction.

For a 13° pitch angle helix the useful bandwidth of operation extends from a frequency at which the circumference measured in free space wavelengths, *ka*, has a value of 0.75, up to a value of approximately 1.25. Over this frequency range the radiation pattern will become slightly narrower as the frequency increases, but otherwise will retain the same general shape. For smaller pitch angles the useful frequency range of operation decreases without having any compensating advantages, while for pitch angles greater than 16° the increased separation between the terms introduces higher sidelobes. Consequently a pitch angle of 13° is about optimum from a pattern and hence a power gain point of view.

As far as the input impedance of the axial mode helical antenna is concerned, this is substantially purely resistive, with a value of approximately $150ka\Omega$, over the frequency range for which the radiation pattern is as previously described. The input impedance becomes of this form at substantially the same frequency, and hence the same value of *ka*, as the pattern takes on its characteristic form and this provides a convenient measurement of the lowest operating frequency of the antenna. But the input impedance remains purely resistive at frequencies well above the frequency when the radiation pattern breaks up, so that there is no corresponding impedance measurement of the upper frequency limit of the antenna.

In addition to the relatively large side lobes associated with the helix the backward lobe may also be large. The size of this lobe is a function of the ground plane size, but if the ground plane is made much greater than 1λ square it is found that ground plane currents also affect the main forward lobe of the antenna. A better solution in this case is to use a conical ground plane, so that the antenna then becomes a horn antenna with a helical feed.

6.6 Analysis of axial mode helical antenna

The helical antenna to be analysed is shown in Fig. 6.5. The variables which would have to be included in an exact analysis would be the radius a of the helix, the pitch angle α, the number of turns N, the radius of wire used for winding the helix, and the shape and size of the planar ground screen. The object of the analysis would be to calculate the distribution of current in magnitude and phase over both the helical conductor and both sides of the ground screen. Given this information it would then be possible to calculate the radiation field at any point in space.

Because this is not possible, a number of simplifications will be made so as to produce a model for the helical antenna which can be analysed, and which produces results in agreement with experiment. Firstly the ground plane will be ignored. Secondly the helix will be taken to be infinitely long in order to be able to calculate the phase velocity of propagation along it. Thirdly the actual wire helix will be replaced by a circular cylindrical surface, as in Fig. 6.6, which conducts only in the helical direction, and in this direction the conductivity is assumed to be infinite. In the orthogonal direction along the surface the conductivity is zero. This model could be constructed from fine insulated wire wound side by side on the cylindrical surface at the required pitch angle, and is known as the sheath helix model.

6.6.1 *Analysis of infinite sheath helix model*

Because it is expected that current will flow in the helical direction on the surface of the sheath helix cylinder, it follows that there will be an electric field in that direction and a magnetic field orthogonal to that direction. There will thus, in general, be axial components of both electric and magnetic field both inside and outside the sheath helix, and consequently scalar wave equations for both E_z and H_z must be satisfied everywhere in space. These equations are

$$\left.\begin{aligned} \nabla^2 E_z + k^2 E_z = 0 \\ \nabla^2 H_z + k^2 H_z = 0 \end{aligned}\right\} \qquad (6.19)$$

Fig. 6.6. Coordinate system for sheath–helix model. (After Maclean and Farvis, 1962.)

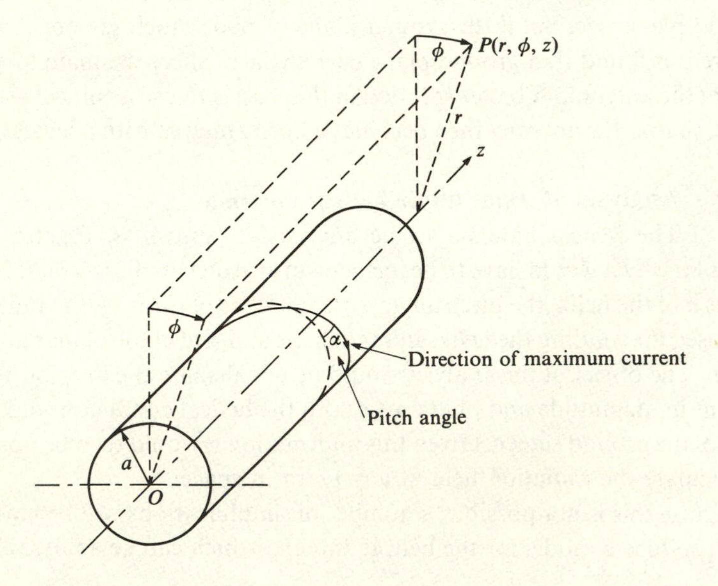

When written in cylindrical coordinates these equations are of the form

$$\frac{1}{r}\frac{\partial}{\partial r}\left(r\frac{\partial\psi}{\partial r}\right)+\frac{1}{r^2}\frac{\partial^2\psi}{\partial\phi^2}+\frac{\partial^2\psi}{\partial z^2}+k^2\psi=0 \tag{6.20}$$

where ψ stands for both E_z and H_z.

Assuming a solution of the form $\psi = f(r)e^{-jm\phi}e^{-j\beta z}$, where β is the axial phase constant and m must be integral if the fields are to be single valued in ϕ, there results

$$r\frac{\partial^2 f}{\partial r^2}+r\frac{\partial f}{\partial r}-[(\beta^2-k^2)r^2+m^2]f=0 \tag{6.21}$$

For slow wave propagation along the helix in the z-direction, $\beta > k$, and the solutions for the axial electric and magnetic fields become

$$E_z=[A_m I_m(\gamma r)+C_m K_m(\gamma r)]e^{-jm\phi}e^{-j\beta z} \tag{6.22a}$$

$$H_z=[B_m I_m(\gamma r)+D_m K_m(\gamma r)]e^{-jm\phi}e^{-j\beta z} \tag{6.22b}$$

where A_m, B_m, C_m, and D_m are constants and

$$\gamma^2=\beta^2-k^2 \tag{6.23}$$

Outside the helix the I_m functions are not suitable since their values become infinite as the radius tends to infinity. Likewise inside the helix the K_m functions must be discarded as these become infinite as the radius tends to zero. Hence the desired solutions become

$$\left.\begin{aligned}
E_z^i &= A_m^i I_m(\gamma r)e^{-jm\phi}e^{-j\beta z} && 0\leqslant r\leqslant a\\
E_z^e &= A_m^e K_m(\gamma r)e^{-jm\phi}e^{-j\beta z} && a\leqslant r\leqslant \infty\\
H_z^i &= B_m^i I_m(\gamma r)e^{-jm\phi}e^{-j\beta z} && 0\leqslant r\leqslant a\\
H_z^e &= B_m^e K_m(\gamma r)e^{-jm\phi}e^{-j\beta z} && a\leqslant r\leqslant \infty
\end{aligned}\right\} \tag{6.24}$$

where a is the radius of the helix and the superscripts i and e refer to regions internal and external to the helical surface respectively.

Using the solutions for the z-directed components of fields, the r- and ϕ-components can be obtained, as outlined in standard texts on electromagnetic theory, from the equations

$$\left.\begin{aligned}
E_r &= \frac{j\beta}{\gamma^2}\frac{\partial E_z}{\partial r}+\frac{j\omega u}{\gamma^2 r}\frac{\partial H_z}{\partial\phi}\\
H_r &= -\frac{j\omega\varepsilon}{\gamma^2 r}\frac{\partial E_z}{\partial\phi}+\frac{j\beta}{\gamma^2}\frac{\partial H_z}{\partial r}\\
E_\phi &= \frac{j\beta}{\gamma^2 r}\frac{\partial E_z}{\partial\phi}-\frac{j\omega\mu}{\gamma^2}\frac{\partial H_z}{\partial r}\\
H_\phi &= \frac{j\omega\varepsilon}{\gamma^2}\frac{\partial E_z}{\partial r}+\frac{j\beta}{\gamma^2 r}\frac{\partial H_z}{\partial\phi}
\end{aligned}\right\} \tag{6.25}$$

where

$$\gamma^2 = \beta^2 - k^2 \tag{6.26}$$

These equations give for the fields inside the helix

$$\left.\begin{aligned}
E_r^i &= \left[\frac{j\beta}{\gamma} A_m^i I'_m(\gamma r) + \frac{\omega\mu m}{\gamma^2 r} B_m^i I_m(\gamma r)\right] e^{-jm\phi} e^{-j\beta z} \\
E_\phi^i &= \left[\frac{m\beta}{\gamma^2 r} A_m^i I_m(\gamma r) - \frac{j\omega\mu}{\gamma} B_m^i I'_m(\gamma r)\right] e^{-jm\phi} e^{-j\beta z} \\
H_r^i &= \left[-\frac{m\omega\varepsilon}{\gamma^2 r} A_m^i I_m(\gamma r) + \frac{j\beta}{\gamma} B_m^i I'_m(\gamma r)\right] e^{-jm\phi} e^{-j\beta z} \\
H_\phi^i &= \left[\frac{j\omega\varepsilon}{\gamma} A_m^i I'_m(\gamma r) + \frac{m\beta}{\gamma^2 r} B_m^i I_m(\gamma r)\right] e^{-jm\phi} e^{-j\beta z}
\end{aligned}\right\} \tag{6.27}$$

Similarly for the external fields

$$\left.\begin{aligned}
E_r^e &= \left[\frac{j\beta}{\gamma} A_m^e K'_m(\gamma r) + \frac{\omega\mu m}{\gamma^2 r} B_m^e K_m(\gamma r)\right] e^{-jm\phi} e^{-j\beta z} \\
E_\phi^e &= \left[\frac{m\beta}{\gamma^2 r} A_m^e K_m(\gamma r) - \frac{j\omega\mu}{\gamma} B_m^e K'_m(\gamma r)\right] e^{-jm\phi} e^{-j\beta z} \\
H_r^e &= \left[-\frac{m\omega\varepsilon}{\gamma^2 r} A_m^e K_m(\gamma r) + \frac{j\beta}{\gamma} B_m^e K'_m(\gamma r)\right] e^{-jm\phi} e^{-j\beta z} \\
H_\phi^e &= \left[\frac{j\omega\varepsilon}{\gamma} A_m^e K'_m(\gamma r) + \frac{m\beta}{\gamma^2 r} B_m^e K_m(\gamma r)\right] e^{-jm\phi} e^{-j\beta z}
\end{aligned}\right\} \tag{6.28}$$

The boundary conditions at the surface $r = a$ require that the tangential electric fields outside and inside the helix are continuous in the ϕ- and z-directions, and must be equated to zero along the directions of the helical winding, together with the requirement that the tangential magnetic field along the helical conductor be continuous. Formally these conditions are expressed by the equations

$$E_z^i = E_z^e \tag{6.29a}$$

$$E_\phi^i = E_\phi^e \tag{6.29b}$$

and, from Fig. 6.6

$$E_z^i \sin\alpha = -E_\phi^i \cos\alpha \tag{6.29c}$$

and

$$H_z^i \sin\alpha + H_\phi^i \cos\alpha = H_z^e \sin\alpha + H_\phi^e \cos\alpha \tag{6.29d}$$

The four eqns (6.29(a)–(d)) enable the four unknown constants A_m^i, A_m^e, B_m^i and B_m^e to be eliminated, leading to the characteristic equation

$$\frac{I'_m(\gamma a)K'_m(\gamma a)}{I_m(\gamma a)K_m(\gamma a)} = -\frac{(\gamma^2 a^2 + m\beta a \cot\alpha)^2}{k^2 a^2 \gamma^2 a^2 \cot^2\alpha} \tag{6.30}$$

6.6.2 *Numerical solution of characteristic equation for sheath helix*

For a given pitch angle α and a specified numeric m the characteristic equation can readily be solved to show the relation between the phase velocity of propagation along the z-axis and the diameter of the helix measured in free space wavelengths. For helical antenna operation m is taken equal to unity since this gives an angular variation of 2π radius when the circumference is approximately one free space wavelength as is customary for such antennas. Then for an arbitrary value of γa equal to 0.25, say, the left-hand side of eqn (6.30) is equal to -17.93. Using eqn (6.30), the right-hand side then leads to a quadratic equation in ka, the solutions to which, for a pitch angle of 13°, are 0.83 and 0.59. Since the phase velocity relative to that of light is given by the ratio $ka/\beta a$ the corresponding phase velocities are 0.96 and 0.92 for these different values of ka. Similarly, by selecting additional values of γa, the numerical solution of eqn (6.30) is obtained, as shown in Fig. 6.7.

6.6.3 *Application of characteristic equation for sheath helix to end-fire helical antenna*

From Fig. 6.7 it is seen that for a pitch angle of 13° four waves can propagate up to a frequency corresponding to ka equal to 0.68, but only two waves for values of ka above this figure. When four waves are present the current distribution is understandably complex. Two of the waves have comparable, but not equal, phase velocities, while the other two waves have much lower velocities over most of the range of ka. A considerable simplification arises, however, when the number of waves reduces to two. The phase velocities are widely separated in this case and from the field eqns (6.27) and (6.28) it can be shown that the power carried by the mode with the higher phase velocity is very much larger than the power associated with the lower phase velocity mode. The experimental results for the radiation pattern agree with those calculated from this higher velocity of propagation mode also, so that the other mode will therefore be ignored.

Considering therefore this higher velocity mode travelling axially at a phase velocity greater than 0.8 times the velocity of light, it is desired to find the radiation pattern of a finite length helical antenna, assuming that the phase velocity for the infinite helix is unaltered when the length of the helix is reduced to a finite value. Referring to Fig. 6.8 which depicts a helical antenna aligned perpendicular to the xy-plane, gives for the cylindrical components of current density associated with the travelling wave of current in the helical direction,

$$\left.\begin{aligned} J_{\phi'} &= J\cos\alpha e^{-j\beta z'}e^{-j\phi'} \\ J_{z'} &= J\sin\alpha e^{-j\beta z'}e^{-j\phi'} \end{aligned}\right\} \qquad (6.31)$$

Fig. 6.7. Velocity ratio as function of circumference for various pitch angles.

$m = +1$	$m + 1$ (backward wave)
(1) $\alpha = 1°$	(a) $\alpha = 1°$
(2) $\alpha = 4°$	(b) $\alpha = 4°$
(3) $\alpha = 7°$	(c) $\alpha = 7°$
(4) $\alpha = 10°$	(d) $\alpha = 10°$
(5) $\alpha = 13°$	(e) $\alpha = 13°$
(6) $\alpha = 16°$	(f) $\alpha = 16°$

(After Maclean and Farvis, 1962.)

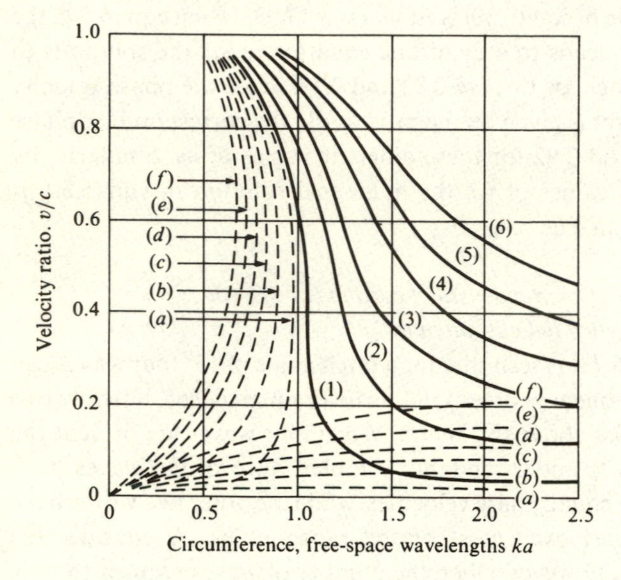

Fig. 6.8. Sheath helical antenna aligned perpendicular to xy-plane.

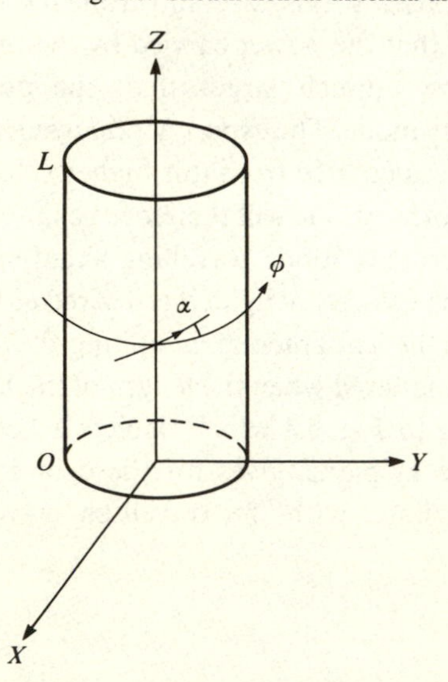

These set up rectangular components of magnetic vector potential at an observation point $P(r, \theta, \phi)$ in the radiation field, given by

$$\left.\begin{aligned}
A_x &= -\frac{\mu}{4\pi}\int_0^L\int_0^{2\pi}\frac{Ja\cos\alpha}{r}\sin\phi' e^{-j\beta z'}e^{-j\phi'}e^{-jkr}\,d\phi'\,dz' \\
A_y &= -\frac{\mu}{4\pi}\int_0^L\int_0^{2\pi}\frac{Ja\cos\alpha}{r}\sin\phi' e^{-j\beta z'}e^{-j\phi'}e^{-jkr}\,d\phi'\,dz' \\
A_z &= \frac{\mu}{4\pi}\int_0^L\int_0^{2\pi}\frac{Ja\sin\alpha}{r}e^{-j\beta z'}e^{-j\phi'}e^{-jkr}\,d\phi'\,dz'
\end{aligned}\right\} \tag{6.32}$$

where, in the radiation field from Fig. 6.9,

$$\begin{aligned}
r \approx r_0 - \rho\cos\psi &= r_0 - \rho\cos\theta\cos\theta' - \rho\sin\theta\sin\theta'\cos\phi\cos\phi' \\
&\quad - \rho\sin\theta\sin\theta'\sin\phi\sin\phi' \\
&= r_0 - x'\sin\theta\cos\phi - y'\sin\theta\sin\phi - z'\cos\theta \\
&= r_0 - a\cos\phi'\sin\theta\cos\phi - a\sin\phi'\sin\theta\sin\phi - z'\cos\theta
\end{aligned} \tag{6.33}$$

Hence in the radiation field

$$A_x = -\frac{\mu Ja\cos\alpha}{4\pi r_0}e^{-jkr_0}\int_0^L\int_0^{2\pi}\sin\phi' e^{-j(\beta z' - kz'\cos\theta)}$$

$$\times\, e^{-j\phi'}e^{jka\sin\theta\cos(\phi-\phi')}\,d\phi'\,dz'$$

Fig. 6.9. Radiation from current on surface of sheath helix.

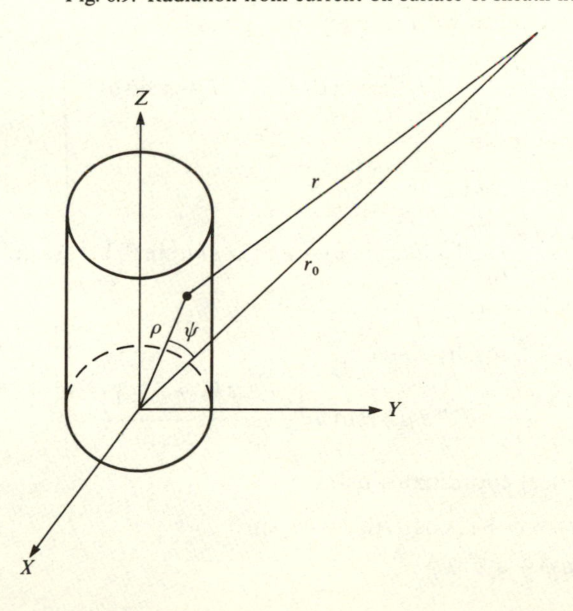

$$A_y = -\frac{\mu J a \cos\alpha}{4\pi r_0} e^{-jkr_0} \int_0^L \int_0^{2\pi} \cos\phi' e^{-j(\beta z' - kz'\cos\theta)} \times e^{-j\phi'} e^{jka\sin\theta\cos(\phi-\phi')} d\phi'\, dz' \tag{6.34}$$

$$A_z = \frac{\mu J a \sin\alpha}{4\pi r_0} e^{-jkr_0} \int_0^L \int_0^{2\pi} e^{-j(\beta z' - kz'\cos\theta)} e^{-j\phi'} e^{jka\sin\theta\cos(\phi-\phi')} d\phi'\, dz'$$

Using

$$\int_0^{2\pi} e^{ju\cos(\phi-\phi')} \begin{matrix}\cos n\phi' \\ \sin n\phi'\end{matrix} d\phi' = 2\pi j^n \begin{matrix}\cos n\phi \\ \sin n\phi\end{matrix} J_n(u)$$

gives

$$A_x = -\frac{\mu J a \cos\alpha}{4\pi r_0} e^{-jkr_0} \int_0^L e^{-jz'(\beta - k\cos\theta)} \times \{-j\pi J_0(ka\sin\theta) - j\pi\cos 2\phi J_2(ka\sin\theta) - \pi\sin 2\phi J_2(ka\sin\theta)\}\, dz'$$

$$A_y = \frac{\mu J a \cos\alpha}{4\pi r_0} e^{-jkr_0} \int_0^L e^{-jz'(\beta - k\cos\theta)} \times \{\pi J_0(ka\sin\theta) - \pi\cos 2\phi J_2(ka\sin\theta) + j\pi\sin 2\phi J_2(ka\sin\theta)\}\, dz' \tag{6.35}$$

$$A_z = \frac{\mu J a \sin\alpha}{4\pi r_0} e^{-jkr_0} \int_0^L e^{-jz'(\beta - k\cos\theta)} \times \{2\pi j\cos\phi J_1(ka\sin\theta) + 2\pi\sin\phi J_1(ka\sin\theta)\}\, dz'$$

Performing the integration with respect to z' gives

$$\left.\begin{aligned} A_x &= -\frac{\mu J a \cos\alpha}{4}\frac{e^{-jkr_0}}{r_0}[J_0(ka\sin\theta) + e^{-j2\phi}J_2(ka\sin\theta)] \times \frac{[e^{-jL(\beta - k\cos\theta)} - 1]}{(\beta - k\cos\theta)} \\ A_y &= j\frac{\mu J a \cos\alpha}{4}\frac{e^{-jkr_0}}{r_0}[J_0(ka\sin\theta) - e^{-j2\phi}J_2(ka\sin\theta)] \times \frac{[e^{-jL(\beta - k\cos\theta)} - 1]}{(\beta - k\cos\theta)} \\ A_z &= \frac{\mu J a \sin\alpha}{2}\frac{e^{-jkr_0}}{r_0} e^{-j\phi} J_1(ka\sin\theta)\frac{[e^{-jL(\beta - k\cos\theta)} - 1]}{(\beta - k\cos\theta)} \end{aligned}\right\} \tag{6.36}$$

Converting the spherical coordinates using

$$a_\theta = a_x\cos\theta\cos\phi + a_y\cos\theta\sin\phi - a_z\sin\theta$$
$$a_\phi = -a_x\sin\phi + a_y\cos\phi$$

gives

$$A_\theta = -\frac{\mu J a}{4}\frac{e^{-jkr_0}}{r_0}\frac{[e^{-jL(\beta - k\cos\theta)} - 1]}{(\beta - k\cos\theta)}\{[J_0(ka\sin\theta)\cos\theta e^{-j\phi} + e^{-j\phi}J_2(ka\sin\theta)\cos\theta]\cos\alpha + 2J_1(ka\sin\theta)\sin\alpha e^{-j\phi}\sin\theta\} \quad (6.37)$$

$$A_\phi = j\frac{\mu J a}{4}\frac{e^{-jkr_0}}{r_0}\frac{[e^{-jL(\beta - k\cos\phi)} - 1]}{(\beta - k\cos\theta)}\{J_0(ka\sin\theta)e^{-j\phi} - e^{-j\phi}J_2(ka\sin\theta)\}\cos\alpha \quad (6.38)$$

Then since in the radiation field

$$E_\theta = \frac{1}{\sqrt{(\mu\varepsilon)}}\frac{\partial A_\theta}{\partial r}$$

and

$$E_\phi = \frac{1}{\sqrt{(\mu\varepsilon)}}\frac{\partial A_\phi}{\partial r}$$

the radiation electric fields become

$$E_\theta = jkZ_0\frac{Ja}{4}\frac{e^{-jkr_0}}{r_0}\frac{[e^{-jL(\beta - k\cos\theta)} - 1]}{(\beta - k\cos\theta)}\{[J_0(ka\sin\theta) + J_2(ka\sin\theta)]e^{-j\phi}\cos\alpha\cos\theta + 2J_1(ka\sin\theta)\sin\alpha\sin\theta e^{-j\phi}\} \quad (6.39)$$

$$E_\phi = kZ_0\frac{Ja}{4}\frac{e^{-jkr_0}}{r_0}\frac{[e^{-jL(\beta - k\cos\theta)} - 1]}{(\beta - k\cos\theta)}\{J_0(ka\sin\theta) - J_2(ka\sin\theta)\}e^{-j\phi}\cos\alpha \quad (6.40)$$

It will be noted that along the forward axis

$$\left.\begin{aligned} E_\theta(\theta = 0°) &= j\frac{kZ_0 Ja}{4}\frac{e^{-jkr_0}}{r_0}\frac{[e^{-j(\beta - k)L} - 1]}{(\beta - k)}e^{-j\phi}\cos\alpha \\ E_\phi(\theta = 0°) &= \frac{kZ_0 Ja}{4}\frac{e^{-jkr_0}}{r_0}\frac{[e^{-j(\beta - k)L} - 1]}{(\beta - k)}e^{-j\phi}\cos\alpha \end{aligned}\right\} \quad (6.41)$$

so that the fields in this direction are circularly polarised. Moreover when $(\beta - k)L$ is equal to π radians, the fields in the forward direction are a maximum. This condition is known as the Hansen–Woodyard condition and applies to every travelling wave end-fire antenna.

6.6.4 *Radiation from single wire axial mode helical antenna*

In the previous analysis the radiating antenna took the form of a sheath helix, which as outlined earlier could take the form of fine insulated wire wound side by side on a cylindrical surface. Although it would be possible in principle to excite such an antenna at microwave frequencies from a circular waveguide, helical antennas are normally used at lower frequencies, and are excited by a coaxial line fed against a ground plane. Along the wire forming the helix it is known from direct measurement of the current along a six turn helix using a loop probe that the current may be expressed, for Fig. 6.10, approximately in the idealised form of a travelling wave, i.e.

$$I(s) = I_0 e^{-j\beta s} \tag{6.42}$$

where s is the coordinate of distance along the arc of the helical conductor.

If L is the length of one turn of the helix which is assumed to begin in the xy-plane at the point with cylindrical coordinates $(a, 0°)$, then

$$s = \frac{\phi'}{2\pi} L \tag{6.43}$$

where ϕ' is the coordinate of angle of the element ds. The rectangular coordinates of the element ds are given in parametric form by the equations

$$\left.\begin{aligned} x &= a\cos\phi' \\ y &= a\sin\phi' \\ z &= \frac{S}{2\pi}\phi' \end{aligned}\right\} \tag{6.44}$$

where S is the pitch of the helix.

Due to the travelling wave of current in an element ds of the helix a magnetic vector potential with the following rectangular components is set

Fig. 6.10. Radiation from single wire helical antenna.

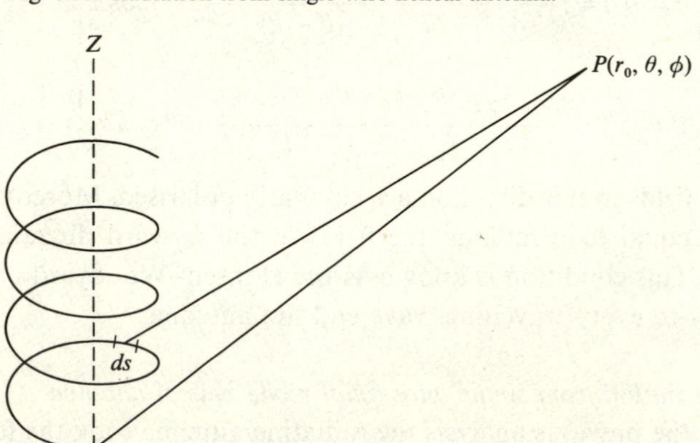

up at a point of observation $P(r_0, \theta, \phi)$;

$$\left.\begin{aligned} dA_x &= \frac{\mu I_0}{4\pi r} e^{-j((\beta L/2\pi)\phi' + kr)} \cdot (-a \sin \phi') d\phi' \\ y &\qquad \cdot (a \cos \phi') d\phi' \\ z &\qquad \cdot \left(\frac{S}{2\pi}\right) d\phi' \end{aligned}\right\} \quad (6.45)$$

where r is the distance from the element ds to P, which is related in the radiation field to the distance r_0 from the origin, by the equation

$$r \approx r_0 - a \cos \phi' \sin \theta \cos \phi - a \sin \phi' \sin \theta \sin \phi - \frac{S}{2\pi} \phi' \cos \theta$$

Hence the total components of vector potential become in the far field.

$$\left.\begin{aligned} A_x &= \frac{\mu I_0}{4\pi} \frac{e^{-jkr_0}}{r_0} \int_0^{\phi'_0} e^{-j((\beta L/2\pi)\phi' - ka \sin\theta \cos(\phi - \phi') - (kS/2\pi)\phi' \cos\theta)} \\ &\quad \times (-a \sin \phi') \, d\phi' \\ A_y &= \frac{\mu I_0}{4\pi} \frac{e^{-jkr_0}}{r_0} \int_0^{\phi'_0} e^{-j((\beta L/2\pi)\phi' - ka \sin\theta \cos(\phi - \phi') - (kS/2\pi)\phi' \cos\theta)} \\ &\quad \times (a \cos \phi') \, d\phi' \\ A_z &= \frac{\mu I_0}{4\pi} \frac{e^{-jkr_0}}{r_0} \int_0^{\phi'_0} e^{-j((\beta L/2\pi)\phi' - ka \sin\theta \cos(\phi - \phi') - (kS/2\pi)\phi' \cos\theta)} \cdot \frac{S}{2\pi} \, d\phi' \end{aligned}\right\} \quad (6.46)$$

Defining

$$H = \frac{1}{2\pi}(\beta L - kS \cos \theta)$$

and

$$u = ka \sin \theta$$

enables these equations to be written more simply as

$$\left.\begin{aligned} A_x &= -\frac{\mu I_0 a}{4\pi} \frac{e^{-jkr_0}}{r_0} \int_0^{\phi'_0} e^{-j(H\phi' - u \cos(\phi - \phi'))} \sin \phi' \, d\phi' \\ A_y &= \frac{\mu I_0 a}{4\pi} \frac{e^{-jkr_0}}{r_0} \int_0^{\phi'_0} e^{-j(H\phi' - u \cos(\phi - \phi'))} \cos \phi' \, d\phi' \\ A_z &= \frac{\mu I_0 S}{8\pi^2} \frac{e^{-jkr_0}}{r_0} \int_0^{\phi'_0} e^{-j(H\phi' - u \cos(\phi - \phi'))} \, d\phi' \end{aligned}\right\} \quad (6.47)$$

It will be noted that, as far as radiation in the forward direction is concerned, for which θ is equal to zero, if βL were, for example, 2.5π radians and kS were 0.5π radians, both of which are typical values for a helical antenna, H would have the value of unity. For this ideal case the integrals become with a single-turn helix for which ϕ_0' is equal to 2π,

$$\left.\begin{aligned} A_x &= j\frac{\mu I_0 a}{8\pi}\frac{e^{-jkr_0}}{r_0}\int_0^{2\pi} e^{ju\cos(\phi-\phi')}(1-\cos 2\phi' + j\sin 2\phi')\,d\phi' \\ A_y &= \frac{\mu I_0 a}{8\pi}\frac{e^{-jkr_0}}{r_0}\int_0^{2\pi} e^{ju\cos(\phi-\phi')}(1+\cos 2\phi' - j\sin 2\phi')\,d\phi' \\ A_z &= \frac{\mu I_0 S}{8\pi^2}\frac{e^{-jkr_0}}{r_0}\int_0^{2\pi} e^{ju\cos(\phi-\phi')}(\cos\phi' - j\sin\phi')\,d\phi' \end{aligned}\right\} \tag{6.48}$$

Using the standard integrals

$$\int_0^{2\pi} e^{ju\cos(\phi-\phi')}\frac{\cos n\phi'}{\sin n\phi'}\,d\phi' = 2\pi j^n \frac{\cos n\phi}{\sin n\phi} J_n(u)$$

gives the results in closed form as

$$\left.\begin{aligned} A_x &= j\frac{\mu I_0 a}{4}\frac{e^{-jkr_0}}{r_0}\{J_0(ka\sin\theta) + e^{-j2\phi}J_2(ka\sin\theta)\} \\ A_y &= \frac{\mu I_0 a}{4}\frac{e^{-jkr_0}}{r_0}\{J_0(ka\sin\theta) - e^{-j2\phi}J_2(ka\sin\theta)\} \\ A_z &= j\frac{\mu I_0 S}{4\pi}\frac{e^{-jkr_0}}{r_0}J_1(ka\sin\theta)e^{-j\phi} \end{aligned}\right\} \tag{6.49}$$

Converting to spherical coordinates using

$$\bar{a}_\theta = \bar{a}_x\cos\theta\cos\phi + \bar{a}_y\cos\theta\sin\phi - \bar{a}_z\sin\theta$$
$$\bar{a}_\phi = -\bar{a}_x\sin\phi + \bar{a}_y\cos\phi$$

gives

$$\begin{aligned} A_\theta &= \frac{\mu I_0 a}{4}\frac{e^{-jkr_0}}{r_0}\Big\{ jJ_0(ka\sin\theta)\cos\theta e^{-j\phi} \\ &\qquad + jJ_2(ka\sin\theta)\cos\theta e^{-j\phi} - j\frac{S}{\pi a}J_1(ka\sin\theta)\sin\theta e^{-j\phi}\Big\} \\ &= \frac{j\mu I_0 a}{4}\frac{e^{-jkr_0}}{r_0}e^{-j\phi}\Big\{\cos\theta[J_0(ka\sin\theta) + J_2(ka\sin\theta)] \\ &\qquad - \frac{S}{\pi a}\sin\theta J_1(ka\sin\theta)\Big\} \\ &= \frac{j\mu I_0 a}{4}\frac{e^{-jkr_0}}{r_0}e^{-j\phi}\Big\{2\cos\theta\frac{J_1(ka\sin\theta)}{ka\sin\theta} \\ &\qquad - \frac{S}{\pi a}\sin\theta J_1(ka\sin\theta)\Big\} \end{aligned} \tag{6.50}$$

$$A_\phi = \frac{\mu I_0 a}{4}\frac{e^{-jkr_0}}{r_0}\{J_0(ka\sin\theta)e^{-j\phi} - J_2(ka\sin\theta)e^{-j\phi}\}$$

$$= \frac{\mu I_0 a}{4}\frac{e^{-jkr_0}}{r_0}e^{-j\phi}\{J_0(ka\sin\phi) - J_2(ka\sin\theta)\}$$

$$= \frac{\mu I_0 a}{2}\frac{e^{-jkr_0}}{r_0}e^{-j\phi}J_1'(ka\sin\theta) \tag{6.51}$$

Then since in the radiation zone

$$E_\theta = \frac{1}{\sqrt{(\mu\varepsilon)}}\frac{\partial A_\theta}{\partial r}$$

and

$$E_\phi = \frac{1}{\sqrt{(\mu\varepsilon)}}\frac{\partial A_\phi}{\partial r}$$

the radiation electric fields become

$$\left.\begin{aligned} E_\theta &= \frac{kZ_0 I_0 a}{4}\frac{e^{-jkr_0}}{r_0}e^{-j\phi} \\ &\quad\times\left\{2\cos\theta\frac{J_1(ka\sin\theta)}{ka\sin\theta} - \frac{S}{\pi a}\sin\theta J_1(ka\sin\theta)\right\} \\ E_\phi &= -\frac{jkZ_0 I_0 a}{2}\frac{e^{-jkr_0}}{r_0}e^{-j\phi}J_1'(ka\sin\theta) \end{aligned}\right\} \tag{6.52}$$

Fig. 6.11. E_θ pattern for 20 turn, 13° helix. (After Maclean and Kouyoumijian, 1959.)

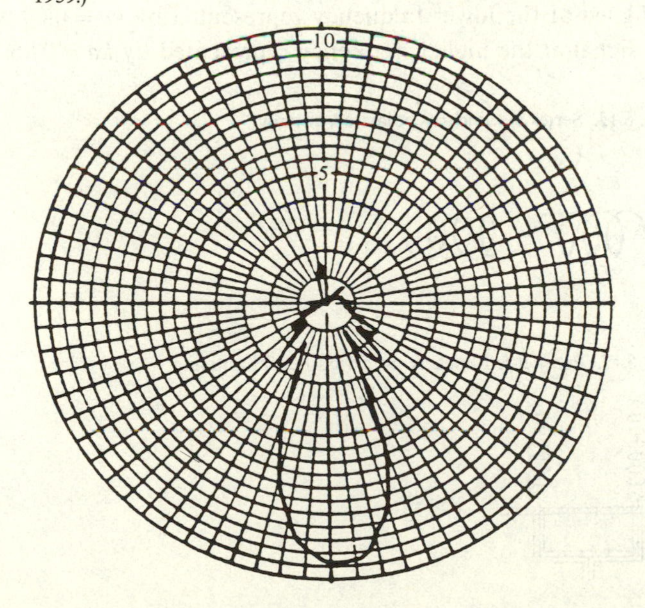

Along the forward axis this gives

$$E_\theta(\theta = 0°) = \frac{kZ_0 I_0 a}{4} \frac{e^{-jkr_0}}{r_0} e^{-j\phi} \tag{6.53}$$

$$E_\phi(\theta = 0°) = -\frac{jkZ_0 I_0 a}{4} \frac{e^{-jkr_0}}{r_0} e^{-j\phi} \tag{6.54}$$

so that the fields in this direction due to a single turn of the helix are circularly polarised.

For a cylindrical helix wound with N turns the total radiation pattern will be given by the above expressions for a single turn, multiplied by the array factor $(\sin N\pi H)/(\sin \pi H)$. For a normal helical antenna this array factor will dominate the pattern of the single turn. A representative pattern for a 20 turn, 13° helix is shown in Fig. 6.11.

6.7 Linearly polarised helical antennas

Linear polarisation is achieved as the resultant of combining oppositely directed circularly polarised travelling waves. These may be produced by combining two oppositely wound helices in series or in parallel.

Series connection of helices, as shown in Fig. 6.12, means that the first circularly polarised wave has to travel through an oppositely wound helix, for which the phase velocity of propagation is unsuitable. The resultant polarisation rotates with frequency, e.g., from 0° at $ka = 0.68$ to 90° at $ka = 0.86$. This is therefore an unsuitable design since such an antenna when correctly aligned at the lower frequency represented by $ka = 0.68$ would pick up no signal at the higher frequency represented by $ka = 0.86$.

Fig. 6.12. Series-fed linearly polarised antenna.

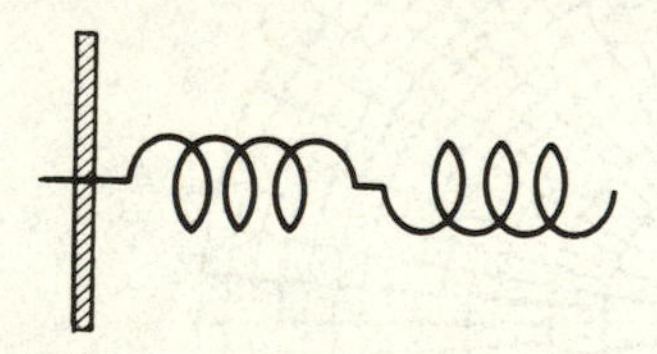

Fig. 6.13. Parallel-fed linearly polarised helical antenna.

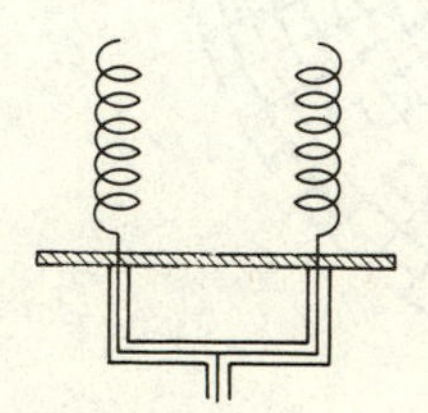

Parallel connection of helices, as shown in Fig. 6.13, results in the same bandwidth of operation for the pair as for each alone if the spacing between them is at least half a wavelength.

Further reading

A.Z. Fradin: *Microwave Antennas*: Pergamon Press, 1961.

J.D. Kraus: *Antennas*: McGraw-Hill, New York, 1950.

T.S.M. Maclean and W.E.J. Farvis: 'The sheath helix approach to the helical aerial': Proc. *IEEE*(C), **109**, 1962. (Copyright C, 1962, *IEEE*.)

T.S.M. Maclean and R.G. Kouyoumjian: 'The bandwidth of helical antennas': *Trans. IRE* (*APG*), **AP-7**, 1959. (Copyright C, 1959, *IRE* (now *IEEE*).)

7

+ – +

Yagi–Uda antennas

The Yagi–Uda antenna is a linearly polarised, travelling-wave, end-fire planar array of linear dipoles, as shown in Fig. 7.1. One of these dipoles only is fed. All the other elements are parasitically excited, with a single reflector element normally, and usually with between one and 20 director elements, although as many as 100 director elements have been used. This antenna is one of the most commonly used antennas for television reception because

(a) it is inexpensive to build; and
(b) it is easy to feed, by comparison with arrays in which separate transmission lines are taken to each radiating element.

Fig. 7.1. Yagi–Uda antenna.

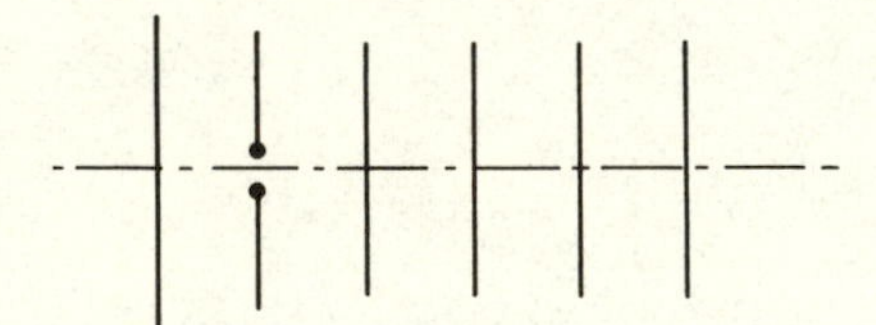

On the other hand it is more difficult to design, than, for example, a corner reflector antenna, which has a power gain in the same range of 10–16 dB as the Yagi–Uda antenna, and is also linearly polarised.

Three different approaches will be used to study this antenna. The first will be a consideration of a two element array with a fed element plus a reflector or director element. The second will treat theoretically a Yagi–Uda array with a continuous array of short director elements, and the third will combine the results of an experimental study of the antenna with the results of the second approach applied to the discrete element case.

7.1 Two dipole array

7.1.1 *Fed element plus single parasitic element – approximate solution*

Referring to Fig. 7.2 the array of two elements can be treated as a two port network which can be described by the equations

$$\left.\begin{aligned} V_1 &= I_1 Z_{11} + I_2 Z_{12} \\ 0 &= I_1 Z_{21} + I_2 Z_{22} \end{aligned}\right\} \tag{7.1}$$

These equations give for the ratio of the currents,

$$\frac{I_2}{I_1} = -\frac{Z_{21}}{Z_{22}} \tag{7.2}$$

and for the input impedance,

$$Z_i = Z_{11} - \frac{Z_{12}^2}{Z_{22}} \tag{7.3}$$

It is clear from eqns (7.2) and (7.3) that both the radiation pattern and the input impedance of the arbitrary, two element array of linear dipoles, depend only on the self and mutual impedances of these dipoles. Since the dipoles are of approximate length $\lambda/2$ this means that their input reactances in particular are highly dependent on the exact departure from the resonant length of each dipole, but the mutual impedance is not greatly dependent on this difference. Consequently it is legitimate in developing an approximate physical understanding of the array to consider each of the two dipoles as being exactly $\lambda/2$ in length as far as their mutual impedance is concerned.

Using the experimental result that a spacing of 0.1λ between the dipoles is a suitable minimum separation gives for the mutual impedance between two such $\lambda/2$ dipoles, from Fig. 3.9,

$$Z_{12} = (67.5 + j10)\ \text{ohms} \tag{7.4}$$

Fig. 7.2. Two-dipole array.

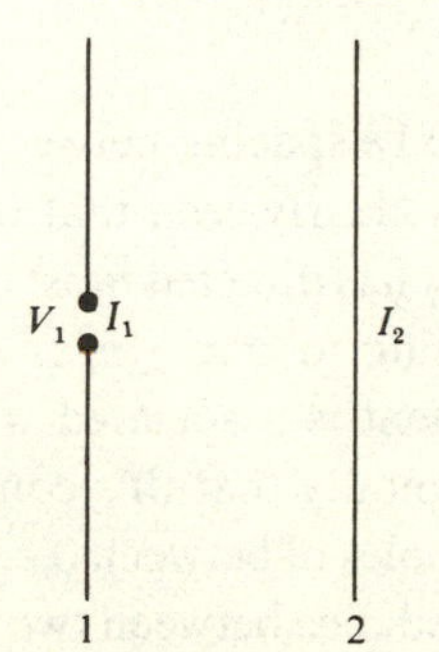

It then remains to choose the lengths of these dipoles more accurately so as to make their self impedance such that the currents flowing in them, and the input impedance of the driven element, provide both the desired field strength in the forward direction and a convenient impedance for matching to the fed dipole.

The variation of input reactance of an isolated dipole with frequency has an incremental slope of approximately 1500 ohms per wavelength centred on the resonant frequency for a typical length/diameter ratio of dipole. The corresponding input resistance variation has an incremental slope of approximately 350 ohms per wavelength. Consequently from eqn (7.2) the ratio of the dipole currents for a parasitic length l_2/λ of 0.47 when the resonant input impedance is $(73+j0)$ ohms is

$$\frac{I_2}{I_1} = -\frac{68.2\angle\ 8.4^\circ}{\left(73+\frac{\Delta l_2}{\lambda}350\right)+j\frac{\Delta l_2}{\lambda}(1500)} \tag{7.5}$$

It is clear from eqn (7.5) that the current I_2 in a parasitic element shorter than the resonant length must lead the current I_1 in the driven element by between 188.4° and 278.4°. The ratio I_2/I_1 in these two extreme cases is 68.2/73.0 and zero. As a particular example for $\Delta l_2/\lambda$ equal to -0.02.

$$\frac{I_2}{I_1} = -\frac{68.2\angle\ 8.4^\circ}{67-j30} = 0.93\angle\ -147.5^\circ \tag{7.6}$$

The associated input impedance of the fed dipole is then, from eqn (7.3)

$$Z_i = Z_{ii} - 47.9 - j41.7$$

If l_1/λ is taken as 0.5 this gives for the input impedance

$$Z_i \approx (30+j0)\ \text{ohms} \tag{7.7}$$

From eqn (7.6) the radiation pattern in the equatorial plane of the two vertical dipoles is given by

$$E(\phi) = C\{e^{-jkd\cos\phi} + 0.93e^{-j147.5^\circ}\} \tag{7.8}$$

where C is a constant and kd is equal to 36° for 0.1λ spacing between the dipoles. Eqn (7.8) is plotted in Fig. 7.3 where it is clearly seen that in the transmitting situation the presence of the element l_2 has directed most of the radiation into the forward half-space corresponding to $\phi < \pm\pi/2$.

Consider now the situation when the fed element is associated with a reflector dipole of length l_2. It is known experimentally that this combination works well with a separation between the dipoles of between 0.2λ and 0.25λ. When the separation is 0.2λ the mutual impedance between two such

$\lambda/2$ dipoles is, from Fig. 3.9,

$$Z_{12} = (51 - j22)\,\text{ohms} \tag{7.9}$$

and since eqns (7.1)–(7.3) still apply the ratio of the parasitic to driven dipole currents for $\Delta l_2/\lambda$ equal to $+0.03$ is

$$\frac{I_2}{I_1} = -\frac{56.1\angle -23^\circ}{83.5 + j45} = 0.59\angle\ 129^\circ \tag{7.10}$$

The associated input impedance of the fed dipole, from eqn (7.3) is

$$Z_i = Z_{11} - 8.9 + j32$$

If l_1/λ is again taken as 0.5λ, this gives for the input impedance

$$Z_i \approx (70 + j74)\,\text{ohms} \tag{7.11}$$

The radiation pattern in the equatorial plane of the two vertical dipoles is given by

$$E(\phi) = C\{e^{-jkd\cos\phi} + 0.59e^{j129^\circ}\} \tag{7.12}$$

where C is again a constant and kd is now equal to 72° for 0.2λ spacing between the dipoles. Eqn (7.12) is plotted in Fig. 7.4 and the main radiation is seen to be along the direction ϕ equal to π since the radiation along the forward direction is reflected by the second parasitic dipole.

7.1.2 *Fed element plus single parasitic element – computer solution*

From Figs. 7.3 and 7.4 it is clear that the presence of a parasitic dipole can provide additional directive gain in either the forward or reverse

Fig. 7.3. Radiation pattern of two dipole array: 1 fed element; 1 director element.

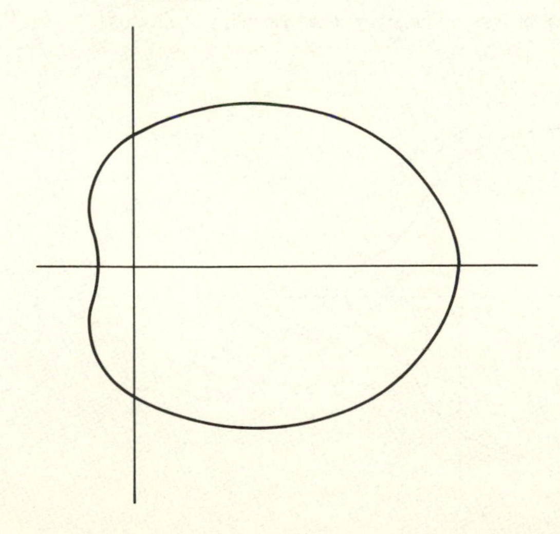

directions depending on the lengths of the dipoles and their separation. For the particular case of a pair of driven and parasitic dipoles, each of length 0.5λ, computer solutions have been obtained of front/back ratio, power gain and half-power beam width for different dipole separations up to 0.25λ. These results are shown in Table 7.1.

In this table the half-angle beamwidth is defined as the angle between the direction of maximum radiation and the direction of the half-power radiation. It will be noted that for the spacing of 0.2λ the front/back ratio of 2.91 in the table compares with 2.80 in the approximate solution represented by Fig. 7.4.

Computer solutions are also available for the input impedance of a two dipole array where the dipoles are of length between 0.40λ and 0.58λ. For the case of dipoles of equal radius 0.006 66λ, the results are shown in Fig. 7.5 for interelement spacing of 0.10λ, 0.15λ, 0.20λ and 0.25λ. From this diagram

Table 7.1. *Pair of driven and parasitic dipoles of equal length* 0.5λ

Dipole radius: $a_1 = a_2 = 0.006\,66\lambda$				
Spacing-d	Front/back ratio	Power gain relative to isotropic source	Half-angle beam width H-plane (°)	E-plane (°)
0.1	2.63	7.24	60.0	33.2
0.15	2.87	6.8	64.0	34.0
0.20	2.91	6.34	70.0	35.1
0.25	2.79	5.75	76.8	36.6

Fig. 7.4. Radiation pattern of two dipole array: 1 fed element; 1 reflector element.

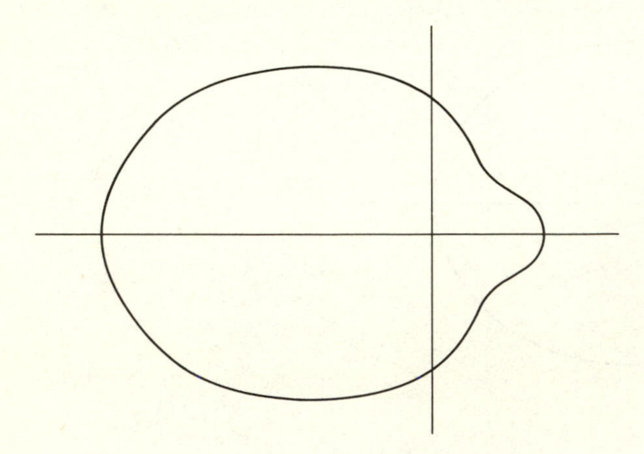

it is seen that both the input resistance and the input reactance increase with spacing between the two dipoles, but the increase of the input resistance with spacing is much more marked. For small separation between the dipoles the input resistance can be very low for element lengths less than 0.48λ, as shown in Fig. 7.5(*a*) and (*b*). Fig. 7.5(*c*) and (*d*) confirm that tuning is less critical at spacing of 0.2λ and 0.25λ, with both spacings readily providing an input resistance of at least 50Ω. In particular, for two $\lambda/2$ dipoles spaced at 0.2λ the input resistance is 70Ω, in agreement with the approximate results obtained in Section 7.1.1.

7.2 Long Yagi–Uda array with continuous distribution of short director elements

In Fig. 7.6 a continuous array of infinitesimally short dipoles aligned parallel to the *y*-axis and lying in the *yz*-plane is shown, with the axis of the array directed along the *z*-axis. Due to an element *dz* of this continuous array, which is carrying a postulated current density J_y amperes per metre that is uniform in magnitude but varies in phase as $\exp(-j\beta z)$ along the *z*-axis, the vector potential at a far field point $P(r_0, \theta, \phi)$ is given by

$$dA_y = \frac{\mu J dy}{4\pi r} e^{-j\beta z} e^{-jkr} \tag{7.13}$$

where *r* is the distance from the source element *dz* to the point of

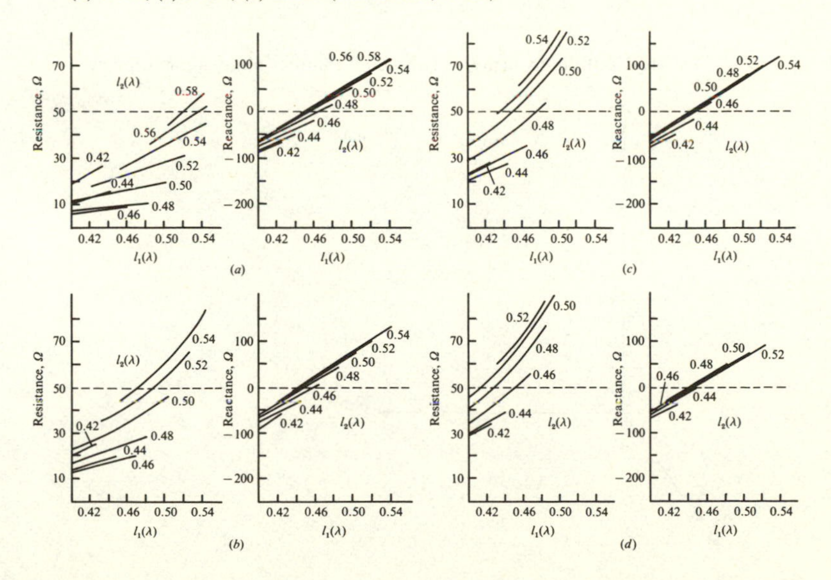

Fig. 7.5. Input impedance of dipole antenna, with driven and parasitic elements of equal radius ($0.006\,66\lambda$), for interelement spacings of (*a*) 0.10λ; (*b*) 0.15λ; (*c*) 0.20λ; (*d*) 0.25λ. (After Neri, 1980.)

observation. In the radiation field this distance r is related to r_0, the distance from the origin, by the equation

$$r \approx r_0 - z\cos\theta$$

The total vector potential at P due to the full continuous array of length l is then given by

$$A_y = \frac{\mu J\,dy}{4\pi r_0} e^{-jkr_0} \int_0^l e^{j(k\cos\theta - \beta)z}\,dz$$

$$= \frac{\mu J\,dy}{2\pi r_0} e^{-jkr_0} e^{j(k\cos\theta - \beta)l/2} \frac{\sin(k\cos\theta - \beta)\frac{l}{2}}{(k\cos\theta - \beta)} \tag{7.14}$$

To find the components of the radiation magnetic field in spherical coordinates we use

$$H_\phi \approx \frac{1}{\mu}\frac{\partial A_\theta}{\partial r}$$

and

$$H_\theta \approx -\frac{1}{\mu}\frac{\partial A_\phi}{\partial r}$$

with

$$A_\theta = A_y \sin\phi\cos\theta$$

and

$$A_\phi = A_y \cos\phi$$

Fig. 7.6. Continuous array of y-directed infinitesimally short dipoles.

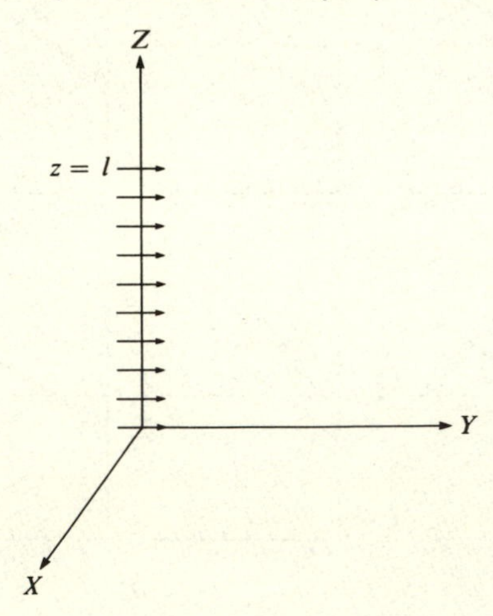

Then

$$H_\phi = -\frac{jkJdy\cos\theta\sin\phi}{2\pi r_0}e^{-jkr_0}e^{j(k\cos\theta-\beta)l/2}\frac{\sin(k\cos\theta-\beta)\frac{l}{2}}{(k\cos\theta-\beta)} \tag{7.15}$$

and

$$H_\theta = \frac{jkJdy\cos\phi}{2\pi r_0}e^{-jkr_0}e^{j(k\cos\theta-\beta)l/2}\frac{\sin(k\cos\theta-\beta)\frac{l}{2}}{(k\cos\theta-\beta)} \tag{7.16}$$

Since the power density in the far field is

$$P_r = (H_\theta H_\theta^* + H_\phi H_\phi^*)\frac{Z_0}{2}$$

in which $*$ denotes complex conjugate, and Z_0 is the impedance of free space, the total power radiated by the continuous array is given by

$$W = \int_0^\pi \int_0^{2\pi} \frac{k^2J^2dy^2(\cos^2\phi + \cos^2\theta\sin^2\phi)}{4\pi^2 r_0^2} \times \frac{\sin^2(k\cos\theta-\beta)\frac{l}{2}}{(k\cos\theta-\beta)^2}\frac{Z_0}{2}r_0^2\sin\theta\, d\theta\, d\phi \tag{7.17}$$

The integral on ϕ is

$$\int_0^{2\pi}\left[1 - \frac{\sin^2\theta(1-\cos 2\phi)}{2}\right]d\phi$$

which is readily evaluated to give $\pi(1+\cos^2\theta)$, so that the radiated power reduces to

$$W = \frac{k^2J^2dy^2Z_0}{8\pi}\int_0^\pi \sin\theta(1+\cos^2\theta)\frac{\sin^2(k\cos\theta-\beta)\frac{l}{2}}{(k\cos\theta-\beta)^2}d\theta \tag{7.18}$$

Using the substitution of variable

$$u = \left(\cos\theta - \frac{\beta}{k}\right)$$

so that

$$du = -\sin\theta d\theta$$

and

$$\cos^2\theta = \left(u + \frac{\beta}{k}\right)^2$$

gives for the radiated power

$$W = -\frac{J^2 dy^2 Z_0}{8\pi} \int_{(1-\beta/k)}^{-(1+\beta/k)} \frac{\left(1 + u^2 + 2\frac{\beta}{k}u + \frac{\beta^2}{k^2}\right)}{u^2} \frac{(1 - \cos klu)}{2} du \tag{7.19}$$

Consider now the following three separate integrals, with each of the limits remaining as in eqn (7.19),

$$I_1 = \int \frac{1 - \cos klu}{2} du$$

$$I_2 = \int \frac{\frac{\beta}{k}(1 - \cos klu)}{u} du$$

$$I_3 = \int \frac{\left(1 + \frac{\beta^2}{k^2}\right)(1 - \cos klu)}{2u^2} du$$

The first integral is straightforward and leads to

$$I_1 = 1 - \frac{\sin kl \cos \beta l}{kl}$$

To evaluate I_2 use is made of the standard Cin function defined by

$$\text{Cin}\, v = \int_0^v \frac{1 - \cos u}{u} du$$

and this leads to

$$I_2 = \frac{\beta}{k}[\text{Cin}(\beta - k)l - \text{Cin}(\beta + k)l]$$

Similarly using the Si function defined as

$$\text{Si}\, v = \int_0^v \frac{\sin u}{u} du$$

the solution to the third integral becomes

$$I_3 = \frac{1 + \left(\frac{\beta}{k}\right)^2}{2} \left\{ \frac{[\cos(\beta - k)l - 1]}{1 - \frac{\beta}{k}} + \frac{[\cos(\beta + k)l - 1]}{1 + \frac{\beta}{k}} - kl\,\text{Si}(\beta - k)l + kl\,\text{Si}(\beta + k)l \right\}$$

Hence the radiated power is

$$W = \frac{(J\,dy)^2 Z_0}{8\pi}\left\{\frac{\left[1+\left(\frac{\beta}{k}\right)^2\right]}{2}\frac{[\cos(\beta-k)l-1]}{1-\frac{\beta}{k}} + \frac{[\cos(\beta+k)l-1]}{1+\frac{\beta}{k}} - kl\,\mathrm{Si}(\beta-k)l + kl\,\mathrm{Si}(\beta+k)l + \frac{\beta}{k}[\mathrm{Cin}(\beta-k)l - \mathrm{Cin}(\beta+k)l] + 1 - \frac{\sin kl\cos\beta l}{kl}\right\} \tag{7.20}$$

The power gain of the aerial is defined as

$$G = \frac{\text{maximum power density}}{\text{average power density}}$$

where the maximum power density occurs along the positive z-axis at (0°, $\pi/2$) and is given by

$$\frac{(kJ\,dy)^2 Z_0}{8\pi^2 r_0^2}\,\frac{\sin^2(k-\beta)\frac{l}{2}}{(k-\beta)^2}$$

The average power density is $W/4\pi r_0^2$, where W is given by eqn (7.20), and hence G can be evaluated.

Fig. 7.7. Power gain as a function of overall length. (After Bennett, 1974.)

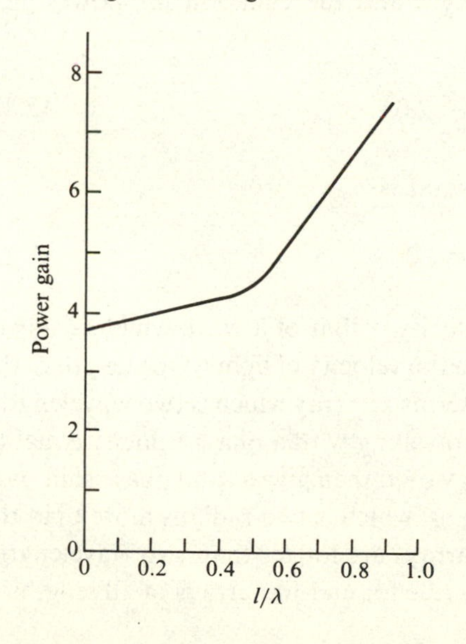

It is thus seen that the power gain of this end-fire array depends on the current density J, the infinitesimal length dy over which J extends, the length of the array l and the phase constant β with which the wave is guided along the surface in the yz-plane. For a fixed length of array l, and a specified current density moment $J dy$ it is therefore of interest to study the variation of the power gain of this antenna as a function of the phase constant β or, equivalently, as a function of the ratio β/k at a fixed frequency. When this is done it is found that for any fixed array length l/λ there is a particular value of β/k which maximises the forward power gain. Fig. 7.7 shows this maximum gain as a function of the array length l/λ. It is seen that the slope of the power gain curve changes at an array length l/λ which is approximately equal to 0.5. Below this array length the maximum power gain is given approximately by

$$G \approx 3.65 + 1.32\frac{l}{\lambda} \qquad \frac{l}{\lambda} < 0.5$$

and the associated relative phase constant is

$$\frac{\beta}{k} \approx 0.96\frac{\lambda}{l} \qquad \frac{l}{\lambda} < 0.5 \tag{7.21}$$

Eqn (7.21) indicates that to achieve such a phase velocity the exciting wave must be slowed down relative to the velocity of light so that the conductors carrying the current density J are required to behave as an artificial dielectric, or must have dielectric loading added.

Above the array length l/λ equal to approximately 0.5 the slope of the power gain curve in Fig. 7.7 changes and the equation for power gain becomes

$$G \approx 1.8 + 5.6\frac{l}{\lambda} \qquad 0.5 < \frac{l}{\lambda} < 2.0 \tag{7.22}$$

The associated relative phase constant is

$$\frac{\beta}{k} \approx 1 + 0.55\frac{\lambda}{l} \qquad 0.5 < \frac{l}{\lambda} < 2.0 \tag{7.23}$$

Again the corresponding phase velocity is that of a wave which is slower than that of light but which tends to the velocity of light as the length of the array tends to infinity. In practical terms an array which is two wavelengths long should be excited by a wave travelling with a phase velocity equal to 0.86 times the velocity of light. This would then give a total phase shift over the length of the array of 14.6 radians, which is two radians more than the free space shift. Many Yagi–Uda arrays are longer than two wavelengths and a useful approximate working rule for end-fire arrays of all lengths is that the total phase shift over the length of the array should be approximately π radians more than the free space phase shift. This is not to say that the above analysis applied to the two wavelength array is in error, but only that because the postulated uniform current density is not achieved in practice the optimum phase velocity is different from that calculated for the above idealised case. An experimentally derived distribution of current along a discrete array is shown in Section 7.3.

7.3 Experimental approach to Yagi–Uda antenna design

An illuminating experimental design study of the Yagi–Uda antenna was carried out by Ehrenspeck and Poehler using near and far field measurements of a transmitting antenna of this kind at a frequency close to 9 GHz. All the measurements were carried out above a ground plane, as shown in Fig. 7.8, with the fed monopole being excited from below. The individual reflector and director elements took the form of brass rods whose height above the ground plane could readily be adjusted and the separation of the elements could also be changed in multiples of 0.1λ. The size of the ground plane was 12 ft × 6 ft so that far field measurements could be taken along the larger dimension which represented a distance of 110 wavelengths.

Fig. 7.8. Yagi–Uda array of monopoles imaged in a ground plane. (After Ehrenspeck and Poehler, 1959.)

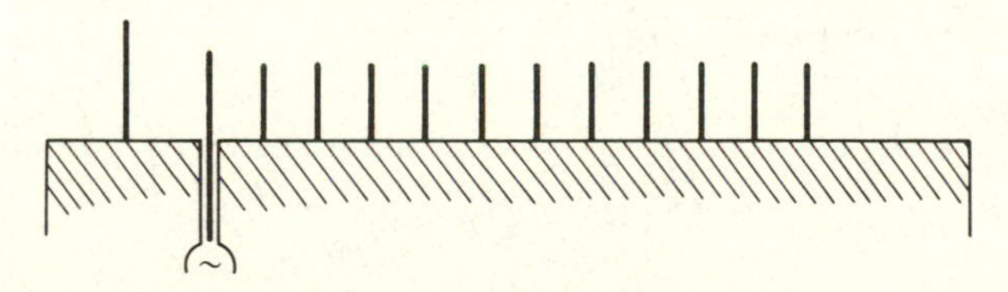

In order to measure the phase velocity along the directors a probe was moved along the line of the array immediately above the director elements. The distance between two successive minima when a reflecting plate was placed at the end of the array then represented one-half wavelength of the exciting wave. The ground plane was surrounded by absorbing material to reduce the effect of reflection at its edges and all measurements were conducted in an anechoic chamber.

From experimental measurements it was found that after adjustment, in the absence of any director elements, of the feed plus reflector element for maximum forward power, no re-adjustment was necessary when the director elements were placed in position. This remained true for all combinations of director elements. Likewise, for a 6λ Yagi–Uda array, measurements of the phase contours along the director elements in the absence of the reflector confirmed that these remained unchanged when the reflector was added, showing that the phase velocity of the exciting wave along the directors was unaltered by the presence of the reflector. Consequently the actions of the reflector and director elements were found to be mutually independent for this array, so that each could be optimised independently for maximum forward gain.

The near fields of the feed plus director elements without a reflector are shown in the contour plots of Fig. 7.9, which provide a phase and amplitude plot out to 17λ in the forward direction with contour spacing at 5 dB intervals. Likewise in Fig. 7.10 the effect on the amplitude contours of adding the reflector is clearly seen, but the phase remains unchanged, with a total phase change of π radians more than that associated with its 6.0λ length for this antenna adjusted for maximum forward field strength.

Fig. 7.9. Nearfield amplitude and phase plot of Yagi–Uda antenna without reflector ($L = 6.0\lambda$; $S_D = 0.20\lambda$; $S_R = 0.25\lambda$). (After Ehrenspeck and Poehler, 1959.)

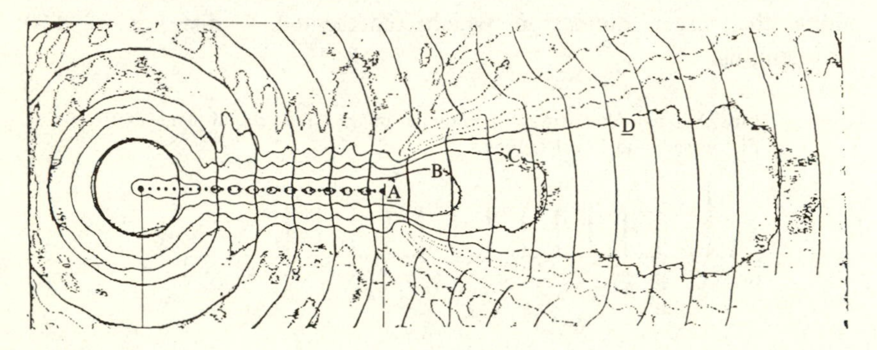

Fig. 7.10. Nearfield amplitude and phase plot of Yagi–Uda antenna with reflector ($L = 6.0\lambda$; $S_D = 0.20\lambda$; $S_R = 0.25\lambda$). (After Ehrenspeck and Poehler, 1959.)

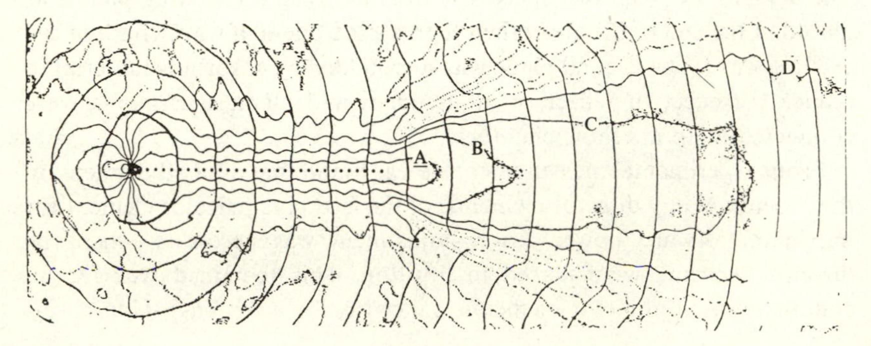

7.3.1 *Optimisation of feed–reflector combination*

The height of the monopole feed was first adjusted for resonance. Then a sequence of different reflectors, with heights from 0.205λ to 0.265λ, and of radius 0.024λ, were added and the power in the far field was measured with each reflector. The input power to the antenna was kept constant in each case. These measurements of radiated power in the forward direction were made for reflector spacing between 0.083λ and 0.313λ. The results obtained showed that a maximum forward power gain of 3.85 dB over the isolated monopole could be achieved using a reflector of height 0.226λ placed at a distance of 0.250λ behind the feed monopole.

7.3.2 *Optimisation of director elements*

With the feed–reflector combination adjusted for maximum forward field strength as described in the previous section, a sequence of experiments involving the variables of director spacing and director height was carried out for two Yagi–Uda antenna of lengths 1.2λ and 6.0λ. It was found that any director spacing of between 0.1λ and 0.3λ could be used with both lengths of antenna to give maximum possible forward field strength, provided that the director heights were increased as the spacing between the director elements increased. In terms of maximum gain being associated with a particular phase constant of propagation along the structure, this is equivalent to saying that a certain minimum volume of metal is required to slow the wave down to the velocity associated with that particular phase constant. The experimental results for power gain in the forward direction above that of the isolated monopole, plotted against director element height for a range of uniform director spacing, are shown in Fig. 7.11 for the antenna lengths of 1.2λ and 6.0λ. For these particular lengths the maximum power gains are seen to be 10 dB and 14 dB respectively, and for other

Fig. 7.11. Power gain of Yagi–Uda array as a function of director height and spacing for array length: (*a*) 1.2λ; (*b*) 6.0λ. (After Ehrenspeck and Poehler, 1959.)

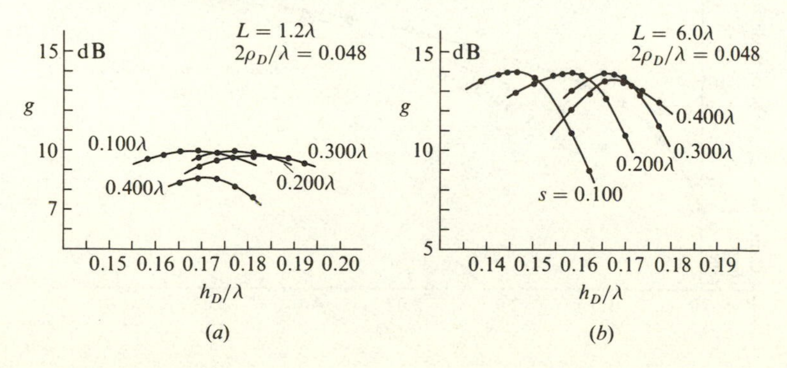

lengths between 0.6λ and 6.0λ the experimental values of maximum forward power gain are shown in Fig. 7.12. Superimposed on this graph is the power gain which would result from the assumption of the phase delay across the antenna being π radians greater than that for the free space phase velocity. This condition is referred to as the Hansen–Woodyard condition, but it must be remembered that the condition assumes equal excitation of all the elements in the antenna.

7.3.3 *Director currents*

The probe measurements taken immediately above the director elements are a measure of the electric fields there. These are proportional to the rate of change of current in the directors, and hence to the current itself at the base of the monopole. The distribution of this measured current for a 3λ antenna is shown in Fig. 7.13. It is clear that while the currents in the director elements are substantially equal over the final 2.5λ, the current in the monopole feed is approximately six times as large as these director currents. Consequently a rigid application of the Hansen–Woodyard condition would be quite wrong in respect of a Yagi–Uda antenna. Nevertheless it provides a useful single criterion of assessment by focussing attention on the one factor of phase velocity, which is affected by each of the variables of director spacing, director height, director radius and also it is found experimentally by the overall length of the length if this is less than approximately three wavelengths.

From Fig. 7.13, showing the currents in the feed and director elements,

Fig. 7.12. Maximum power gain of Yagi–Uda array as a function of array length. (After Ehrenspeak and Poehler, 1959.)

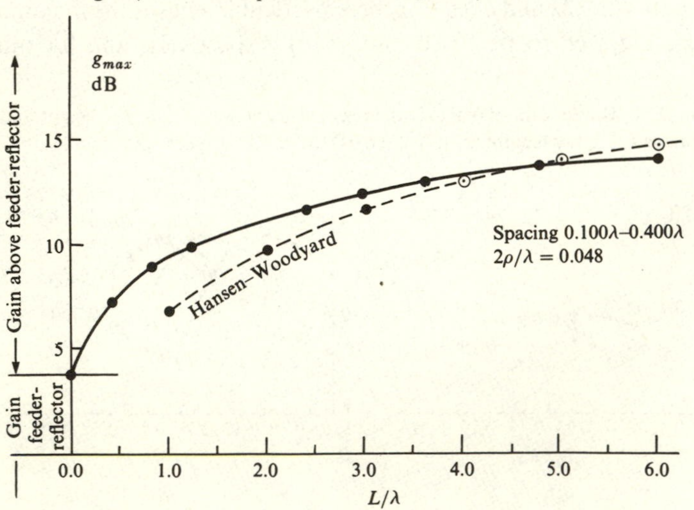

the forward field strength can be calculated as the sum of two components. The first term is the radiation field from the feed, which is derived from the vector potential due to a current of magnitude $6I_D$ where I_D is the current in a director element. The second term is the summation of 15 components associated with the director elements in the array, each carrying approximately an equal current I_D but with a phase delay between successive elements which may be calculated from the measured phase velocity curve shown in Fig. 7.14. When these two components are combined with the

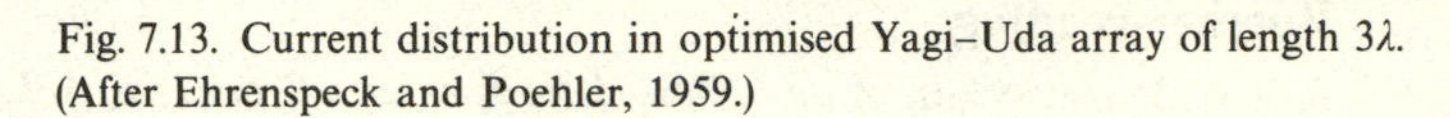
Fig. 7.13. Current distribution in optimised Yagi–Uda array of length 3λ. (After Ehrenspeck and Poehler, 1959.)

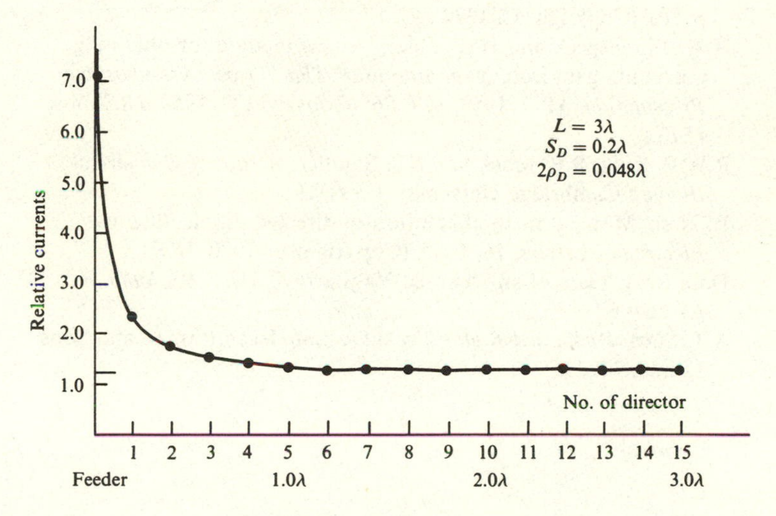

Fig. 7.14. Phase velocity for obtaining maximum power gain as a function of array length. (After Ehrenspeck and Poehler, 1959.)

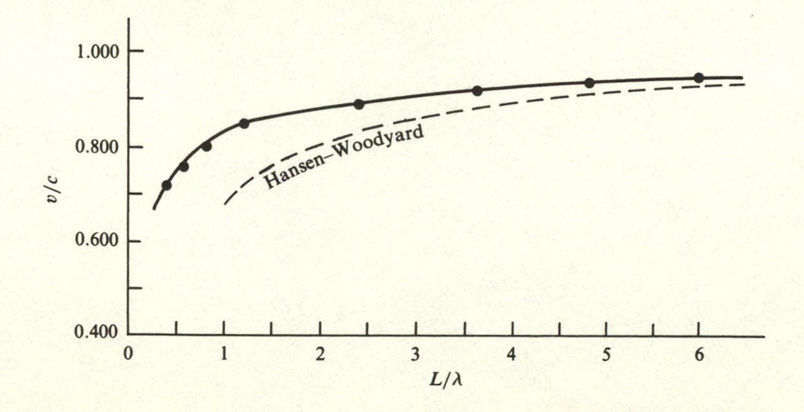

appropriate phase between them the total forward field strength is known, and hence the power gain due to the feed and director elements can be calculated since the input resistance of the antennas can also be measured. The total power gain of the antenna is then obtained by adding the power gain improvement of 3.85 dB associated with the reflector element. Power gains of 16 dB are not uncommon for Yagi–Uda arrays, and in the laboratory a power gain of 23 dB has been achieved using an array of length 40 wavelengths.

Further reading

R.G.T. Bennet: 'Gain of an idealised Yagi array': *Proc. IEE*, **121**, 1974, p. 116. (Copyright C, 1974 *IEE*.)

H.W. Ehrenspeck and H. Poehler: 'A new method for obtaining maximum gain from Yagi antennas': *IRE Trans. Antennas and Propagation*, **AP-7**, 1959, 379–86. (Copyright C, 1959, *IRE* (now *IEE*).)

R.W.P. King, R.B. Mack and S.S. Sandler: *Arrays of Cylindrical Dipoles*: Cambridge University Press, 1968.

R. Neri: 'Moment method solution of directed dipole antennas': *Electronics Letters*, **16**, 1980. (Copyright C, 1980, *IEE*.)

D.G. Reid: 'Gain of an idealised Yagi array': *JIEE*, **93**, 1946, Part III A), 564–6.

A.T. Starr: *Radio and Radar Technique*: Sir Isaac Pitman and Sons, London, 1953.

8

+−+

Frequency independent and logarithmically periodic antennas

Linear dipole antennas have radiation patterns which change little with frequency from very low frequencies up to the frequency when the dipole is one-half wavelength long. Over this frequency range, which may span several octaves, the pattern of a vertical dipole remains omnidirectional in azimuth and has a beamwidth between half-power angles in elevation which changes from 90° to 78° with increasing frequency. On the other hand the input impedance of such a dipole changes from a negligible input resistance associated with a large capacitive reactance to approximately $(73 + j42)$ ohms as the frequency increases.

In contrast to this, the input impedance of an axial mode helix or a rhombic antenna changes relatively little with frequency by comparison with the change in its radiation pattern. One of the reasons for this difference in performance is that the current distribution in both the rhombic and helical antennas is of travelling-wave form. Hence the physical dimension of antenna length does not enter into its input impedance calculation, unlike the cases of antennas which have a current distribution of resonant standing wave form.

Any structure excited in such a way that it produces no reflections will satisfy the condition of constant input impedance with frequency since this impedance is simply the ratio of incident voltage/incident current at the origin. As an example, an infinitely long bicone will satisfy this condition. It should be noted that infinite length is not necessarily associated with progressively decreasing current, i.e. the essential boundary condition of tangential electric field being zero on the surface of a perfect conductor can be satisfied by a current of constant magnitude along the surface. This is true whether or not there are reflections along it.

Since any practical antenna must be of finite size it is desirable that the incident current distribution along it should fall towards zero value before the end of the antenna is reached, unless the reflected wave is attenuated more rapidly than the incident wave. A progressive fall of the incident current with distance may be achieved by increasing the characteristic impedance of the antenna. Alternatively if the incident wave is directed through a resonant region it may radiate sufficient power from this region for the current continuing beyond it to be highly attenuated. This is the technique which has found most application in practice.

If a travelling wave of current on one of these finite structures is set up in some such way, there still remains the question of the value of the resistive input impedance, which is equal to the characteristic impedance offered to the wave. By considering a planar structure, such as bifin in Fig. 8.1, which will be assumed to be loaded in some way that only an incident wave exists on it, it will be clear that as the included angle of the fin increases so will the input impedance decrease. Thus when the included angle tends to π radians the impedance must tend to zero, and when the included angle tends to zero the impedance is proportional to the ratio of two elliptic integrals, which tends to a very high value. But when the included angle is equal to $\pi/2$ the characteristic impedance is equal to $Z_0/2$ where Z_0 is the impedance of free space. In this case the bifin is congruent to its complement, where the complement is defined as the region of the plane not occupied by the bifin, and so the bifin with the included angle of $\pi/2$ is referred to as a self complementary structure. An infinite number of examples of one particular type of self complementary structure may be generated by drawing any non-overlapping curve from an origin to infinity and then drawing the same curve rotated through 90°, 180° and 270°. Alternate sections are then made perfectly conducting.

Fig. 8.1. Bifin antenna.

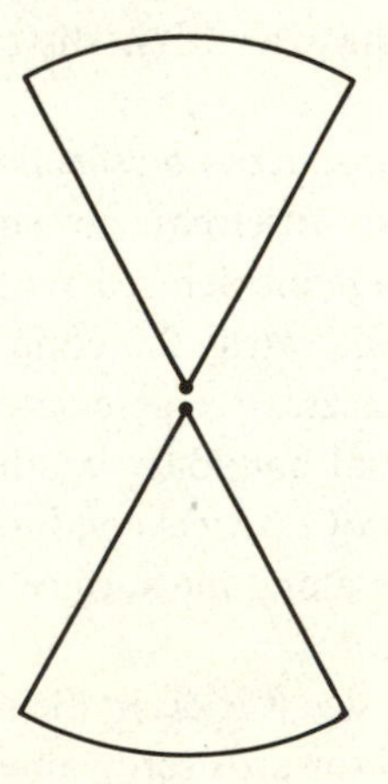

Since all such examples are congruent their input impedances are equal for each alternate pair and these impedances are equal to $(Z_0/2 + j0)$ ohms for all frequencies. Although this applies strictly to an infinite structure only, it will also apply in practice to a structure where the current has effectively fallen to zero before the structure is truncated. In the measurement of such impedances it is clearly necessary that balanced excitation of the structure should be ensured, which would not be the case if direct connection by a coaxial cable were made without the presence of an intervening balanced-to-unbalanced transformer.

As an example of a complementary structure which is not congruent consider the combination of a planar dipole and an identical slot cut in an infinite perfectly conducting sheet. For both this and all other complementary structures in which it is possible to locate a pair of terminals not bridged by a short circuit, which in the dipole–slot combination implies both symmetric and asymmetric excitation of the dipole and slot, the product of the input impedances equals $Z_0^2/4$. Thus for the dipole–slot combination

$$Z_d \cdot Z_s = \frac{Z_0^2}{4} \tag{8.1}$$

For the case of a resonant dipole of input impedance 73 ohms, the input impedance of the complementary slot will therefore be $(487 + j0)$ ohms.

8.1 Planar equiangular spiral antenna

The planar equianglar spiral antenna shown in Fig. 8.2 was proposed by V.H. Rumsey as a structure which was defined essentially by angles. Its shape is described by the equation, shown in Fig. 8.3,

$$r = r_0 e^{a\phi} \tag{8.2}$$

where r_0 is the starting radius at the angle ϕ equal to zero. The form of this curve depends only on the constant a and is independent of the constant r_0. Although mathematically the radius would go on increasing as the angle ϕ increases, it must be terminated in practice after a relatively few revolutions. One of the main reasons for the success of this antenna is that the current is found to attenuate rapidly after the length of one turn becomes equal to the free space wavelength. Because of this the equiangular spiral may then be truncated without affecting its operation at frequencies above this resonant wavelength. The high frequency limit is then set approximately by the perimeter of the first turn, and in practice it may be desirable to adopt a printed circuit form of construction when this upper frequency is in the gigahertz range. The maximum ratio of useful upper to lower frequency

Fig. 8.2. Equiangular spiral antenna.

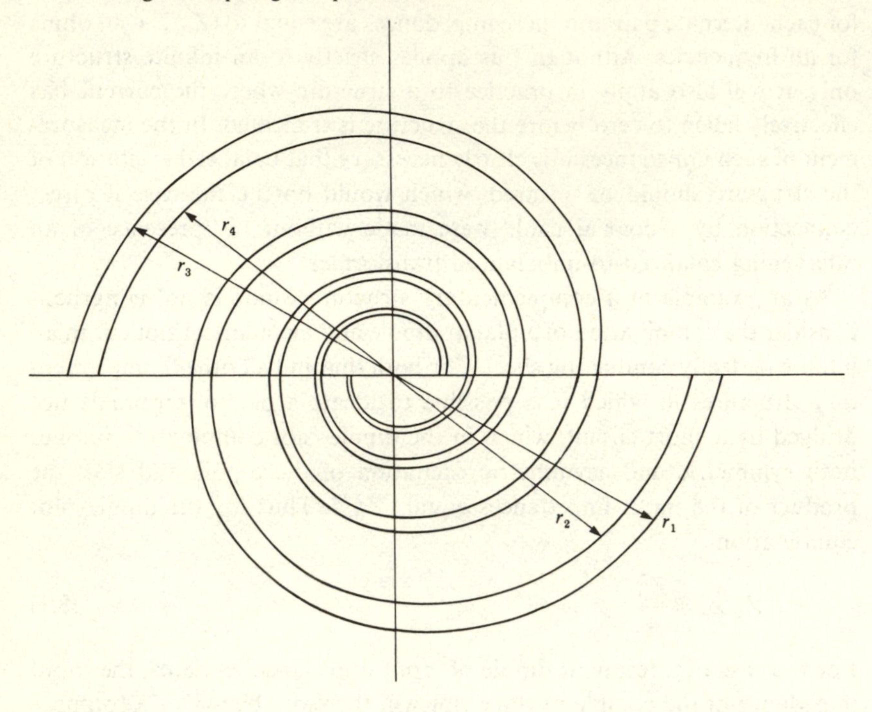

Fig. 8.3. Planar equiangular spiral: $r = y_0 \exp(a\phi); r_0 = 0.2; a = 0.15$.

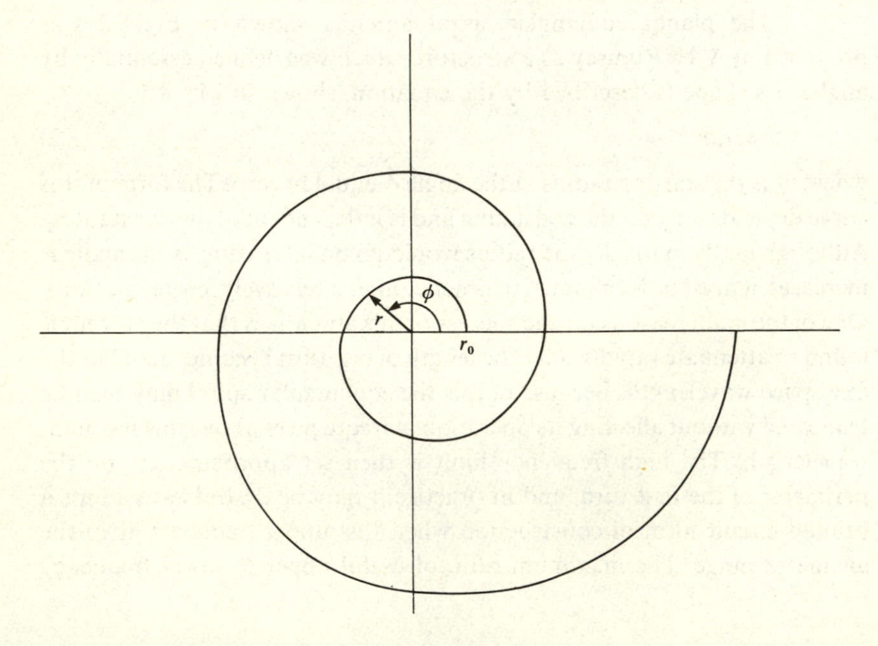

limit which has been achieved with this antenna is of the order of 40:1, which represents an enormous improvement in bandwidth over any antenna made previously. Although both impedance and radiation pattern are required to be broadband it has been found that the radiation pattern is the requirement which limits the bandwidth in practice. The impedance requirement can, if desired, be met more easily by constructing the equiangular spiral so as to be self complementary.

Since it is known that the finite cone or fin is not broadband it is clear that the rapid attenuation of current in the finite equiangular spiral must be associated with the curvature of the lines of current flow. But there are many other curves which also possess this characteristic of curved lines of current flow and which do not exhibit the phenomenal bandwidth of the equiangular spiral. To see why this should be the case note that along any one single curve, out of the four equiangular spiral curves which constitute the antenna, the ratios of successive radii at a fixed angle ϕ are given by

$$\frac{r_1}{r_2} = \frac{r_2}{r_3} = \frac{r_3}{r_4} = \cdots = e^{-a2\pi} = \tau \tag{8.3}$$

where τ is a constant. Thus if at a given frequency f_1 the radius r_n corresponds to a perimeter centred on r_n of one wavelength, then at a lower frequency τf_1 the radius r_{n+1} will correspond to a perimeter centred on r_{n+1} of one wavelength of τf_1. Because of the attenuation of the current in each such resonant turn of 20–30 dB it may be assumed that the radiation pattern is due largely to the current in a single turn, and consequently the radiation patterns will be identical at the frequencies

$$f_1, \tau f_1, \tau^2 f_1, \ldots, \tau^n f_1$$

This identity of pattern may be written more succinctly as occurring at the frequencies

$$f_{n+1} = \tau f_n$$

so that

$$\ln f_{n+1} = \ln \tau + \ln f_n \tag{8.4}$$

or the patterns are identical on a log-periodic basis. Often the variation in pattern is so slight within a period that it is said to be frequency independent, and there is one situation in which this statement is precise. Consider the equation of the equiangular spiral expressed as a radius normalised to the wavelength of operation, i.e. as

$$\left.\begin{aligned} \frac{r}{\lambda} &= \frac{r_0}{\lambda} e^{a\phi} \\ \text{or} \quad & \\ r' &= r'_0 e^{a\phi} \end{aligned}\right\} \tag{8.5}$$

Taking logarithms of both sides gives

$$\ln r' = \ln r'_0 + a\phi$$

which may be expressed as

$$\left.\begin{aligned} r' &= e^{a[\phi + (1/a)\ln(r_0/\lambda)]} \\ \text{or} \quad & \\ r' &= e^{a[\phi + \phi_0]} \end{aligned}\right\} \tag{8.6}$$

where ϕ_0 is equal to the constant $(1/a)\ln(r_0/\lambda)$. Thus a change of frequency results only in a rotation of the spiral and hence rotation of its pattern. In particular when ϕ is equal to 2π,

$$\left.\begin{aligned} r' &= e^{a2\pi} \cdot e^{a\phi_0} \\ \text{or} \quad & \\ \tau r' &= e^{a\phi_0} = \text{constant} \end{aligned}\right\} \tag{8.7}$$

so that the spirals, expressed in wavelengths, and the radiation patterns are identical at frequencies multiplied by the constant τ.

8.2 Practical equiangular spiral antenna

Referring to Fig. 8.2 the equiangular spiral is completely defined by the equations for the four radial arms:

$$\left.\begin{aligned} r_1 &= r_0 e^{a\phi} \\ r_2 &= K r_1 \\ r_3 &= -r_1 \\ r_4 &= K r_3 \end{aligned}\right\} \tag{8.8}$$

where in practice values of a given by $0.2 \leqslant a \leqslant 1.2$ have been used and, similarly, values of K within the range $0.375 \leqslant K \leqslant 0.97$. If the outside diameter of the spiral is regarded as being fixed and maximum bandwidth is wanted then the value chosed for a should be small so that more turns can be included. But this necessarily implies a smaller value of the width factor K and there is a practical limit to this since the metal arms must remain sufficiently wide to carry the coaxial feeding cable which is soldered to the arms or ground plane. The inner of the coaxial cable is connected to the other half of the antenna and a length of dummy cable is sometimes soldered to this half in order to maintain symmetry in the radiation pattern.

Two kinds of structure may be used. The spiral may be constructed of planar metallic arms which may have to be supported mechanically or, alternatively, the spiral may take the form of slots cut out of a metallic sheet. If the spirals have at least $1\frac{1}{2}$ turns the radiation patterns are bidirectional and substantially circularly polarised even at large angles from the normal to the planar sheet.

Measurements of the fields along the spiral arms have been carried out and these indicate a decay of approximately 20 dB in the first wavelength, irrespective of the frequency. Consequently the active portion of the spiral decreases as the frequency rises, but an effectively constant aperture expressed in wavelengths is maintained so that the beamwidth tends to remain moderately constant. As an example a spiral constructed at the University of Illinois with the following constants:

$a = 0.303$
$K = 0.75$
$r_0 = 0.2''$
spiral arm length 13″

gave a 3 dB beamwidth of between approximately 50° and 100° over the frequency range 2–5.18 GHz. Useful patterns were, however, obtained over the frequency range 594 MHz–12 GHz. Over 80% of the bandwidth between 1 and 12 GHz the radiation was circularly polarised up to 40° off axis, where the criterion for circular polarisation is an axial ratio of less than 2:1. Typical patterns are shown in Fig. 8.4.

The input impedance remains substantially constant provided the arm length of the spiral is longer than one wavelength, and when the spiral is fed from a 50 ohm transmission line the VSWR is usually better than 2:1.

Fig. 8.4. Typical radiation patterns for E_θ, E_ϕ from 595 MHz to 12 GHz. (After Dyson, 1959.)

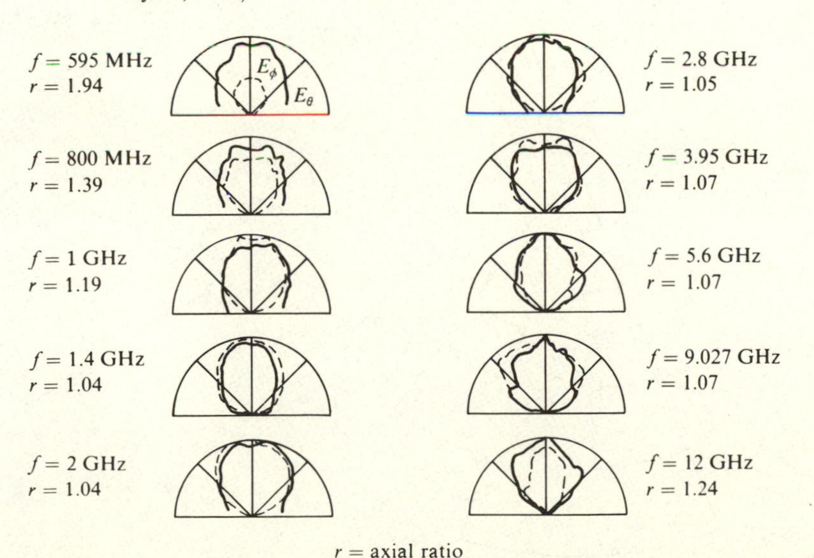

8.3 Linearly polarised frequency independent antenna

It has already been mentioned that the infinite bifin satisfies the first requirement of a frequency independent antenna, namely that it should be defined in terms of angles alone. When it is made finite, however, by truncating it at a radius r, it fails to satisfy the second requirement that the incident current should fall to zero at or before the truncation. The proposal was therefore made by R.H. DuHamel that a series of discontinuities be introduced, as in Fig. 8.5, in the bifin to see if they would accelerate the decay of current along the structure. In order to minimise the number of characteristic lengths introduced he proposed that these discontinuities should exist between two angles, the angle of the bifin and an inner angle. Nevertheless the radial distances at which they would be introduced remained to be chosen, and working from the equiangular spiral example it

Fig. 8.5. Log-periodic antenna with circumferential teeth.

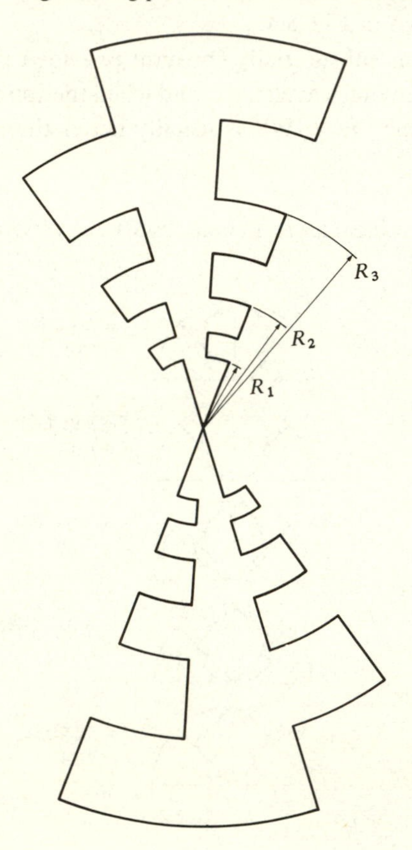

was proposed that they should occur at radii

$$\frac{R_1}{R_2} = \frac{R_2}{R_3} = \cdots = \frac{R_n}{R_{n+1}} = \tau \tag{8.9}$$

Then provided the zero current condition at the truncation is satisfactory, the radiation pattern and impedance should be identical at frequencies

$$f_{n+1} = \tau f_n \tag{8.10}$$

as for the equianglar spiral. But unlike that case a change of frequency does not result in an effective rotation of the bifin or its radiation pattern.

The discontinuities introduced originally took the form of circumferential teeth within a self complementary structure. Experiments showed that when the teeth were large enough the current did fall to zero before the truncation and the variation of the pattern between the log-periodic frequencies was acceptable. The polarisation was linear, perpendicular to what it would be for the bifin with no teeth, i.e. the polarisation was dominated by currents flowing along the teeth. Experimentally it was found that the circumferential teeth could be replaced by trapezoidal teeth, as in Fig. 8.6. Radiation from either structure is bidirectional, normal to its plane, as for the equianglar spiral.

Fig. 8.6. Log-periodic antenna with trapezoidal teeth.

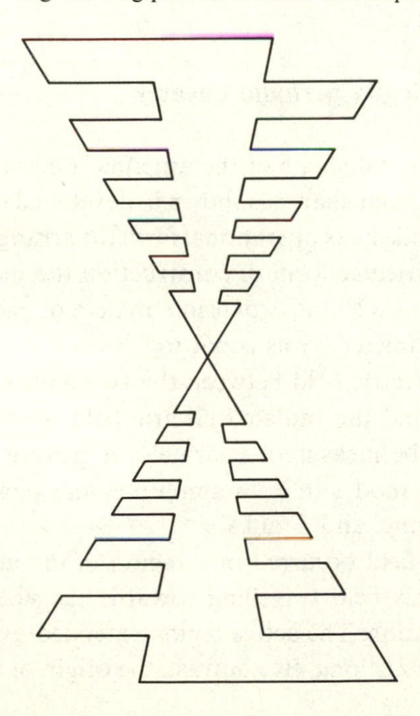

8.4 Unidirectional linearly polarised frequency independent antenna

For many applications the bidirectional radiation from the frequency independent antennas previously discussed is a disadvantage. But if a planar linearly polarised frequency independent antenna is imagined to be hinged at its apex then the two halves may be inclined towards each other at an angle ψ. In its original form ψ is equal to π radians. The operation of reducing the angle ψ affects both the input impedance and the radiation pattern. The effect on the input impedance is clearly that as the angle is decreased so is the input impedance, tending towards zero as the spacing between the planes tends to zero. The variation in the radiation pattern has been found to increase as the angle ψ is decreased, though the effect was improved by using straight rather than circular teeth. But the radiation pattern is now unidirectional, with the maximum field being radiated symmetrically off the apex of the structure. Typical practical values of the angle ψ are within the range 30°–60°. The polarisation is in the direction of the teeth and the currents in the two displaced halves give rise to fields which add in phase off the apex. The E-plane beamwidths are typically 60°–90° and for the H-plane 40°–155°.

In practice although conducting sheet is convenient for applications in the gigahertz frequency range it is unsuitable for low frequency use and in this range the sheet is replaced by wire forming the outline of the toothed sheet.

8.5 Triangular logarithmically periodic linearly polarised antenna

In this form the trapezoidal shape of the antenna used at lower frequencies is replaced by a triangular shape, as shown in developed form in Fig. 8.7. When viewed from the side in its operational form the arrangement is shown in Fig. 8.8. In one particular form of construction the included angle between the two halves was 32°, the expansion angle α of each half was 28° and the expansion parameter τ was equal to 0.85.

Since the transmission line electric field between the two halves of the structure is in the z-direction and the radiated electric field is in the x-direction, these two fields can be measured separately. A picture of the transmission line and radiation modes in both amplitude and phase has thus been built up by Bell, Elfving, and Franks.

The origin of the E_x-electric field occurred in a region of the antenna known as the active region, this field travelling towards the apex and beyond, with only small attenuation. The active region extended over 3–4 teeth which were approximately $\lambda/2$ long. By contrast, the origin of the E_z-

electric field naturally occurred at the apex, and its amplitude diminished rapidly throughout the active region, tending to a negligible value beyond it.

Measurements of current along the triangular teeth revealed approximately sinusoidal standing waves over the active region. The maximum of each standing wave occurred at the mid-points of the teeth where connection was made to a supporting conducting boom, and the currents at the teeth discontinuities were almost zero. The input current to the active region was much lower than the current maxima, and beyond the active region it fell almost to zero.

Fig. 8.7. Developed form of triangular log-periodic antenna.

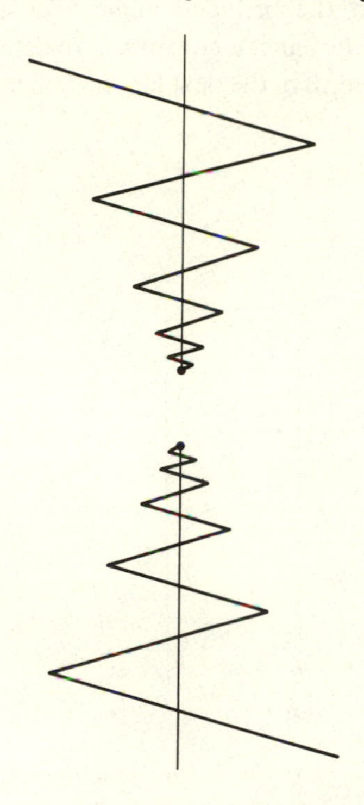

Fig. 8.8. Side-view of triangular log-periodic antenna.

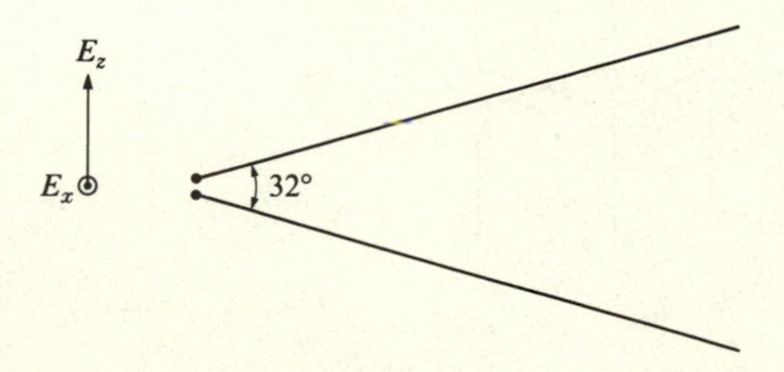

8.6 Logarithmically periodic dipole array

The log-periodic dipole array is a planar end-fire array of individually fed dipoles as shown in Fig. 8.9(a), with the ratio of successive radii from the apex chosen to be a constant τ, as in the equiangular spiral, i.e.

$$\frac{R_1}{R_2} = \frac{R_2}{R_3} = \frac{R_3}{R_4} = \cdots = \tau \tag{8.11}$$

Then by similar triangles it follows that τ is also equal to the ratio of consecutive dipole element lengths, i.e.

$$\tau = \frac{L_1}{L_2} = \frac{L_2}{L_3} = \cdots \tag{8.12}$$

The combination of this ratio τ and the included angle α define the geometrical structure of the antenna, although it is convenient to define also the ratio of element spacing d_n to the length of the next larger element L_{n+1}

Fig. 8.9. Log-periodic dipole array.

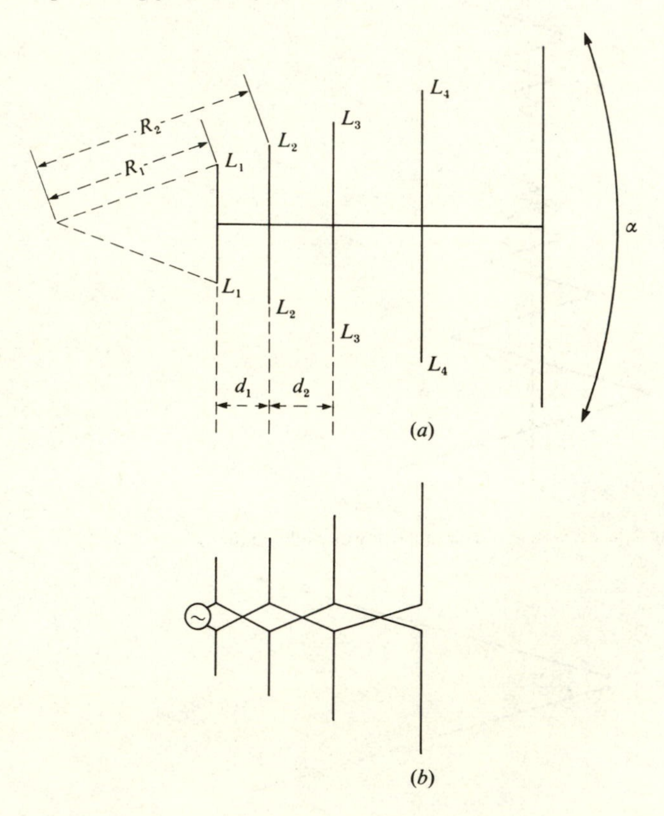

by the equation

$$\sigma = \frac{d_1}{L_2} = \frac{d_2}{L_3} = \frac{d_n}{L_{n+1}} \tag{8.13}$$

Then the relation between the three constants α, τ and σ is given by

$$\tan\frac{\alpha}{2} = \frac{L_2 - L_1}{2d_1} = \frac{1-\tau}{2\sigma} = \cdots \tag{8.14}$$

As shown in Fig. 8.9(*b*) the feeds to successive dipoles are switched and this switching is conveniently arranged in practice by bringing out successive dipoles in opposite directions from both sides of the supporting transmission line, as show in Fig. 8.10. The reason for this switching is to obtain the desired phase velocity of propagation along the transmission line loaded by the radiating dipoles.

Thus since the main beam of radiation from the array would be expected to come from the shortest dipole in the direction opposite to the direction of the larger dipole elements, it is necessary that the sum of the forward transmission and backward radiation phase delays per segment should be a multiple of 2π radians. Thus in the absence of phase switching if β_t is the phase constant of the loaded transmission line and β_r the phase constant of the loaded radiation path towards the shortest dipole, then for cophasal summation of the fields,

$$(\beta_t + \beta_r)d = 2\pi \tag{8.15}$$

But if phase switching is used then the corresponding equation becomes

$$(\beta_t + \beta_r)d = \pi \tag{8.16}$$

since a phase shift of π radians is provided by the switching of the transmission line. Using an experimentally determined value of d of 0.1λ means that for the unswitched case the average phase velocity expressed as

Fig. 8.10. Switching by bringing out successive arms in opposite directions.

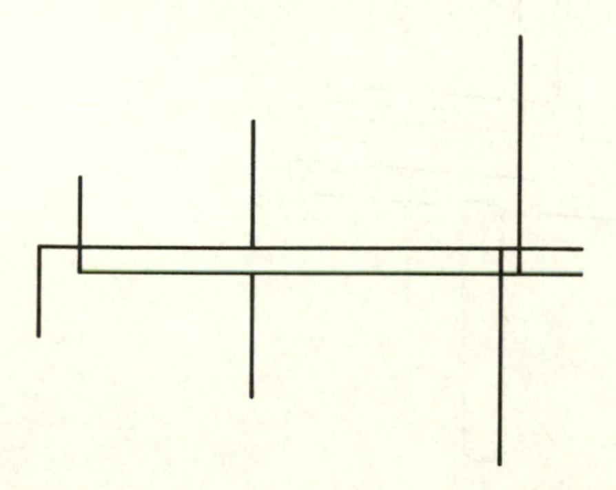

a fraction of the velocity of light, over the combined transmission and radiation paths, is

$$\frac{v_{av \cdot u}}{c} = \frac{k}{\beta_{av}} = \frac{2k}{(\beta_t + \beta_r)} = 0.2 \tag{8.17}$$

By comparison for the switched case the corresponding average phase velocity expressed as a fraction of the velocity of light is

$$\frac{v_{av \cdot s}}{c} = 0.4 \tag{8.18}$$

It is clearly much easier to slow a wave to 0.4c rather than to 0.2c. Experimental measurement of phase velocity for an antenna with the above value of d gave an average velocity of 0.46c.

The feed to the antenna may be made by a balanced line to the smallest dipole, with the transmission line then being continued to successive dipoles. Alternatively an unbalanced feed may be used by running a coaxial cable inside one of the transmission lines shown in Fig. 8.11 from the long dipole end, and connecting its inner conductor to the other line at the small dipole end.

As far as a more complete understanding of the antenna is concerned a study will next be carried out in network terms of a transmission line loaded by a uniformly spaced array of dipoles. The object of this analysis is to determine the complex phase of propagation along the structure so that this result can be applied as an approximation to the single section of a log-periodic dipole array which has this uniform spacing.

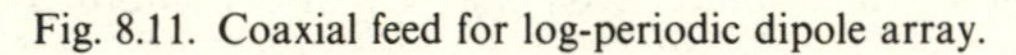
Fig. 8.11. Coaxial feed for log-periodic dipole array.

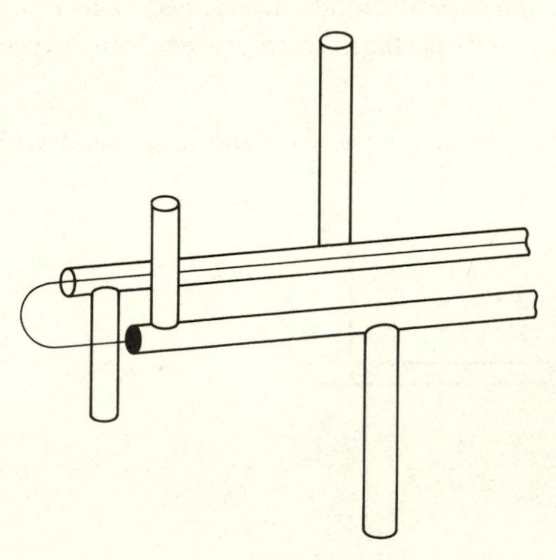

8.7 Network analysis of uniformly loaded periodic structure

The structure to be analysed is one period of a transmission line which is loaded regularly with dipole antennas of constant height $2H$, spaced distance d apart, as shown in Fig. 8.12(a). Consider first a section of length d of an unloaded transmission line, as in Fig. 8.12(b). This gives for the input and output currents,

$$\left.\begin{aligned} I_1 &= Y_{11}V_1 + Y_{12}V_2 \\ I_2 &= Y_{21}V_1 + Y_{22}V_2 \end{aligned}\right\} \tag{8.19}$$

where

$$Y_{11} = \left(\frac{I_1}{V_1}\right)_{V_2=0} = -jY_0 \cot kd = Y_{22}$$

and

$$Y_{12} = \left(\frac{I_1}{V_2}\right)_{V_1=0} = jY_0 \operatorname{cosec} kd = Y_{21}$$

For sections to be connected in cascade it is convenient to take the positive direction of current flow in such networks always to the right, so that the equations at the input and output ports are

$$\left.\begin{aligned} I_1 &= Y_{11}V_1 + Y_{12}V_2 \\ I_2 &= -Y_{21}V_1 - Y_{22}V_2 \end{aligned}\right\} \tag{8.20}$$

For an infinitely long lossless transmission line $V_2 = V_1 e^{-j\beta d}$ and $I_2 = I_1 e^{-j\beta d}$ so that

$$\left.\begin{aligned} I_1 &= V_1(Y_{11} + Y_{12}e^{-j\beta d}) \\ I_1 &= -V_1(Y_{21}e^{j\beta d} + Y_{22}) \end{aligned}\right\} \tag{8.21}$$

Fig. 8.12. Transmission line uniformly loaded with dipole antennas.

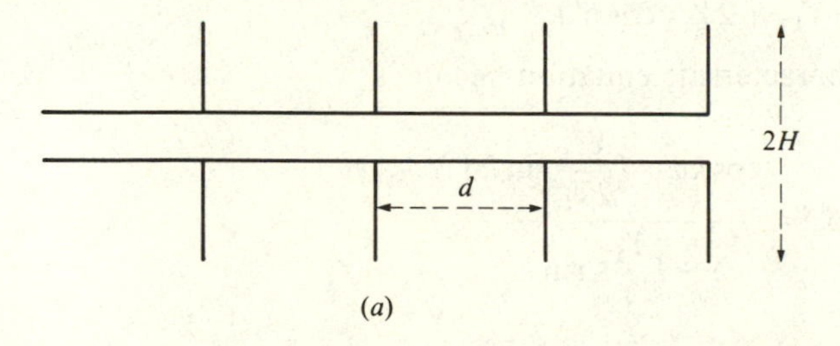

(a)

I_1 d I_2

V_1 V_2

(b)

Hence

$$\cos kd = -\frac{(Y_{11} + Y_{22})}{2Y_{12}} \tag{8.22}$$

Now let an admittance Y_i be added across the line at the junction of each segment of length d so that the input current to the first segment becomes

$$I'_1 = I_1 + Y_i V_1 = (Y_{11} + Y_i)V_1 + Y_{12}V_2 \tag{8.23}$$

Hence every Y_{11} in the above analysis must be replaced by $(Y_{11} + Y_i)$, and eqn (8.22) becomes

$$\cos \beta' d = \cos kd - \frac{Y_i}{2Y_{12}} \tag{8.24}$$

where β' is now complex, equal to $(\beta + j\alpha)$, i.e.

$$\cos \beta' d = \cos kd - \frac{Y_i}{j2Y_0} \sin kd \tag{8.25}$$

In eqn (8.25) Y_i is the input admittance of a single dipole in the presence of all the other dipoles in the array. As an approximation it will be assumed that the neighbouring two dipoles, one on each side of the single dipole being considered, will shield it from the influence of all the other dipoles in the array, so that the input admittance to this single dipole becomes

$$Y_i = Y_{11} + Y_{12}\frac{V_2}{V_1} + Y_{1(-1)}\frac{V_{-1}}{V_1}$$

Since

$$V_2 = V_1 e^{-j\beta' d}$$

and

$$V_{-1} = V_1 e^{j\beta' d}$$

this gives for the input admittance

$$Y_i = Y_{11} + 2Y_{12} \cos \beta' d$$

Hence the characteristic equation becomes

$$\cos \beta' d = \frac{\cos kd + j\dfrac{Y_{11}}{2Y_0} \sin kd}{1 - j\dfrac{Y_{12}}{Y_0} \sin kd} \tag{8.26}$$

In carrying out numerical computations it is desirable to begin with dipole impedances rather than admittances, since the self and mutual impedances of dipole elements are available and are independent of the presence of other elements in the array. The formulae for these impedances for equal length dipoles of length $2H$ spaced distance d apart, and based on assumed trigonometric current distributions on these dipoles, are available in the literature. From these formulae it is therefore possible to write down an impedance matrix for the complete array of identical dipoles. Denoting the dipole at which it is desired to calculate the input admittance Y_i by the number two let the dipole array be assumed to extend from $-\infty$ to $+\infty$ along the transmission line, but consider coupling to exist with dipole numbers one and three only. Then this approximate matrix of three elements is

$$[Z] = \begin{bmatrix} Z_{11} & Z_{12} & 0 \\ Z_{12} & Z_{11} & Z_{12} \\ 0 & Z_{12} & Z_{11} \end{bmatrix} \tag{8.27}$$

The corresponding admittance matrix is

$$[Y] = \begin{bmatrix} Y_{11} & Y_{12} & 0 \\ Y_{12} & Y_{11} & Y_{12} \\ 0 & Y_{12} & Y_{11} \end{bmatrix} \tag{8.28}$$

Since the admittance matrix is the inverse of the impedance matrix Z

$$[Z][Y] = [I]$$

Using these approximate matrices with nine elements each, multiplication of the matrices using the formula

$$\sum_{i=1}^{5} Z_{ri} Y_{ic}$$

where r, c stand for row and column respectively, gives

$$\begin{bmatrix} (Z_{11}Y_{11} + Z_{12}Y_{12}) & (Z_{11}Y_{12} + Z_{12}Y_{11}) & Z_{12}Y_{12} \\ (Z_{12}Y_{11} + Z_{11}Y_{12}) & (Z_{11}Y_{11} + 2Z_{12}Y_{12}) & (Z_{11}Y_{12} + Z_{12}Y_{11}) \\ Z_{12}Y_{12} & (Z_{12}Y_{11} + Z_{11}Y_{12}) & (Z_{11}Y_{11} + Z_{12}Y_{12}) \end{bmatrix}$$

Although it is clear that this product does not form a unit matrix this is the result of both the original impedance and admittance matrices being approximate only. A closer approximation to the unit matrix is obtained by taking five elements in each of the Z and Y matrices while still retaining only the self and two adjacent terms in each row. Hence the following approximate equations are obtained from the equality with the unit matrix

$$Z_{11}Y_{12} + Z_{12}Y_{11} = 0 \tag{8.29}$$

$$Z_{11}Y_{11} + 2Z_{12}Y_{12} = 1 \tag{8.30}$$

From eqns (8.29) and (8.30) Y_{11} and Y_{12} are obtained using calculated values of Z_{11} and Z_{12}. Hence the complex value of β' may be found from eqn (8.26) for a given spacing d between the dipoles.

Calculations of β' have been carried out by Mittra and Jones for

$d = 6.3\,\text{cm}$, $H = 28\,\text{cm}$ and $H/a = 88$ where a is the radius of the dipole. The results are shown in Fig. 8.13, along with experimental results obtained at the University of Illinois for the case of switched excitation, for which the mutual impedances are given by

$$Z_{1n}^{S} = (-1)^{n+1} Z_{1n}$$

Fig. 8.13. Complex propagation constant β' vs frequency. (After Mittra and Jones, 1963.)

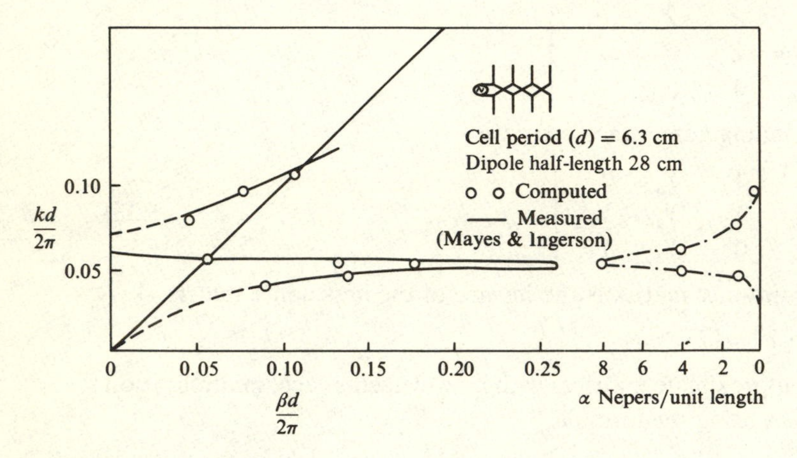

Fig. 8.14. Transmission line voltage along log-periodic dipole array. (After Carrel, 1961.)

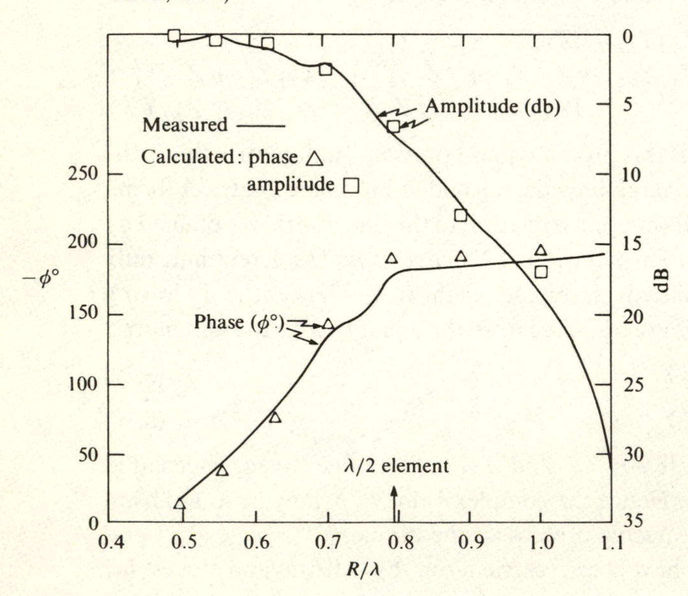

8.8 Application of uniformly loaded periodic structure to logarithmically periodic dipole array

In order to apply the results of the previous section to a LPD Array it is sufficient to divide the dipole array into a sequence of cells in which the separation distance d progressively increases from the feed end. For each such cell the complex value of the phase constant β' is calculated as though the array were uniform. Results have been computed on this basis for an actual LP dipole and compared with measurements made by Carrel at Illinois. The agreement, shown in Fig. 8.14, for the resultant transmission line voltage in magnitude and phase is quite remarkable considering the approximations which have been made in the analysis.

Further reading

C.A. Balanis: Antenna Theory: *Analysis and Design*: Harper and Row, New York, 1982.

R.L. Bell, C.T. Elfving and R.E. Franks: 'Near-field measurements on a logarithmically periodic antenna': *IRE Trans. Antennas and Propagation*, 1960, 559–67.

R. Carrel: 'The design of log-periodic dipole antennas': IRE *International Convention Record*, 1961, pp. 61–75. (Copyright C, 1961, *IRE* (now *IEEE*).)

J.D. Dyson: 'The equiangular spiral antenna': *Trans. IRE* (*APG*), **A-P7**, 1959. (Copyright C, 1959, *IRE* (now *IEEE*).)

R. Mittra and K.E. Jones: 'Theoretical Brillouin ($k - \beta$) diagram for monopole and dipole arrays, and their application to log-periodic antennas: *IRE International Convention Record*, 1963, pp. 118–28. (Copyright C, 1963, *IEEE*.)

H. Jasik: *Antenna Engineering Handbook*: McGraw-Hill, New York, 1961.

W.L. Weeks: *Antenna Engineering*: McGraw-Hill, New York, 1968.

9

Noise power delivered by wire antennas

Any antenna which is used in a receiving situation will pick up not only the desired signal and other coherent interfering signals, but also random fluctuations of voltage which are due to incoming noise fields, some of which come from outer space. In order to quantify such noise fields we will consider first an infinitesimal dipole which is placed at the centre of a large sphere at a uniform temperature TK, as shown in Fig. 9.1.

Fig. 9.1. Infinitesimal dipole at centre of large sphere, temperature T.

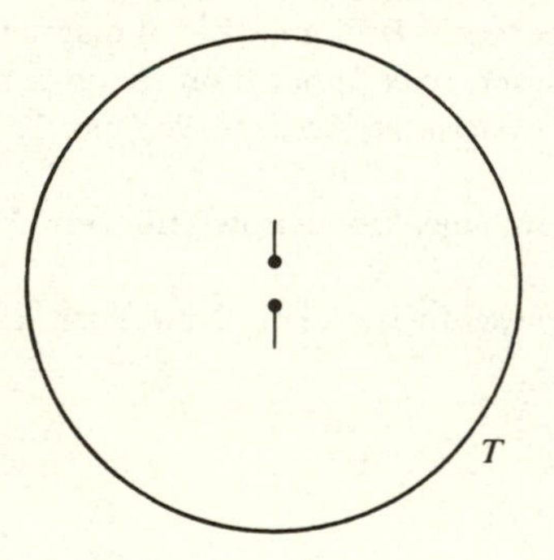

9.1 Noise power delivered to matched load of infinitesimal dipole in isotropic noise field

Within this large sphere, from Planck's law, the energy contained per unit volume per unit bandwidth is given by

$$W = \frac{8\pi f^3 h}{c^3(\exp(hf/KT) - 1)} \tag{9.1}$$

where h is Planck's constant, f is the frequency, c is the velocity of light in vacuo, K is Boltzmann's constant and T is the temperature of the sphere in degrees Kelvin, the sphere being treated as a black body on the surface of which all the thermal noise sources are uncorrelated.

In the vicinity of the dipole the total incoming energy may be assumed to be contained in a summation of uniform plane waves travelling in all radial directions from the spherical surface towards the dipole. Such plane waves possess energy associated with both their electric and magnetic fields, but since these fields are related by the impedance of free space the total energy can be described in terms of either field alone. The electric field will be used in this discussion. Since the energy stored per unit volume at a frequency f, due to a mean square electric field $e^2(f)$, is $\varepsilon_0 e^2(f)$, where ε_0 is the permittivity of free space, this gives from eqn (9.1) for the total mean square electric field along a single coordinate direction

$$e^2(f) = \frac{8\pi h}{3\varepsilon_0 \lambda^3(\exp(hf/KT) - 1)} df \tag{9.2}$$

The mean square applied voltage across the short length dl of the infinitesimal dipole is then $e^2(f)dl^2$, i.e.

$$v^2(f) = \frac{8\pi h dl^2}{3\varepsilon_0 \lambda^3(\exp(hf/KT) - 1)}$$

In the normal radiofrequency range the exponential term can be replaced by the first two terms in its power series, giving

$$v^2(f) = \frac{8\pi}{3} \frac{KTZ_0 dl^2}{\lambda^2} df \tag{9.3}$$

in which Z_0 is the impedance of free space. It will be noted that this mean square voltage is proportional to the temperature of the circumscribing sphere and proportional also to the square of the frequency.

But this applied voltage across the ends of the infinitesimal dipole of length dl must not be confused with the open-circuit voltage which would be obtained from terminals at the middle of the dipole. As shown in Chapter 3 these two voltages are related by the equation.

$$v_{oc} = \tfrac{1}{2} v_{applied}$$

so that the mean square open-circuited voltage is, from eqn (9.3),

$$v_{oc}^2 = \frac{2\pi}{3} \frac{KTZ_0 dl^2}{\lambda^2} df \tag{9.4}$$

Since the input resistance of a lossless infinitesimal dipole is given by

$$R_i = 5k^2 dl^2$$

this gives for the noise power available from this dipole

$$P_a = \frac{v_{oc}^2}{4R_i} = \frac{\pi}{3} \frac{KTZ_0}{10.4\pi^2} df = KTdf \tag{9.5}$$

and for the open-circuit mean square noise voltage

$$v_{oc}^2 = 4KTR_i df \tag{9.6}$$

Thus the noise power, delivered to the matched load of the infinitesimal dipole, is proportional to the temperature of the circumscribing sphere and the band width df

9.2 Noise power delivered to matched load of any antenna in isotropic noise field

Eqn (9.2) gives the mean square electric field along a single coordinate direction at any point within a large spherical surface maintained at a temperature TK. At a specified point of observation there will be three orthogonal components of the total mean square electric field e_T^2, which is thus given by

$$e_T^2 = \frac{8\pi h}{\varepsilon_0 \lambda^3 (\exp(hf/KT) - 1)} df \tag{9.7}$$

In the normal radiofrequency range this reduces to

$$e_T^2 = \frac{8\pi KTZ_0}{\lambda^2} df$$

so that the total power density becomes

$$P_T = \frac{8\pi KT}{\lambda^2} df \tag{9.8}$$

This total incoming power density will now be considered as the integral of an isotropic power density S_T per unit solid angle $\sin\theta\, d\theta\, d\phi$, so that

$$P_T = \int_0^{\pi} \int_0^{2\pi} S_T \sin\theta\, d\theta\, d\phi$$

where for isotropic noise S_T is independent of θ and ϕ, so that

$$P_T = 4\pi S_T$$

Hence the isotropic power density is from eqn (9.8)

$$S_T = \frac{2KT}{\lambda^2} df$$

But of this total power density S_T, only one-half, on average, will actuate any antenna, whether it is a linearly or circularly polarised antenna. Hence the effective total power density becomes one-half of S_T, i.e.

$$S = \frac{KTdf}{\lambda^2}$$

where the units of S are watts per unit solid angle.

Consider now any antenna, with an effective receiving area, as a function of angle $A_{eff}(\theta, \phi)$, so that the power delivered to a matched load is given by

$$P_{del} = \int_0^{\pi} \int_0^{2\pi} S A_{eff}(\theta, \phi) \sin\theta \, d\theta \, d\phi$$

But since the effective area is related to the power gain by the equation,

$$G_P = \frac{4\pi}{\lambda^2} A_{eff}$$

this gives for the power delivered to the matched load,

$$P_{del} = \frac{S\lambda^2}{4\pi} \int_0^{\pi} \int_0^{2\pi} G_P(\theta, \phi) \sin\theta \, d\theta \, d\phi$$

Writing the power gain as the product of efficiency and directional gain gives

$$P_{del} = \frac{\eta S\lambda^2}{4\pi} \int_0^{\pi} \int_0^{2\pi} G_D(\theta, \phi) \sin\theta \, d\theta \, d\phi$$

But since directional gain is defined in the transmitting situation as

$$G_D(\theta, \phi) = \frac{\text{radiation intensity in direction } (\theta, \phi)}{\text{average radiation intensity}}$$

the power delivered to a matched load by any antenna becomes

$$P_{del} = \frac{\eta S\lambda^2 4\pi r^2}{4\pi P_{rad}} \int_0^{\pi} \int_0^{2\pi} \frac{|E_T|^2}{2Z_0} \sin\theta \, d\theta \, d\phi$$

where E_T is the transmitted radiation field in direction (θ, ϕ). Since this double integral is equal to the radiated power P_{rad}/r^2, this gives

$$\begin{aligned} P_{del} &= \eta S\lambda^2 \\ &= \eta KTdf \end{aligned} \tag{9.9}$$

Because this noise power is a constant for a given temperature of spherical enclosure, irrespective of the size of receiving antenna, the use of a large receiving antenna improves the signal level only, and thereby the signal/noise ratio.

Since the noise power given by eqn (9.9) can also be written as

$$P_{del} = \frac{v_{oc}^2}{4R_L}$$

where R_L is the matched load resistance connected to the receiving antenna, and since also the antenna efficiency is the ratio of radiation resistance to input resistance, then

$$v_{oc}^2 = 4KTR_r df \tag{9.10}$$

that is, the radiation resistance is seen, at the antenna receiving terminals, to be at the same temperature as the large circumscribing sphere, as far as its opencircuited mean square noise terminal voltage is concerned.

The same result may be obtained more simply but in a possibly less convincing way by considering the equivalent circuit of a receiving antenna as shown in Fig. 9.2. The antenna is loaded with an impedance $Z_L = (R_r + R_l + jX)$ where X is an arbitrary load reactance, and R_r, R_l are the radiation and loss resistances of the antenna, such that

$$R_L = R_r + R_l$$

The loss resistance of the antenna will be understood to produce a mean square noise current

$$i_n^2 = \frac{4KTR_l df}{(4(R_l + R_r)^2 + (X + Xa)^2)}$$

where T is the temperature associated with both R_l and R_r. This noise current will produce a radiated noise power given by

$$P_r = i_n^2 R_r = \frac{4KTR_r R_l df}{(4(R_l + R_r)^2 + (X + Xa)^2)} \tag{9.11}$$

Under thermal equilibrium conditions this radiated power due to the loss resistance R_l must be balanced by an incoming noise power

$$P_i = \frac{v_{oc} R_l}{(4(R_l + R_r)^2 + (X + Xa)^2)} \tag{9.12}$$

Fig. 9.2. Equivalent circuit of receiving antenna.

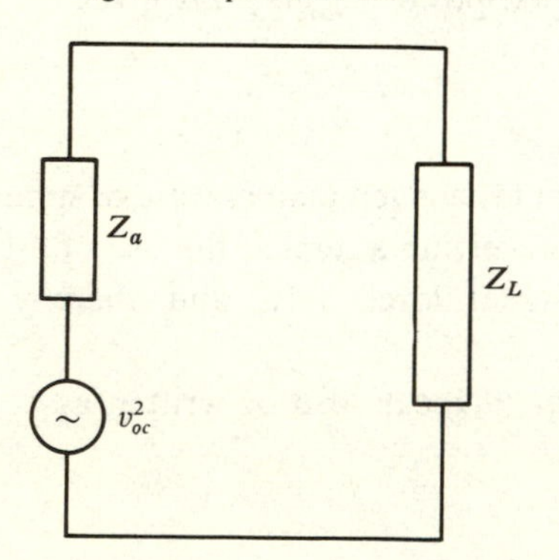

since R_l is the only physically real resistance in the antenna. Hence equating the radiated and incoming powers gives as in eqn (9.10)

$$v_{oc}^2 = 4KTR_r df \tag{9.13}$$

9.3 Network representation of noise power delivered by wire antennas

9.3.1 *Mean square output noise current in short-circuited wire antenna*

Consider a short-circuited wire antenna of arbitrary length which is conceptually considered to be divided into a number N of equal length segments, as shown in Fig. 9.3, with the segments being sufficiently short in terms of a wavelength for the current in each to be considered constant in amplitude and phase. This antenna is now considered to be placed at the centre of a large sphere at temperature TK on the surface of which all the

Fig. 9.3. Short-circuited dipole divided into N equal segments.

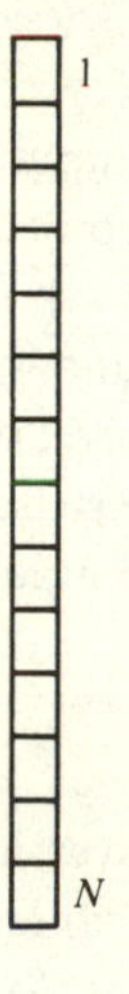

Fig. 9.4. Noise equivalent circuit of antenna divided into N segments.

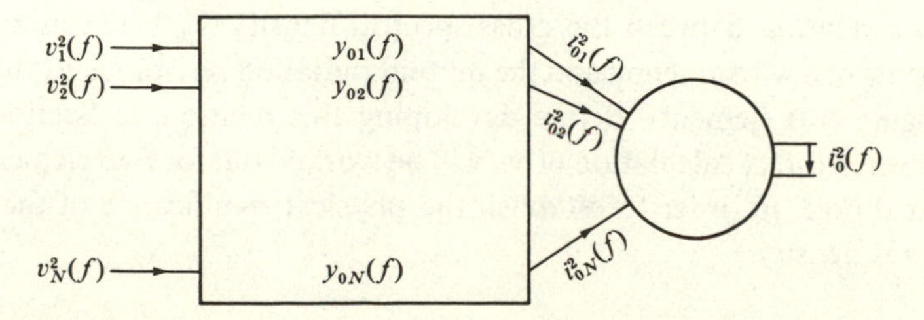

noise sources are taken to be uncorrelated. As shown in Section 9.1 the mean square applied voltage across any one of the equal length segments is given by eqn (9.3), i.e.

$$v_j^2(f) = \frac{8\pi}{3} \frac{KTZ_0 dl^2}{\lambda^2} df \tag{9.14}$$

The network representation of the antenna, as shown in Fig. 9.4, then has N noise input signals across the N segments of the antenna and a single noise output. Although the noise fields coming from the constant temperature enclosure are uncorrelated as far as the fields coming from different directions are concerned, a correlation exists between the noise fields arriving from a single direction. Consequently the voltages due to these fields coming from a single direction, which are applied across the different segments of the antenna, are also correlated. For such correlated mean square input voltages $v_j^2(f)$, the short-circuited mean square output noise current is given by

$$i_0^2(f) = \sum_{j=1}^{N} \sum_{k=1}^{N} Y_{0j}^*(f) Y_{0k}(f) S_{jk}(f, \theta, \phi) v_j^2(f) \tag{9.15}$$

where $Y_{0j,k}(f)$ is the mutual admittance between the j, kth input segment and the output segment; the asterisk denotes complex conjugate and $S_{jk}(f, \theta, \phi)$ is the cross-spectral density between the jth and kth segments which is the Fourier transform of the cross-correlation function of the voltages applied across these segments. For the case of uncorrelated input voltages, S_{jk} is equal to unity for j equal to k, and zero otherwise, so that eqn (9.14) reduces to the standard equation for uncorrelated input signals,

$$i_0^2(f) = \sum_{j=1}^{N} |Y_{0j}|^2 v_j^2(f)$$

The mutual admittances in eqn (9.15) are given by the inversion of the impedance matrix representing the N segments of the antenna, the evaluation of which was considered in Chapter 4. The algebraic form taken by the cross-spectral density function $S_{jk}(f, \theta, \phi)$ depends, like the mutual impedance elements, on the geometry of the wire antenna involved. There is a close relation between the cross-spectral density S_{jk} between any two elements of a wire antenna and the mutual radiation resistance R_{jk} between the same two elements. Before developing this relation, in Section 9.3.4, however, a direct calculation of S_{jk} will be worked out for two elements of a linear dipole, in order to establish the physical significance of the cross-spectral density.

9.3.2 *Evaluation of cross-spectral density S_{jk} between two elements of linear dipole*

Fig. 9.5 represents an incoming planar noise wavefront arriving from the direction (θ, ϕ) with two perpendicular electric field components E_θ and E_ϕ. Only the E_θ-field has a component along the segments j, k of the dipole lying along the z-axis and separated by a distance d. Since, on the average, E_θ will be equal to E_ϕ, this means that the mean square noise voltage applied across element j, due to this single incoming noise wave, is given by

$$v_j^2(t) = \tfrac{1}{2} e_T^2 dl^2 \sin^2 \theta$$

where the total mean square electric field e_T^2 is such that when integration of $v_j^2(t)$ is carried out over the whole sphere, $v_j^2(t)$ will be one-half of $v_j^2(f)$ because of the two orthogonal electric field components. Hence

$$e_T^2 = \frac{2KTZ_0}{\lambda^2} \sin \theta \, d\theta \, d\phi \, df \tag{9.16}$$

Integrating over all frequencies and all angles θ, ϕ gives the total mean square noise voltage across element j as

$$v_j^2(t) = \int_{-\infty}^{+\infty} \int_0^{\pi} \int_0^{2\pi} \frac{KTZ_0}{\lambda^2} dl^2 \sin^3 \theta \, d\theta \, d\phi \, df \tag{9.17}$$

Over the frequency range df this equation reduces to eqn (9.3). Likewise the mean square voltage across segment k has the same value.

Fig. 9.5. Incoming planar noise wavefront with E_θ and E_ϕ components.

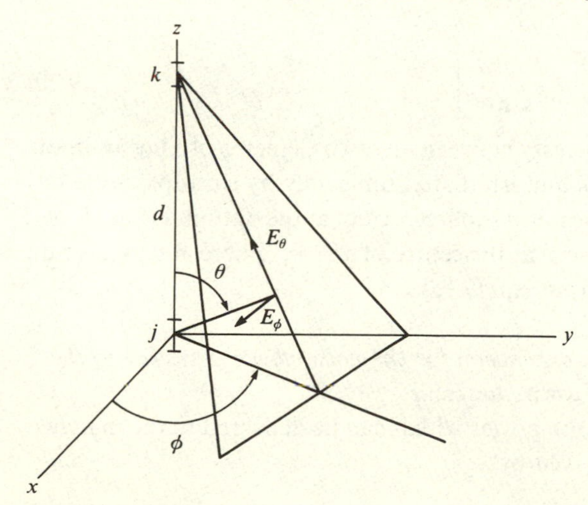

The cross-correlation function $R_{jk}(\tau)$ between these mean square voltages is obtained from

$$R_{jk}(\tau) = E(v_j^{2*}(t)v_k^2(t+\tau))^{\frac{1}{2}}$$
$$= \int_{-\infty}^{+\infty}\int_0^{\pi}\int_0^{2\pi} e^{j\omega(\tau + d\cos\theta/c)}\frac{KTZ_0}{2}dl^2 \sin^3\theta\, d\theta\, d\phi\, df \tag{9.18}$$

since only the same signal at the same frequency arriving from the same angle (θ, ϕ) is wanted. This gives delta functions for frequency and (θ, ϕ) so that the expression remains as a triple integral, i.e.

$$R_{jk}(\tau) = 2\pi\int_{-\infty}^{+\infty} e^{j\omega\tau}\int_0^{\pi} e^{jkd\cos\theta}\sin^3\theta\frac{KTZ_0}{2}dl^2\, d\theta\, df$$

The substitution of variable

$$u = \cos\theta$$

then gives

$$R_{jk}(\tau) = 2\pi\int_{-\infty}^{+\infty} e^{j2\pi f\tau}\frac{4KTZ_0}{\lambda^2}\left[\frac{\sin kd}{k^3d^3} - \frac{\cos kd}{k^2d^2}\right]dl^2\, df \tag{9.19}$$

With the definition

$$V_j^2(f) = \frac{8\pi}{3}\frac{KTZ_0}{\lambda^2}dl^2$$

this gives

$$R_{jk}(\tau) = \int_{-\infty}^{+\infty} e^{j2\pi f\tau}S_{jk}(f)V_j^2(f)\, df$$

where

$$S_{jk}(f) = 3\left[\frac{\sin kd}{k^3d^3} - \frac{\cos kd}{k^2d^2}\right] \tag{9.20}$$

Thus the cross-spectral density between any two elements of a linear dipole separated by a distance d and irradiated uniformly by isotropic noise has been evaluated. Hence the mean square current at the output terminals of a short-circuited dipole placed at the centre of a large sphere at temperature TK can now be found from eqn (9.15).

9.3.3 *General network expression for the noise power delivered to the matched load of a wire antenna*

The available noise power which can be delivered to the matched load of any antenna is given by

$$P_a = \frac{v_{oc}^2(f)}{4R_L}$$

where R_L is the resistive component of the matched load. The mean square open-circuit voltage $v_{oc}^2(f)$ is related to the mean square short-circuit current at the output terminals by the equation

$$v_{oc}^2(f) = i_0^2(f)|Z_i(f)|^2 \tag{9.21}$$

where $Z_i(f)$ is the input impedance of the antenna at frequency f and $i_0^2(f)$ is given by eqn (9.15). Hence the available noise power becomes for any wire antenna,

$$P_a = \frac{|Z_i(f)|^2}{4R_L} \operatorname{Re} \sum_{j=1}^{N} \sum_{k=1}^{N} Y_{0j}^*(f) Y_{0k}(f) S_{jk}(f, \theta, \phi) v_j^2(f) \tag{9.22}$$

which can be readily evaluated by computer and is numerically equal to $\eta KTdf$.

9.3.4 *Reduction of the general network expression for the noise power for a linear dipole antenna*

It is desirable to show that the general expression given by eqn (9.22) is analytically as well as numerically equal to $\eta KTdf$. This can be done readily and will be illustrated for the case of a linear dipole. From eqns (9.20) and (9.3) the product of the cross-spectral density and the mean square voltage applied across each segment is

$$S_{jk}(f, \theta, \phi) v_j^2(f) = \frac{8\pi KTZ_0 dl^2}{\lambda^2} \left(\frac{\sin kd}{k^3 d^3} - \frac{\cos kd}{k^2 d^2} \right) df \tag{9.23}$$

But the mutual resistance between two collinear elementary dipoles is given by

$$R_{jk}(f) = 60k^2 dl^2 \left(\frac{\sin kd}{k^3 d^3} - \frac{\cos kd}{k^2 d^2} \right)$$

so that eqn (9.23) simplifies to

$$S_{jk}(f, \theta, \phi) v_j^2(f) = 4KTR_{jk}(f) df \tag{9.24}$$

Now consider the same linear dipole in a transmitting situation where the radiated power is given by

$$P_{rad} = \frac{1}{2} \sum_{j=1}^{N} \sum_{k=1}^{N} |I_j||I_k| R_{jk}(f) \cos v_{jk} \tag{9.25}$$

in which I_j and I_k are the transmitting currents in the jth and kth segments, and v_{jk} is the phase angle between the currents in these segments. If these currents are written in terms of the applied voltage at the previous output segment '0' and the mutual admittance between this segment '0' and the elements j, k, the expression for the radiated power becomes

$$P_{rad} = \frac{V_{app}^2}{2} \sum_{j=1}^{N} \sum_{k=1}^{N} |Y_{0j}(f)||Y_{0k}(f)| R_{jk}(f) \cos v_{jk}$$

But since

$$|Y_{0j}(f)||Y_{0k}(f)|\cos v_{jk} = \mathrm{Re}\, Y^*_{0j}(f) Y_{0k}(f)$$

this gives

$$P_{rad} = \frac{V^2_{app}}{2} \mathrm{Re} \sum_{j=1}^{N} \sum_{k=1}^{N} Y^*_{0j}(f) Y_{0k}(f) R_{jk}(f) \tag{9.26}$$

Thus

$$\mathrm{Re} \sum_{j=1}^{N} \sum_{k=1}^{N} Y^*_{0j}(f) Y_{0k}(f) R_{jk}(f) = \frac{2P_{rad}}{V^2_{app}} = \frac{R_r}{|Z_i|}2 \tag{9.27}$$

where R_r is the radiation resistance of the dipole.

Hence eqn (9.22), together with eqns (9.24) and (9.27), gives for the noise power delivered to the matched load of a linear dipole,

$$P_a = KTdf \frac{R_r}{R_L} = \eta KTdf \tag{9.28}$$

a result which is completely general for all antennas.

9.3.5 *Isotropic cross-spectral density between current elements expressed in terms of radiation resistance*

In Section 9.3.2 a direct calculation of the cross-spectral density between two current elements spaced a distance d apart, and forming part of a linear dipole irradiated by isotropic noise, was worked out from first principles. It will now be shown that for any wire antenna a simple relationship exists between the cross-spectral density between any two current elements, and the self and mutual radiation resistances between them.

From eqns (9.14), (9.15) and (9.21) the open-circuit mean square noise voltage at the terminals of the antenna is given by

$$v^2_{oc} = \frac{4KTdf}{Y_0(f) Y^*_0(f)} 20k^2 dl^2$$
$$\times \sum_{j=1}^{N} \sum_{k=1}^{N} Y^*_{0j}(f) Y_{0k}(f) S_{jk}(f) \tag{9.29}$$

where $Y_0(f) Y^*_0(f)$ is the square of the modulus of the input admittance of the antenna. But since this open-circuit mean square noise voltage is also equal to $4KTR_i df$, where R_i is the input resistance of the antenna, then the value of R_i derived from this receiving approach is equal to

$$R_i = \frac{20k^2 dl^2}{Y_0(f) Y^*_0(f)} \sum_{j=1}^{N} \sum_{k=1}^{N} Y^*_{0j}(f) Y_{0k}(f) S_{jk}(f) \tag{9.30}$$

But this input resistance can also be expressed in terms of a double summation involving the mutual resistance between the elements j and k in a transmitting situation as follows. Let currents I_j and I_k flow in segments j and k respectively, with a phase angle γ_{jk} between these currents. Then the total power radiated by the antenna is

$$P = \frac{1}{2} \sum_{j=1}^{N} \sum_{k=1}^{N} |I_j I_k| R_{jk}(f) \cos \gamma_{jk}$$

where $R_{jk}(f)$ is the mutual resistance between segments j and k. It is convenient to write this equation as

$$P = \frac{1}{4} \sum_{j=1}^{N} \sum_{k=1}^{N} (I_j I_k^* + I_k I_j^*) R_{jk}(f)$$

But since this radiated power is equal to $I_0 I_0^* R_i/2$ where I_0 is the input current and R_i is the antenna input resistance, it follows that an alternative expression for this input resistance is

$$R_i = \frac{1}{2I_0 I_0^*} \sum_{j=1}^{N} \sum_{k=1}^{N} (I_j I_k^* + I_k I_j^*) R_{jk}(f) \tag{9.31}$$

In eqn (9.31) the currents I_j and I_k may be written as VY_{0j} and VY_{0k}, where V is the voltage applied across the input terminals and Y_{0j} and Y_{0k} are the mutual admittances between the input segment and j and k segments, respectively, so that the input resistance of the antenna then becomes

$$R_i = \frac{1}{2Y_0(f)Y_0^*(f)} \sum_{j=1}^{N} \sum_{k=1}^{N} \times [Y_{0j}(f)Y_{0k}^*(f) + Y_{0k}(f)Y_{0j}^*(f)] R_{jk}(f) \tag{9.32}$$

Eqn (9.32) is purely real because the quadrature terms associated with the separate admittance products cancel when they are summed. Because also R_{jk} is equal to R_{kj} grouping of terms in the expanded double summation leads to the result that the summation remains real when eqn (9.32) is written as

$$R_i = \frac{1}{Y_0(f)Y_0^*(f)} \sum_{j=1}^{N} \sum_{k=1}^{N} Y_{0j}^*(f) Y_{0k}(f) R_{jk}(f) \tag{9.33}$$

without the use of the prefix Re on the right-hand side of this equation, which would normally be necessary.

From eqns (9.30) and (9.33) it then follows that the cross-spectral density S_{jk} between two current elements is related to the mutual radiation resistance between them by the simple equation

$$S_{jk}(f) = \frac{R_{jk}(f)}{R_{jj}(f)} \tag{9.34}$$

since R_{jj} for a current element is equal to $20k^2dl^2$. The evaluation of a double integral for S_{jk}, as in Section 9.3.2, can thus be avoided for every wire antenna, since formulae for R_{jk} for two general current elements can readily be worked out without involving any integrations. Such formulae are available in the literature.

Thus it is possible to describe mean square noise currents and voltages in receiving antennas either in terms of radiation resistance or in terms of cross-spectral densities, with the relation between them given by eqn (9.34). The algebra associated with Section 9.3 is particularly associated with computer solutions of the noise properties of wire antennas, and thus complements the computer solutions for coherent signals in wire antennas given earlier.

Further reading

S.A. Schelkunoff and H.T. Friis: *Antennas, Theory and Practice*: Wiley and Sons, 1952.

10

+–+

Aperture antennas

Most radiation takes place from the flow of electric current in metallic conductors. In the case of wire antennas, the distribution of this current can be found with good accuracy for antennas up to a few wavelengths in length by means of the moment method. But when a radiator takes the form of a surface on which there may be a transverse as well as a longitudinal variation of current, the linear size of the matrix increases as the area of the radiating surface instead of as its length. Consequently the moment method has not been able to provide quantitative data on the distribution of such surface currents because of limitation on the size of matrix which can be handled by most computers.

In some cases it is possible to estimate the distribution of current on a metallic surface by using the approximation that at every point the surface is locally plane. That is to say, it is assumed that the radius of curvature is infinite and that there are no edges. For large paraboloids of long focal length this approximation is found to be good in the forward direction. This is, firstly, because the curvature is small if the ratio of focal length to diameter is greater than, say, 0.35 and, secondly, because the area over which the effect of the edges has a significant effect on the current distribution is a small percentage of the total area of the paraboloid. But although this gives good results for the main forward field in particular, it is less accurate for determining the far out side lobes and back lobes. One reason for this is that it neglects current flowing on the back surface of the paraboloid. Another is that no account is taken of scattering by the structure used to support the feed.

Consider the question of what information is required to enable the fields anywhere in space to be found. Where metallic conductors only are involved the answer is the same as in the case of wire antennas. That is, provided the electric currents over the whole of the surfaces are known, the potentials and fields can be calculated everywhere. In practice these currents have to be estimated. But there is an alternative approach to the problem. It is often possible to consider a planar surface through which the main body of the radiation can be visualised to come. For example, in the case of the paraboloid this is the surface in the aperture plane. It should be emphasised that this is an approximation. It replaces the front conducting surface only of the paraboloid and neglects the radiation from the back surface currents. Since there are no electric currents flowing in the aperture plane consideration must be given to the question of what fields in this plane are able to provide the forward radiation through it. The answer to this question is found in the equivalence theorem, which for a closed surface provides an exact solution to the radiation through the surface.

10.1 Field equivalence theorem

Consider any closed surface S surrounding a source, or sources, of electromagnetic fields. Let the fields immediately outside this surface at a particular point of observation be $\mathbf{E}_1, \mathbf{H}_1$. Now let the original source or sources inside S be removed, while postulating that the fields $\mathbf{E}_1, \mathbf{H}_1$, just outside the surface remain unchanged, and that just inside the surface the fields at the virtually congruent point are $\mathbf{E}_2, \mathbf{H}_2$. Previously the fields at this point were $\mathbf{E}_1, \mathbf{H}_1$, if the surface S consists of an air boundary there.

The tangential components of fields at the surface where the total fields have been specified are given, on the outside, by

$$\mathbf{n} \times \mathbf{E}_1 \quad \text{and} \quad \mathbf{n} \times \mathbf{H}_1$$

and on the inside of S by

$$-(\mathbf{n} \times \mathbf{E}_2) \quad \text{and} \quad -(\mathbf{n} \times \mathbf{H}_2)$$

where the directions of $\mathbf{n}$ on the two sides are reversed. Taking $\mathbf{n}$ to be that belonging to the outside of S gives a discontinuity of tangential fields

$$\mathbf{n} \times (\mathbf{E}_1 - \mathbf{E}_2) \quad \text{and} \quad \mathbf{n} \times (\mathbf{H}_1 - \mathbf{H}_2)$$

But a discontinuity of tangential magnetic field can only be accounted for by the existence of a surface electric current density on S given by

$$\mathbf{J} = \mathbf{n} \times (\mathbf{H}_1 - \mathbf{H}_2) \tag{10.1}$$

Likewise the discontinuity in tangential electric field could be caused by the presence of a surface magnetic current density on S given by

$$\mathbf{J}_m = -\mathbf{n} \times (\mathbf{E}_1 - \mathbf{E}_2) \tag{10.2}$$

Hence these two current densities $\mathbf{J}$ and $\mathbf{J}_m$ alone are responsible for the total fields external and internal to S and in particular, for producing the fields $\mathbf{E}_1, \mathbf{H}_1$ and $\mathbf{E}_2, \mathbf{H}_2$.

The fields external to S are thus as for the original source or sources, but the fields internal to S have been chosen arbitrarily. They can have any value including zero and, for this case,

$$\mathbf{J} = \mathbf{n} \times \mathbf{H}_1 \tag{10.3}$$

and

$$\mathbf{J}_m = -\mathbf{n} \times \mathbf{E}_1 \tag{10.4}$$

This is Love's form of the equivalence principle, which says that, provided the tangential components of electric and magnetic field are known over any closed surface S, the fields outside S can be obtained from these equivalent currents alone, with the original source or sources removed. Clearly the same argument allows the fields inside S to be found from equivalent sources on the surface, replacing original sources outside S.

The form of these equivalent sources given by eqns (10.3) and (10.4) is extremely simple. Because it is important to realise that they give identical fields to the original source, this will be illustrated by taking the example of a current element as such a source, and by replacing it by equivalent currents on the surface of a sphere surrounding it, and centred on the current element.

10.2 Equivalent currents over spherical surface of radius *a* for original current element

In Fig. 10.1 a current element aligned along the z-axis is positioned at the origin of a rectangular–spherical coordinate system (x, y, z) and (r, θ, ϕ). For such a radiator the electromagnetic fields at any point of

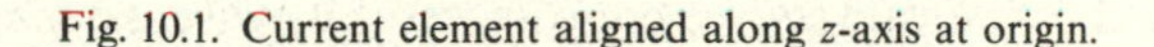

Fig. 10.1. Current element aligned along z-axis at origin.

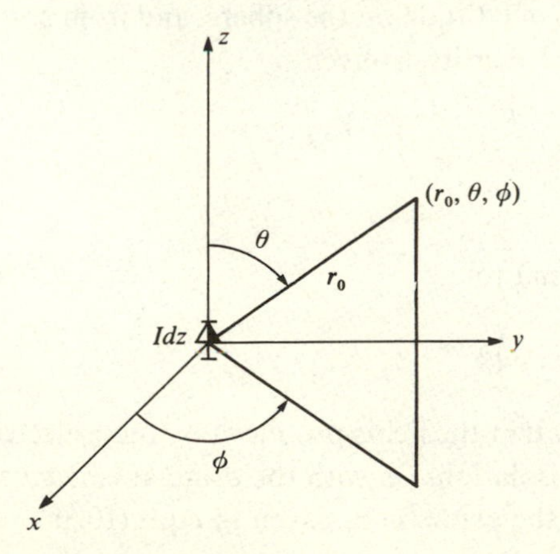

observation are known to be given by

$$\left.\begin{aligned} H_\phi &= \frac{Idz}{4\pi}\left(\frac{jk}{r_0}+\frac{1}{r_0^2}\right)e^{-jkr_0}\sin\theta \\ E_\theta &= \frac{Idz}{4\pi\omega\varepsilon}\left(\frac{jk^2}{r_0}+\frac{k}{r_0^2}-\frac{j}{r_0^3}\right)e^{-jkr_0}\sin\theta \\ E_r &= \frac{Idz}{2\pi\omega\varepsilon}\left(\frac{k}{r_0^2}-\frac{j}{r_0^3}\right)e^{-jkr_0}\cos\theta \end{aligned}\right\} \tag{10.5}$$

The equivalence theorem now asserts that identical fields to these are produced by a combination of electric and magnetic currents distributed over any closed surface such that the current densities over this surface are given by

$$\mathbf{J} = \mathbf{n} \times \mathbf{H}^i \tag{10.6}$$

$$\mathbf{J}_m = -\mathbf{n} \times \mathbf{E}^i \tag{10.7}$$

where $\mathbf{E}^i, \mathbf{H}^i$ are the incident fields obtained from eqns (10.5), and $\mathbf{n}$ is an outwardly directed unit vector normal to the surface. When the closed surface is chosen to be spherical the equations for E_θ and H_ϕ only, from eqns (10.5), contribute to these currents since $\mathbf{n}$ is then a radial unit vector.

For such a spherical surface of radius a the electric current lines of flow are lines of longitude on the sphere, and from eqn (10.6) the magnitude of this current density is given by

$$\begin{aligned} J_{\theta'} &= -\frac{Idz}{4\pi}\left(\frac{jk}{a}+\frac{1}{a^2}\right)e^{-jka}\sin\theta' \\ &= C_1 \sin\theta' \end{aligned} \tag{10.8}$$

where the constant C_1 is equal to $(-Idz/4\pi)[(jk/a)+(1/a^2)]e^{-jka}$, and the prime is introduced to indicate source coordinates. Likewise the magnetic current lines of flow are lines of latitude on the sphere, and from eqn (10.7) the magnitude of this current density is given by

$$\begin{aligned} J_{m\phi'} &= -\frac{Idz}{4\pi\omega\varepsilon}\left(\frac{jk^2}{a}+\frac{k}{a^2}-\frac{j}{a^3}\right)e^{-jka}\sin\theta' \\ &= C_2 \sin\theta' \end{aligned} \tag{10.9}$$

where the constant C_2 is equal to

$$-\frac{Idz}{4\pi\omega\varepsilon}\left(\frac{jk^2}{a}+\frac{k}{a^2}-\frac{j}{a^3}\right)e^{-jka}$$

It is now required to show that the fields produced by these electric and magnetic currents acting in isolation, i.e. with the original current source removed, are identical with the expressions given in eqns (10.5).

10.2.1 *Magnetic vector potential of longitudinal electric current on spherical surface of radius a*

Referring to Fig. 10.2 the longitudinal electric current density $J_{\theta'}$ at a source point (a, θ', ϕ') on the surface of the sphere can be resolved into its three rectangular components

$$J_{x'} = J_{\theta'} \cos\theta' \cos\phi'; \quad J_{y'} = J_{\theta'} \cos\theta' \sin\phi';$$
$$J_{z'} = -J_{\theta'} \sin\theta'$$

where $J_{\theta'}$ is given by eqn (10.8). These three surface current densities, flowing in an elemental area $a^2 \sin\theta' d\theta' d\phi'$, produce at the observation point (r_0, θ, ϕ) components of magnetic vector potential given by

$$\left.\begin{aligned} dA_x &= \frac{\mu C_1 a^2}{4\pi} \sin^2\theta' \cos\theta' \cos\phi' \frac{e^{-jkr}}{r} d\theta' d\phi' \\ dA_y &= dA_x \tan\phi' \\ dA_z &= -\frac{\mu C_1 a^2}{4\pi} \sin^3\theta' \frac{e^{-jkr}}{r} d\theta' d\phi' \end{aligned}\right\} \qquad (10.10)$$

where r is the distance from the source element to the point of observation, with

$$r^2 = r_0^2 + a^2 - 2ar_0 \cos\psi \qquad (10.11)$$

and ψ is the angle between the radii from the origin to the source and observation points respectively.

In eqn (10.10) it is convenient to write e^{-jkr}/r using the source and

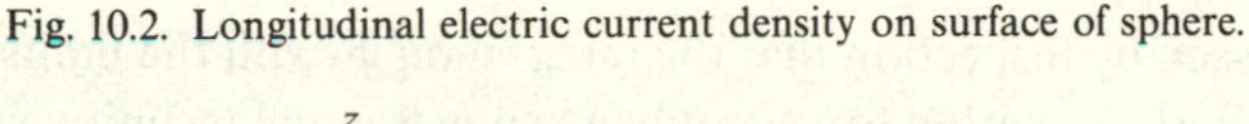

Fig. 10.2. Longitudinal electric current density on surface of sphere.

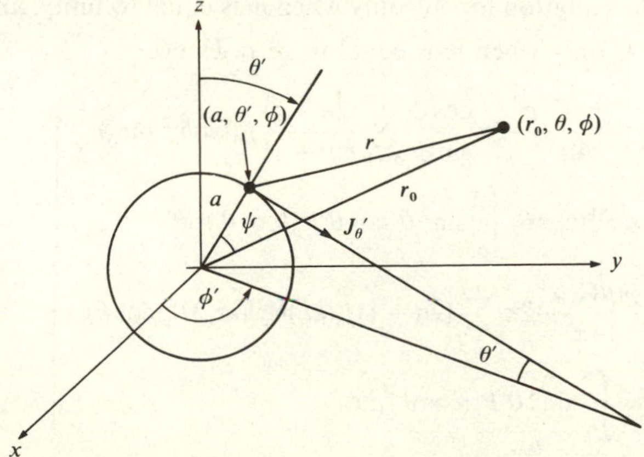

observation angles (θ', ϕ') and (θ, ϕ) as

$$\frac{e^{-jkr}}{r} = -jk \sum_{n=0}^{\infty} (2n+1) j_n(ka) h_n^{(2)}(kr_0) \Bigg\{ P_n(\cos\theta) P_n \cos\theta' + 2 \sum_{m=1}^{n} \frac{(n-m)!}{(n+m)!} P_n^m(\cos\theta) P_n^m(\cos\theta') \cos m(\phi - \phi') \Bigg\} \tag{10.12}$$

Substituting this equation into eqns (10.10) and integrating over the whole sphere gives the components of magnetic vector potential due to the longitudinal electric currents as

$$\left.\begin{aligned}
A_{\substack{x\\y}} &= \frac{-jk\mu C_1 a^2}{4\pi} \int_0^{\pi} \int_0^{2\pi} \sin^2\theta' \cos\theta' \begin{matrix}\cos\phi' \\ \sin\phi'\end{matrix} \\
&\quad \times \sum_{n=0}^{\infty} (2n+1) j_n(ka) h_n^{(2)}(kr_0) \Bigg\{ P_n(\cos\theta) P_n(\cos\theta') \\
&\quad + 2 \sum_{m=1}^{n} \frac{(n-m)!}{(n+m)!} P_n^m(\cos\theta) P_n^m(\cos\theta') \cos m(\phi - \phi') \Bigg\} d\theta'\, d\phi' \\
A_z &= \frac{jk\mu C_1 a^2}{4\pi} \int_0^{\pi} \int_0^{2\pi} \sin^3\theta' \sum_{n=0}^{\infty} (2n+1) j_n(ka) h_n^{(2)}(kr_0) \\
&\quad \times \Bigg\{ P_n(\cos\theta) P_n(\cos\theta') + 2 \sum_{m=1}^{n} \frac{(n-m)!}{(n+m)!} P_n^m(\cos\theta) P_n^m(\cos\theta') \\
&\quad \times \cos m(\phi - \phi') \Bigg\} d\theta'\, d\phi'
\end{aligned}\right\} \tag{10.13}$$

It can be seen by inspection that the integral on ϕ' with the limits $(0, 2\pi)$ is non-zero in the equation for $A_{\substack{x\\y}}$, only when m is equal to unity, and in the equation for A_z only when m is equal to zero. Hence

$$\left.\begin{aligned}
A_{\substack{x\\y}} &= \frac{-jk\mu C_1 a^2}{4\pi} 2\pi \begin{matrix}\cos\phi \\ \sin\phi\end{matrix} \sum_{n=0}^{\infty} \frac{(2n+1)}{n(n+1)} j_n(ka) h_n^{(2)}(kr_0) \\
&\quad \times P_n^1(\cos\theta) \int_0^{\pi} \sin^2\theta' \cos\theta' P_n^1(\cos\theta')\, d\theta' \\
A_z &= \frac{jk\mu C_1 a^2}{4\pi} 2\pi \sum_{n=0}^{\infty} (2n+1) j_n(ka) h_n^{(2)}(kr_0) P_n(\cos\theta) \\
&\quad \times \int_0^{\pi} \sin^3\theta' P_n(\cos\theta')\, d\theta'
\end{aligned}\right\} \tag{10.14}$$

It may be shown that an extension of a particular integral by Barnes,

$$\int_0^1 x(1-x^2)^{\frac{1}{2}} P_n^1(x)\,dx = -\frac{\Gamma(1)\Gamma(\frac{3}{2})\Gamma(2+n)}{4\Gamma(n)\Gamma\left(2-\frac{n}{2}\right)\Gamma\left(\frac{5+n}{2}\right)}$$

where $\Gamma(x)$ is the gamma function of argument x, leads to the result required for the evaluation of $A_{\substack{x\\y}}$, giving

$$\left.\begin{aligned}\int_0^\pi \sin^2\theta' \cos\theta' P_n^1(\cos\theta')\,d\theta' &= -\frac{4}{5} \quad \cdots n=2\\ &= \ \ 0 \quad \cdots n\neq 2\end{aligned}\right\} \tag{10.15}$$

where the definition of $P_n^1(x)$ is that of Bateman *et al.* and not that of Stratton. Similarly the listed integral

$$\int_0^\pi \sin^3(\theta')P_n(\cos\theta')\,d\theta'$$

$$= \frac{\pi\Gamma(2)\Gamma(2)}{\Gamma\left(\frac{n+5}{2}\right)\Gamma\left(2-\frac{n}{2}\right)\Gamma\left(1+\frac{n}{2}\right)\Gamma\left(\frac{1-n}{2}\right)}$$

leads to the particular results required for evaluating A_z, namely

$$\int_0^\pi \sin^3\theta' P_n(\cos\theta')\,d\theta' = \left.\begin{matrix} \frac{4}{3} & \cdots & n=0\\ -\frac{4}{15} & \cdots & n=2\\ 0 & \cdots & n\neq 0,2\end{matrix}\right\} \tag{10.16}$$

Hence the expressions for the components of the total magnetic vector potential, set up by the longitudinal electric currents on the sphere of radius a, at any point of observation (r_0, θ, ϕ), with $r_0 > a$, are given by

$$\left.\begin{aligned} A_x &= \frac{jk\mu I dz}{12\pi} e^{-jka}(1+jka)\, j_2(ka) h_2^{(2)}(kr_0) P_2^1(\cos\theta)\cos\phi\\ A_y &= A_x \tan\phi\\ A_z &= -\frac{jk\mu I dz}{6\pi} e^{-jka}(1+jka)\{ j_0(ka) h_0^{(2)}(kr_0) P_0(\cos\theta)\\ &\quad - j_2(ka) h_2^{(2)}(kr_0) P_2(\cos\theta)\}\end{aligned}\right\} \tag{10.17}$$

Since the original expressions for the fields of the current element are given in spherical coordinates, and are in the form of trigonometric functions of θ and exponential and algebraic forms of r, eqns (10.17) will be transformed through the use of the equations

$$A_r = A_x \sin\theta\cos\phi + A_y \sin\theta\sin\phi + A_z\cos\theta$$
$$A_\theta = A_x \cos\theta\cos\phi + A_y\cos\theta\sin\phi - A_z\sin\theta$$

and

$$j_0(ka) = \frac{\sin ka}{ka};$$

$$j_2(ka) = \left(\frac{3}{k^3a^3} - \frac{1}{ka}\right)\sin ka - \frac{3}{k^2a^2}\cos ka$$

$$h_0^{(2)}(kr_0) = \frac{j}{kr_0}e^{-jkr_0};$$

$$h_2^{(2)}(kr_0) = -j\frac{e^{-jkr_0}}{kr_0}\left(1 - \frac{j3}{kr_0} - \frac{3}{k^2r_0^2}\right)$$

$$P_0(\cos\theta) = 1; \quad P_2(\cos\theta) = (1 + 3\cos 2\theta)/4$$

Using these equations in eqns (10.17) and then taking the curl of **A** in the ϕ direction leads to the magnetic intensity H_ϕ which exists for $r_0 > a$ due to the electric current on the sphere of radius a. This result is

$$H_{\phi_1} = \frac{Idz}{4\pi}\frac{e^{-jk(a+r_0)}}{r_0}(1 + jka) \times \left[\frac{2}{ka}j_1(ka) - j_2(ka)\right]\left(jk + \frac{1}{r_0}\right)\sin\theta \quad (10.18)$$

10.2.2 *Electric vector potential of latitudinal magnetic current on spherical surface of radius a*

Referring to Fig. 10.3 the latitudinal magnetic current density $J_{m\phi'}$ at a source point (a, θ', ϕ') on the surface of the sphere can be resolved into its three rectangular components

$$J_{mx'} = -J_{m\phi'}\sin\phi'; \quad J_{my'} = J_{m\phi'}\cos\phi'; \quad J_{mz'} = 0$$

where $J_{m\phi'}$ is given by eqn (10.9). These magnetic current densities flowing in an elemental area $a^2\sin\theta' d\theta' d\phi'$ produce at the observation point (r_0, θ, ϕ)

Fig. 10.3. Latitudinal magnetic current density on surface of sphere.

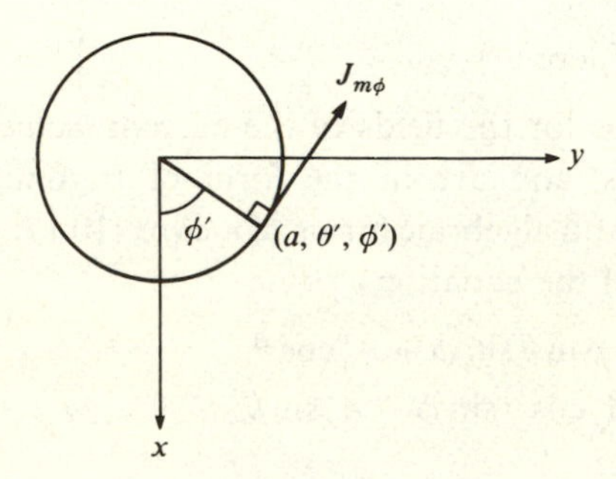

components of electric vector potential given by

$$\left.\begin{aligned} dF_x &= \frac{\varepsilon C_2 a^2}{4\pi}\sin^2\theta' \sin\phi' \frac{e^{-jkr}}{r} d\theta' d\phi' \\ dF_y &= -dF_x \cot\phi' \end{aligned}\right\} \tag{10.19}$$

where r is defined from eqn (10.11). Substituting eqn (10.12) into eqns (10.19) and integrating over the whole sphere gives the components of electric vector potential due to the latitudinal magnetic currents as

$$\begin{aligned} F_{\substack{x\\y}} = \mp\frac{jkC_2 a^2 \varepsilon}{4\pi}\int_0^{\pi}\int_0^{2\pi} \sin^2\theta' \begin{matrix}\sin\phi'\\ \cos\phi'\end{matrix} \\ \times \sum_{n=0}^{\infty}(2n+1) j_n(ka) h_n^{(2)}(kr_0)\Bigg\{ P_n(\cos\theta)P_n(\cos\theta') \\ + 2\sum_{m=1}^{n}\frac{(n-m)!}{(n+m)!} P_n^m(\cos\theta)P_n^m(\cos\theta')\cos m(\phi-\phi')\Bigg\} d\theta'\, d\phi' \end{aligned} \tag{10.20}$$

Again it can be seen by inspection that the integral on ϕ' with the limits $(0, 2\pi)$ is non-zero only when m is equal to unity, for both F_x and F_y. Hence

$$\begin{aligned} F_{\substack{x\\y}} = \mp\frac{jkC_2 a^2 \varepsilon}{4\pi} 2\pi \begin{matrix}\sin\phi\\ \cos\phi\end{matrix} \sum_{n=0}^{\infty}\frac{(2n+1)}{n(n+1)} j_n(ka) h_n^{(2)}(kr_0) \\ \times P_n^1(\cos\theta)\int_0^{\pi}\sin^2\theta' P_n^1(\cos\theta')\, d\theta' \end{aligned} \tag{10.21}$$

The integral on θ' is a standard integral given by

$$\int_0^{\pi}\sin^2\theta' P_n^1(\cos\theta')\, d\theta' = \frac{2\pi\Gamma(1)\Gamma(2)}{\Gamma\left(\frac{4+n}{2}\right)\Gamma\left(\frac{3-n}{2}\right)\Gamma\left(\frac{1+n}{2}\right)\Gamma\left(-\frac{n}{2}\right)}$$

This is zero for all values of n except n equal to unity, for which the answer is $-4/3$. Hence

$$\begin{aligned} F_{\substack{x\\y}} = \pm\frac{j\omega\mu\varepsilon I dz}{4\pi}\begin{matrix}\sin\phi\\ \cos\phi\end{matrix} \\ \times\left(1 + jka - \frac{j}{ka}\right) j_1(ka) h_1^{(2)}(kr_0) P_1^1(\cos\theta) \end{aligned} \tag{10.22}$$

Since the original expressions for the fields of the current element are in spherical coordinates and are in the form of trigonometric functions of θ and exponential and algebraic forms of r, eqns (10.22) will be transformed through the use of the equations

$$F_\phi = -F_x \sin\phi + F_y \cos\phi$$

and

$$j_1(ka) = \frac{\sin ka}{k^2a^2} - \frac{\cos ka}{ka}$$

$$h_1^{(2)}(kr_0) = -\frac{e^{-jkr_0}}{kr_0}\left(1 + \frac{1}{jkr_0}\right)$$

$$P_1^1(\cos\theta) = -\sin\theta$$

This gives for the only component of electric vector potential

$$F_\phi = \frac{-jkIdz}{4\pi\omega}\frac{e^{-jk(a+r_0)}}{r_0}\left(1 + jka - \frac{j}{ka}\right)j_1(ka)\left(1 + \frac{1}{jkr}\right)\sin\theta \tag{10.23}$$

The associated electric field components E_r, E_θ are derived from the equation

$$\mathbf{E} = -\frac{1}{\varepsilon}\nabla \times \mathbf{F}$$

The results are

$$E_r = -\frac{jkIdz}{2\pi\omega\varepsilon}\frac{e^{-jk(a+r_0)}}{r_0^2}\left(1 + jka - \frac{j}{ka}\right)j_1(ka)\left(1 + \frac{1}{jkr}\right)\cos\theta$$

$$E_\theta = -\frac{k^2Idz}{4\pi\omega\varepsilon}\frac{e^{-jk(a+r_0)}}{r_0}\left(1 + jka - \frac{j}{ka}\right)j_1(ka)$$

$$\times\left(1 + \frac{1}{jkr} - \frac{1}{k^2r^2}\right)\sin\theta$$

Hence from Maxwell's equations the magnetic intensity $H_{\phi 2}$ which results from the latitudinal magnetic current on the sphere of radius a is found to be given by

$$H_{\phi 2} = \frac{jIdz}{4\pi}\frac{e^{-jk(a+r_0)}}{r_0}\left(1 + jka - \frac{j}{ka}\right)j_1(ka)\left(jk + \frac{1}{r_0}\right)\sin\theta \tag{10.24}$$

Adding eqns (10.18) and (10.24) then gives the total H_ϕ-field due to the combined effect of the electric and magnetic currents on the sphere,

$$H_\phi = \frac{Idz}{4\pi}\left(\frac{jk}{r_0} + \frac{1}{r_0^2}\right)e^{-jkr_0}\sin\theta \tag{10.25}$$

which agrees with eqn (10.5). Since this is the only **H**-field component, the associated electric fields follow from Maxwell's equations and it is therefore unnecessary to evaluate them separately.

10.3 Application and simplification of the equivalence theorem applied to aperture antennas

The most common application of the equivalence theorem to aperture antennas involves the use of a planar surface S which can be regarded as being closed at plus and minus infinity. If an original source exists on the left-hand side of S, then it may be replaced, as far as the calculation of fields on the right-hand side is concerned, by equivalent sources distributed over the plane between plus and minus infinity, provided the incident fields produced by the original sources over this plane are known. Likewise if the sources were on the right-hand side of the plane then the fields everywhere in the left-hand side region could be found similarly.

A simplification to the equivalence principle is possible by making use of the result of the uniqueness theorem. This states that if the tangential component of $\mathbf{E}$ or $\mathbf{H}$ alone is known over a closed surface S the resulting fields are unique both within and outside S. Hence the equivalence principle in its above form is specifying redundant information by requiring that tangential $\mathbf{E}$ and tangential $\mathbf{H}$ both be known. To eliminate this redundancy in the case of Love's form of the principle, a perfectly conducting surface can be placed just inside S and everywhere tangential to it. In the case of the planar surface S the skin of perfectly conducting material is placed on the source side of S. One effect of this perfect conductor for the planar surface is to cancel the radiation from the electric current density $\mathbf{J}$, which produces a negative image superimposed on itself. The remaining magnetic current density $\mathbf{J}_m$ is now constrained to radiate in the presence of this perfect conductor. This radiation can only be calculated when the surface S lined with the perfect conductor takes the form of an infinite plane. Then using the boundary condition that the normal component of $\mathbf{H}$ at this surface must be zero gives the result that the magnetic current density is doubled because of the presence of the perfect conductor. Hence the radiation from it is calculated in the appropriate half-space as though the skin were absent and the original $\mathbf{J}_m$ were changed to twice this value radiating into free space.

Alternatively, returning to the two current form of Love's equivalence principle, consider the effect of introducing a perfect conductor of magnetism just inside the surface S. The boundary condition to be satisfied at the conducting surface may be put either in the form of tangential $\mathbf{H}$ or normal $\mathbf{E}$ being zero. Taking the case of an infinite planar surface S as before shows that the magnetic current density $\mathbf{J}_m$ is now nullified and the electric current density $\mathbf{J}$ is doubled by the presence of the perfect magnetic conductor. Hence the resultant radiation in the appropriate half-space is that of 2**J** radiating into free space.

Hence for a planar surface it is possible to use

(a) $\mathbf{J} = \mathbf{n} \times \mathbf{H}$

$\mathbf{J}_m = -\mathbf{n} \times \mathbf{E}$

or

(b) $\mathbf{J}_m = -2\mathbf{n} \times \mathbf{E}$

or

(c) $\mathbf{J} = 2\mathbf{n} \times \mathbf{H}$

all radiating into free space as far as the calculation of the fields is concerned, but the fields only exist in half-space. Where the fields over the surface *S* have to be obtained by measurement it is clearly advantageous to use (b) or (c) over the finite range where these are significant.

The above three techniques of calculating forward radiation from an aperture, are, to some extent, paralleled by the situation of finding equivalent sources at two terminals of a transmission line, which can replace the actual generator some distance away. Thus in Fig. 10.4(*a*), the original source of excitation can be replaced by a voltage generator combined with a short circuit, as shown in Fig. 10.4(*b*). Alternatively the superposition of the voltage generator from Fig. 10.4(*b*) with a current generator from Fig. 10.4(*c*) enables the short circuits associated with both

Fig. 10.4. Equivalent sources for transmission line.

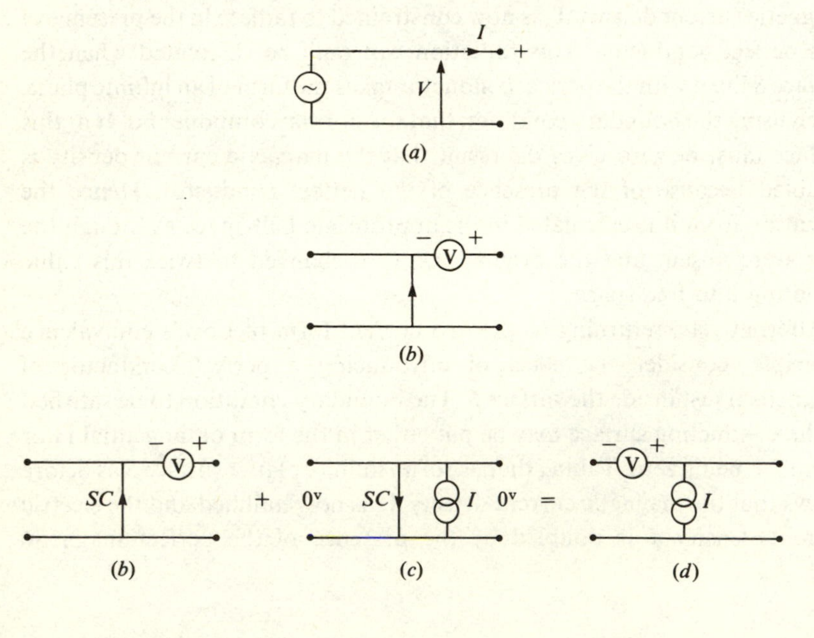

generators to be dispensed with, as shown in Fig. 10.4(*d*). Thus Fig. 10.4(*b*) is the transmission line form of eqn (b) above and Fig. 10.4(*d*) the transmission line form of eqn (a) above. Because transmission lines do not carry magnetic current it is not possible to find an equivalent of eqn (c).

10.4 Further comparison of three aperture field techniques

Each of the three aperture field techniques as given above is exact in the case of an infinite plane where either tangential **E** or **H** or both these fields is specified. Moreover the technique using both fields is clearly exact over any closed surface whereas the other two methods are not. Since these two methods radiate in front of perfectly conducting electric or magnetic screens it is clear that they can produce no fields behind these screens. But the method using both tangential fields does not use such a screen, and care must be taken not to assume that calculated fields in the backward direction apply to the problem being analysed when the aperture is finite. For all three cases the fields available apply to the forward direction in front of the infinite plane only. Moreover the three methods are bound to give identical results provided the specified tangential fields are both known exactly over the full extent of the infinite planar surface, and provided the radiated fields can be calculated analytically. Unfortunately this is not possible in practice and the consequence is that the approximate values of tangential fields, which are used over a finite section only of the infinite plane, do not agree exactly with each other. The difference is most marked when the finite planar aperture being analysed is small and examples will be given later to illustrate these differences quantitatively, but when the aperture is several wavelengths in extent all three methods give virtually identical answers.

For such large apertures it is therefore a matter of convenience which of the three methods is chosen, but in practice method (c) using the tangential magnetic field is not frequently used. This is understandable because where comparison with experiment is required it would not be possible to construct an aperture which was backed by an infinite perfect magnetic conductor, as is postulated in that method. Method (b), the tangential electric field method, is the simplest and, where measured values of field are required, the easiest to implement experimentally. It will, however, be strictly more applicable to planar aperture antennas which make use of a coplanar conducting sheet outside the aperture, than to an isolated aperture. But when the aperture is large there is less radiation close to its plane, so that the presence of this conducting sheet then has little overall effect. The radiation from a planar tangential electric field, i.e. an elemental area of magnetic current will therefore now be investigated.

10.5 Radiation due to elemental area of magnetic current in infinite ground plane

Consider an elemental aperture $dx'dy'$ at the origin immediately in front of a perfectly conducting ground plane situated in the xy-plane, as shown in Fig. 10.5. Let the tangential electric field in this aperture be

$$\mathbf{E} = \mathbf{a}_x E_x + \mathbf{a}_y E_y \tag{10.26}$$

It has previously been shown that the tangential electric field at a surface can be replaced by a magnetic current flowing in the surface, given by

$$\mathbf{J}_m = -\mathbf{n} \times \mathbf{E}$$

Since a perfectly conducting planar sheet has been placed immediately behind the surface this magnetic current density is doubled as far as radiation in the forward half-space is concerned, giving

$$\mathbf{J}_m = \mathbf{a}_x 2E_y - \mathbf{a}_y 2E_x \tag{10.27}$$

Due to this magnetic current density an electric vector potential is set up at a point of observation $P(r, \theta, \phi)$ given by

$$\mathbf{F} = \frac{\varepsilon}{2\pi}(\mathbf{a}_x E_y - \mathbf{a}_y E_x)\frac{e^{-jkr}}{r}dx'dy' \tag{10.28}$$

These cartesian components of vector potential are transformed to spherical coordinates using

$$\left.\begin{aligned} F_r &= F_x \sin\theta\cos\phi + F_y \sin\theta\sin\phi + F_z\cos\theta \\ F_\theta &= F_x\cos\theta\cos\phi + F_y\cos\theta\sin\phi - F_z\sin\theta \\ F_\phi &= -F_x\sin\phi + F_y\cos\phi \end{aligned}\right\} \tag{10.29}$$

Fig. 10.5. Elemental area of magnetic current at origin in ground plane.

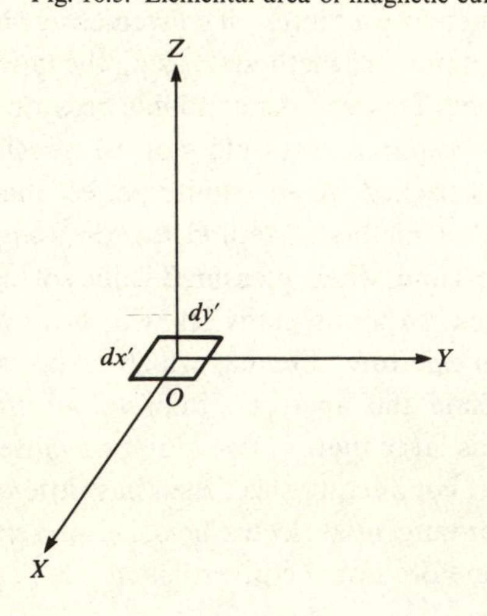

Then using

$$\mathbf{E} = -\frac{1}{\varepsilon}\nabla \times \mathbf{F}$$

gives, for the electric field components,

$$\left.\begin{aligned} E_r &= -\frac{1}{\varepsilon}\left[\frac{1}{r}\frac{\partial F_\phi}{\partial \theta} + \frac{F_\phi}{r}\cot\theta - \frac{1}{r\sin\theta}\frac{\partial F_\theta}{\partial \phi}\right] = 0 \\ E_\theta &= -\frac{1}{\varepsilon}\left[\frac{1}{r\sin\theta}\frac{\partial F_r}{\partial \phi} - \frac{\partial F_\phi}{\partial r} - \frac{F_\phi}{r}\right] \\ &= \frac{1}{2\pi}(E_x\cos\phi + E_y\sin\phi)\left(\frac{1}{r} + jk\right)\frac{e^{-jkr}}{r}dx'dy' \\ \text{and} \\ E_\phi &= -\frac{1}{\varepsilon}\left[\frac{\partial F_\theta}{\partial r} + \frac{F_\theta}{r} - \frac{1}{r}\frac{\partial F_r}{\partial \theta}\right] \\ &= \frac{\cos\theta}{2\pi}(-E_x\sin\phi + E_y\cos\phi)\left(\frac{1}{r} + jk\right)\frac{e^{-jkr}}{r}dx'dy' \end{aligned}\right\} \quad (10.30)$$

It will be noted that in the xy-plane E_ϕ is identically zero since it is tangential to the perfect conductor there. Likewise in the xz-plane the electric field due to E_x which produces the magnetic current density J_y takes the form of semi-circles centred on the y-axis. Similarly the electric field in the yz-plane due to the source field E_y takes the form of semi-circles centred on the x-axis.

10.6 Radiation fields due to finite distribution of magnetic current in infinite ground plane

The results for the elemental area of magnetic current can be extended to deal with a planar aperture of arbitrary shape over which the distribution of tangential electric field is specified. With reference to Fig. 10.6(a) the distance r from an arbitrary source element $dx'dy'$ in the xy-plane to a point of observation $P(r_0, \theta, \phi)$ in the radiation zone is given approximately by

$$r \approx r_0 - \rho' \cos\psi \quad (10.31)$$

where ρ' is the polar distance from the origin to $dx'dy'$ and ψ is the angle between the vectors r_0 and ρ'. But in general the cosine of the angle between two vectors at angles (θ, ϕ) and (θ', ϕ') in spherical coordinates is given by

$$\cos\psi = \cos\theta\cos\theta' + \sin\theta\sin\theta'\cos(\phi - \phi')$$

which reduces since θ' is $\pi/2$ to

$$\cos\psi = \sin\theta\cos(\phi - \phi')$$

If the aperture is circular of radius a and centred on the origin, it is appropriate in Fig. 10.6(b) to replace the elemental area $dx'dy'$ by $\rho' d\rho' d\phi'$, so that the radiation field components become

$$\left.\begin{aligned} E_\theta &= \frac{j}{\lambda}\frac{e^{-jkr_0}}{r_0}\int_0^a\int_0^{2\pi}[E_x(\rho',\phi')\cos\phi + E_y(\rho',\phi')\sin\phi] \\ &\quad\times e^{jk\rho'\sin\theta\cos(\phi-\phi')}\rho'\,d\rho'\,d\phi' \\ &\text{and} \\ E_\phi &= -\frac{j}{\lambda}\frac{e^{-jkr_0}}{r_0}\cos\theta\int_0^a\int_0^{2\pi} \\ &\quad\times[E_x(\rho',\phi')\sin\phi - E_y(\rho',\phi')\cos\phi] \\ &\quad\times e^{jk\rho'\sin\theta\cos(\phi-\phi')}\rho'\,d\rho'\,d\phi' \end{aligned}\right\} \quad (10.32)$$

Likewise if the aperture is symmetrically rectangular of area ab, as shown in Fig. 10.6(c), it is appropriate to write the exponent of the exponential

Fig. 10.6. (a) elemented area of magnetic current arbitrarily located in xy-plane; (b) elemental area of magnetic current in circular aperture; (c) rectangular aperture.

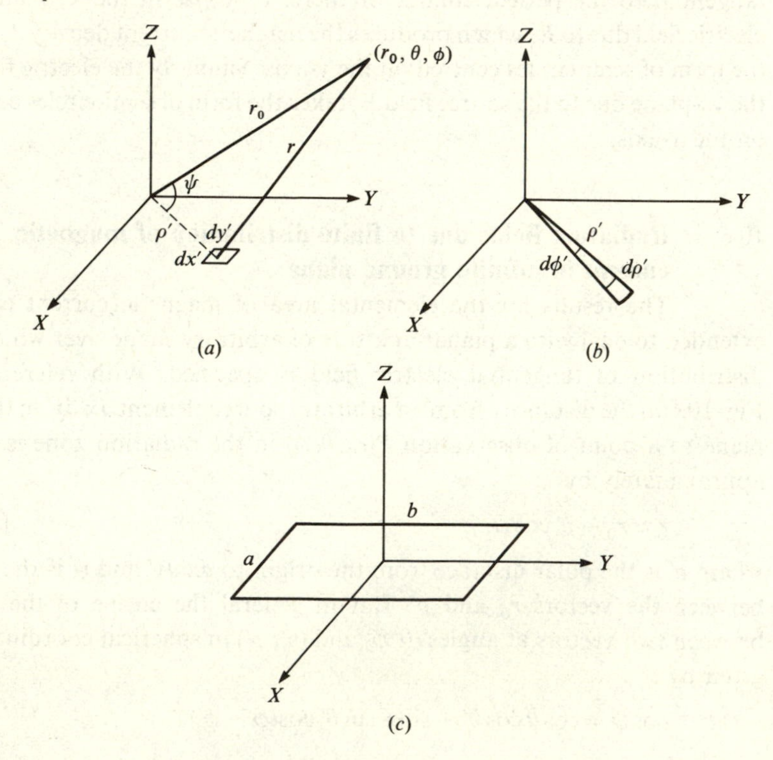

term as

$$jk\rho' \sin\theta \cos(\phi - \phi') = jk(x' \sin\theta \cos\phi + y' \sin\theta \sin\phi)$$

so that the components of the radiation field can be written

$$\left.\begin{aligned}
E_\theta &= \frac{j}{\lambda}\frac{e^{-jkr_0}}{r_0}\int_{-a/2}^{+a/2}\int_{-b/2}^{+b/2} \\
&\quad\times [E_x(x', y')\cos\phi + E_y(x', y')\sin\phi] \\
&\quad\times e^{jk(x'\sin\theta\cos\phi + y'\sin\theta\sin\phi)}\,dx'\,dy' \\
\text{and} \qquad & \\
E_\phi &= -\frac{j}{\lambda}\frac{e^{-jkr_0}}{r_0}\cos\theta\int_{-a/2}^{+a/2}\int_{-b/2}^{+b/2} \\
&\quad\times [E_x(x', y')\sin\phi - E_y(x', y')\cos\phi] \\
&\quad\times e^{jk(x'\sin\theta\cos\phi + y'\sin\theta\sin\phi)}\,dx'\,dy'
\end{aligned}\right\} \qquad (10.33)$$

These formulae are basic for the calculation of radiation patterns using only the tangential electric fields in the aperture of the antenna. Outside this aperture but still in the aperture plane these fields are necessarily zero because of the infinite ground plane. For the purpose of analytic comparison, however, the corresponding expressions based on the knowledge of tangential magnetic field in the aperture of the antenna will now be derived.

10.7 Radiation due to elemental area of electric current in infinite magnetic ground plane

Consider as before an elemental aperture $dx'dy'$ at the origin, immediately in front of a perfectly conducting magnetic ground plane

Fig. 10.7. Elemental area of electric current at origin in ground plane.

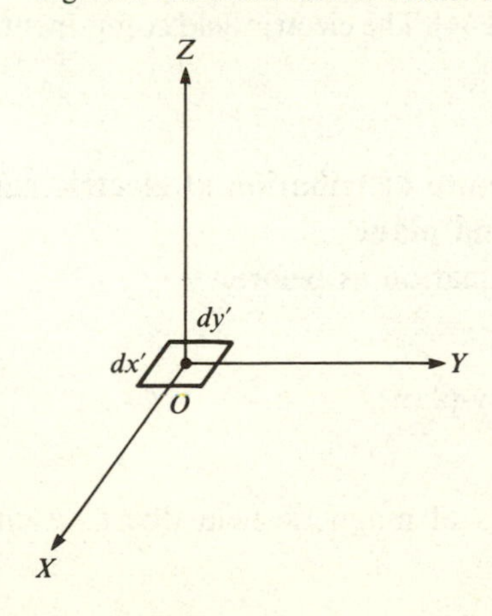

situated in the xy-plane, as shown in Fig. 10.7. Let the tangential magnetic field in this aperture before the ground plane is placed there be

$$\mathbf{H} = \mathbf{a}_x H_x + \mathbf{a}_y H_y \tag{10.34}$$

Consequently, by the equivalence theorem this field can be replaced by an electric current density $\bar{J}$ given by

$$\mathbf{J} = \mathbf{n} \times \mathbf{H}$$

which after taking account of the doubling action of the perfect magnetic conductor gives

$$\mathbf{J} = -\mathbf{a}_x 2H_y + \mathbf{a}_y 2H_x$$

Because of this current flow the magnetic vector potential is given by

$$\mathbf{A} = \frac{\mu}{2\pi}(-\mathbf{a}_x H_y + \mathbf{a}_y H_x)\frac{e^{-jkr}}{r} dx' dy' \tag{10.35}$$

Transforming to spherical coordinates as was done with the electric vector potential, and using the relationship,

$$\mathbf{H} = \frac{1}{\mu} \nabla \times \mathbf{A}$$

gives, for the magnetic field components,

$$\left.\begin{aligned} H_r &= 0 \\ H_\theta &= \frac{1}{2\pi}(H_x \cos\phi + H_y \sin\phi)\left(\frac{1}{r} + jk\right)\frac{e^{-jkr}}{r} dx' dy' \\ &\text{and} \\ H_\phi &= \frac{1}{2\pi}(-H_x \sin\phi + H_y \cos\phi)\cos\theta\left(\frac{1}{r} + jk\right)\frac{e^{-jkr}}{r} dx' dy' \end{aligned}\right\} \tag{10.36}$$

in agreement with the results previously derived for current elements along orthogonal axes in eqns (2.61) and (2.64). The electric field components then follow from Maxwell's equations.

10.8 Radiation fields due to finite distribution of electric current in infinite magnetic ground plane

Using the far field approximation as before,

$$r \approx r_0 - \rho' \cos\psi$$

where, for source elements in the xy-plane,

$$\cos\psi = \sin\theta \cos(\phi - \phi')$$

gives for the radiation components of magnetic field due to a circular aperture of radius a

$$\left.\begin{aligned} H_\theta &= \frac{j}{\lambda}\frac{e^{-jkr_0}}{r_0}\int_0^a\int_0^{2\pi}[H_x(\rho',\phi')\cos\phi + H_y(\rho',\phi')\sin\phi] \\ &\quad\times e^{jk\rho'\sin\theta\cos(\phi-\phi')}\rho'\,d\rho'\,d\phi' \\ \text{and} \\ H_\phi &= \frac{j}{\lambda}\frac{e^{-jkr_0}}{r_0}\cos\theta\int_0^a\int_0^{2\pi} \\ &\quad\times[-H_x(\rho',\phi')\sin\phi + H_y(\rho',\phi')\cos\phi] \\ &\quad\times e^{jk\rho'\sin\theta\cos(\phi-\phi')}\rho'\,d\rho'\,d\phi' \end{aligned}\right\} \tag{10.37}$$

The corresponding electric field components then follow from

$$E_\theta = Z_0 H_\phi \quad \text{and} \quad E_\phi = -Z_0 H_\theta \tag{10.38}$$

Similarly for a rectangular aperture of area ab symmetrically located about the origin, the radiation fields are

$$\left.\begin{aligned} H_\theta &= \frac{j}{\lambda}\frac{e^{-jkr_0}}{r_0}\int_{-a/2}^{a/2}\int_{-b/2}^{b/2} \\ &\quad\times[H_x(x',y')\cos\phi + H_y(x',y')\sin\phi] \\ &\quad\times e^{jk(x'\sin\theta\cos\phi + y'\sin\theta\sin\phi)}\,dx'\,dy' \\ H_\phi &= \frac{j}{\lambda}\frac{e^{-jkr_0}}{r_0}\cos\theta\int_{-a/2}^{a/2}\int_{-b/2}^{b/2} \\ &\quad\times[-H_x(x',y')\sin\phi + H_y(x',y')\cos\phi] \\ &\quad\times e^{jk(x'\sin\theta\cos\phi + y'\sin\theta\sin\phi)}\,dx'\,dy' \end{aligned}\right\} \tag{10.39}$$

with corresponding expressions for E_θ and E_ϕ. It will be noted that H_ϕ is zero in the xy-plane because of the presence of the perfect magnetic conductor there, and the tangential source fields must be zero outside the aperture for the same reason.

10.9 Radiation fields due to combined magnetic and electric currents

When both magnetic and electric source currents are used to calculate the fields in the forward direction outside an aperture in the xy-plane, it is important to note that the range of integration must now extend from the origin to infinity and that the sizes of the currents must be halved. This is because there is now no longer a perfect conductor of electricity in this plane, which would prevent a tangential electric field existing there in the case of the magnetic current model, and likewise a perfect conductor of magnetism in the electric current model. Thus for a circular aperture of radius a centred at the origin, the total radiation components of electric fields are given by the sum of eqns (10.32) and (10.38) with the appropriate change of limit, as

$$E_\theta = \frac{j}{2\lambda}\frac{e^{-jkr_0}}{r_0}\int_0^\infty\int_0^{2\pi} \times [E_x \cos\phi - Z_0 H_x \cos\theta \sin\phi + E_y \sin\phi + Z_0 H_y \cos\theta \cos\phi] e^{jk\rho' \sin\theta \cos(\phi - \phi')}\rho'\, d\rho'\, d\phi'$$

and

$$E_\phi = -\frac{j}{2\lambda}\frac{e^{-jkr_0}}{r_0}\int_0^\infty\int_0^{2\pi} \times [E_x \cos\theta \sin\phi + Z_0 H_x \cos\phi - E_y \cos\theta \cos\phi + Z_0 H_y \sin\phi] \times e^{jk\rho' \sin\theta \cos(\phi - \phi')}\rho'\, d\rho'\, d\phi' \tag{10.40}$$

Similarly for the rectangular aperture previously considered the resultant electric fields are given by

$$E_\theta = \frac{j}{2\lambda}\frac{e^{-jkr_0}}{r_0}\int_{-\infty}^{+\infty}\int_{-\infty}^{+\infty} \times [E_x \cos\phi - Z_0 H_x \cos\theta \sin\phi + E_y \sin\phi + Z_0 H_y \cos\theta \cos\phi] \times e^{jk(x' \sin\theta \cos\phi + y' \sin\theta \sin\phi)}\, dx'\, dy'$$

and

$$E_\phi = -\frac{j}{2\lambda}\frac{e^{-jkr_0}}{r_0}\int_{-\infty}^{+\infty}\int_{-\infty}^{+\infty} \times [E_x \cos\theta \sin\phi + Z_0 H_x \cos\phi - E_y \cos\theta \cos\phi + Z_0 H_y \sin\phi] \times e^{jk(x' \sin\theta \cos\phi + y' \sin\theta \sin\phi)}\, dx'\, dy' \tag{10.41}$$

The infinite limits of integration suggest a Fourier integral relationship between the radiation pattern and the tangential fields in the infinite plane in which the aperture lies, and this matter will be discussed further in the following chapter. It will be noted that the two final equations require both the tangential electric and magnetic fields to be specified over the complete infinite plane before the radiation fields can be found. This is a highly demanding requirement since these same fields themselves form part of the radiation pattern in the θ equal to $\pi/2$ plane. It thus appears that an iterative type of solution would be required to solve this problem exactly, assuming that such a solution were convergent, as the physical solution must be.

An alternative approach to the problem of radiation from apertures will, however, next be considered. This has the advantage that it takes no account of the existence of fictitious magnetic currents, but concentrates attention on uniform plane waves travelling at all angles to the plane of the aperture. Although simple in concept, however, it is rather more lengthy to implement, and for this reason the treatment will be divided into two sections, the first dealing with the one-dimensional problem, before treating the identical problem in two dimensions, of a finite aperture in an infinite plane. The result to be obtained, however, is identical with that given above.

Further reading

R.E. Collin and F.J. Zucker: *Antenna Theory*: McGraw-Hill, 1969.

A.Z. Fradin: *Microwave Antennas*: Pergamon Press, 1961.

R.C. Hansen: *Microwave Scanning Antennas Vol. 1*, Academic Press, New York, 1969.

W.L. Stutzman and G.A. Thiele: *Antenna Theory and Design*: Wiley and Sons, 1981.

11

+−+

Angular spectrum of plane waves

11.1 Angular spectrum for one-dimensional slot

Consider a uniform plane wave travelling at angle θ to the polar axis. Any such wave may be considered to be the superposition of two separate waves, one with its magnetic field horizontal, as shown in Fig. 11.1(*a*), and the other with its electric field horizontal, as shown in Fig. 11.1(*b*). These waves may be described mathematically by the equations

$$\left.\begin{aligned} \mathbf{H} &= \mathbf{a}_x \frac{E}{Z_0} e^{-jk(y\sin\theta + z\cos\theta)} \\ \mathbf{E} &= (-\mathbf{a}_y E\cos\theta + \mathbf{a}_z E\sin\theta) e^{-jk(y\sin\theta + z\cos\theta)} \end{aligned}\right\} \mathbf{H}\text{ horizontal} \tag{11.1}$$

$$\left.\begin{aligned} \mathbf{E} &= -\mathbf{a}_x Z_0 H e^{-jk(y\sin\theta + z\cos\theta)} \\ \mathbf{H} &= (-\mathbf{a}_y H\cos\theta + \mathbf{a}_z H\sin\theta) e^{-jk(y\sin\theta + z\cos\theta)} \end{aligned}\right\} \mathbf{E}\text{ horizontal} \tag{11.2}$$

where E and H are their respective electric and magnetic fields.

Fig. 11.1 (*a*) uniform plane wave with magnetic field horizontal along x-axis; (*b*) uniform plane wave with electric field horizontal along x-axis.

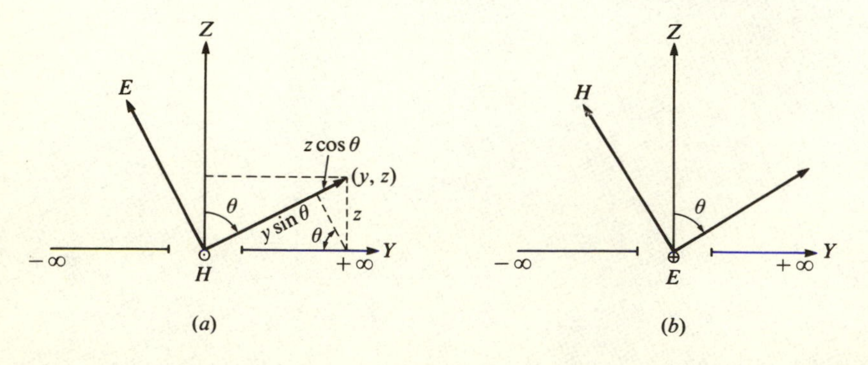

If the xy-plane is now considered as an aperture plane for any field that exists in the region $z \geqslant 0$, it is reasonable to postulate that any arbitrary field within this region can be synthesised from a superposition of an infinite number of such uniform plane waves, each with its own specified amplitude and phase at the origin, and travelling through the plane at all angles θ to the origin. This is analogous to the synthesis of an arbitrary waveform from its Fourier components. For θ between 0° and $\pm\pi/2$ all these waves carry power in the usual way. But in the event of a physical surface existing in the xy-plane it is possible for such a surface to have associated with it stored energy which hugs the surface, and it is not immediately apparent that such fields can be described in this way. Consider, however, the meaning to be attached to a wave which travels along the surface at an angle $(\pm\pi/2 \pm j\alpha)$ where α is a real variable. The real part $\pm\pi/2$ indicates that it is travelling along the surface, and the complete expression $e^{-jk(y\sin(\pm\pi/2\pm j\alpha)+z\cos(\pm\pi/2\pm j\alpha))}$ can be simplified to $e^{-kz\sinh\alpha}e^{\mp jky\cosh\alpha}$. This indicates an exponential decay of amplitude normal to the postulated surface in the xy-plane, together with a progressive phase lag as the wave travels either to the right or to the left from the origin.

Hence the range of all physically possible values of θ is from

$$\left(-\frac{\pi}{2}-j\infty\right) \quad \text{to} \quad \left(-\frac{\pi}{2}-j0\right)$$

$$-\frac{\pi}{2} \quad \text{to} \quad 0 \quad \text{to} \quad +\frac{\pi}{2}$$

$$\left(\frac{\pi}{2}+j0\right) \quad \text{to} \quad \left(\frac{\pi}{2}+j\infty\right)$$

The corresponding range of values of $\sin\theta$ is from

$$-\infty \quad \text{to} \quad -1$$

$$-1 \quad \text{to} \quad +1$$

$$+1 \quad \text{to} \quad +\infty$$

Thus any field for $z \geqslant 0$ can be built up from the integrals

$$\left.\begin{aligned} \mathbf{H}(y,z) &= \mathbf{a}_x \int_{(-\pi/2-j\infty)}^{(\pi/2+j\infty)} \frac{E(\theta)}{Z_0} e^{-jk(y\sin\theta+z\cos\theta)}\,d\theta \\ &\text{and} \\ \mathbf{E}(y,z) &= \int_{(-\pi/2-j\infty)}^{(\pi/2+j\infty)} E(\theta)(-\mathbf{a}_y\cos\theta+\mathbf{a}_z\sin\theta)e^{-jk(y\sin\theta+z\cos\theta)}\,d\theta \end{aligned}\right\} \tag{11.3}$$

for a field with horizontal **H**, and

$$\left.\begin{aligned}\mathbf{E}(y,z) &= -\mathbf{a}_x \int_{(-\pi/2-j\infty)}^{(\pi/2+j\infty)} Z_0 H(\theta) e^{-jk(y\sin\theta + z\cos\theta)}\,d\theta \\ \mathbf{H}(y,z) &= \int_{(-\pi/2-j\infty)}^{(\pi/2+j\infty)} H(\theta)(-\mathbf{a}_y\cos\theta + \mathbf{a}_z\sin\theta) e^{-jk(y\sin\theta + z\cos\theta)}\,d\theta\end{aligned}\right\} \tag{11.4}$$

for a field with horizontal **E**.

These equations can be expressed more succinctly by changing the variable of integration from θ to $k_y = k\sin\theta$, giving, for the horizontal magnetic field case,

$$\left.\begin{aligned}\mathbf{H}(y,z) &= \mathbf{a}_x \int_{-\infty}^{+\infty} \frac{E(k_y)}{Z_0\sqrt{(k^2-k_y^2)}} e^{-j(k_y y + k_z z)}\,dk_y \\ \mathbf{E}(y,z) &= \frac{1}{k}\int_{-\infty}^{+\infty} E(k_y)\left(-\mathbf{a}_y + \mathbf{a}_z \frac{k_y}{\sqrt{(k^2-k_y^2)}}\right) e^{-j(k_y y + k_z z)}\,dk_y\end{aligned}\right\} \tag{11.5}$$

where $dk_y = k\cos\theta d\theta = [k^2 - k_y^2]^{\frac{1}{2}} d\theta = k_z d\theta$, and $E(k_y) = E(\theta)$. Similarly for horizontal **E** fields,

$$\left.\begin{aligned}\mathbf{E}(y,z) &= -\mathbf{a}_x \int_{-\infty}^{+\infty} \frac{Z_0 H(k_y)}{\{[k^2-k_y^2]\}} e^{-j(k_y y + k_z z)} dk_y \\ \mathbf{H}(y,z) &= \frac{1}{k}\int_{-\infty}^{+\infty} H(k_y)\left(-\mathbf{a}_y + \mathbf{a}_z \frac{k_y}{\sqrt{(k^2-k_y^2)}}\right) e^{-j(k_y y + k_z z)}\,dk_y\end{aligned}\right\} \tag{11.6}$$

Since these four equations are valid in the aperture itself, for which $z = 0$, this gives for the tangential y components of $\mathbf{E}(y,z)$ and $\mathbf{H}(y,z)$ in the aperture

$$E_y(y,0) = -\frac{1}{k}\int_{-\infty}^{+\infty} E(k_y) e^{-jk_y y}\,dk_y \tag{11.7}$$

for the wave with only horizontal **H** and

$$H_y(y,0) = -\frac{1}{k}\int_{-\infty}^{+\infty} H(k_y) e^{-jk_y y}\,dk_y \tag{11.8}$$

for the horizontally polarised wave with **E** parallel to the xy-plane. These are Fourier transform relationships which enable the amplitudes and phases of the uniform plane waves passing through the origin for all angles of travel θ to be found. Thus a horizontal electric field in the aperture plane, this field being directed in the y-direction, gives rise to a spectrum of plane waves each of which has its magnetic field horizontally directed at right angles to this in the x-direction. Similarly a y-directed magnetic field in the aperture plane sets up a spectrum of plane waves with electric field components in the x-direction.

The Fourier transform relationships, eqns (11.7) and (11.8), allow the magnitude and phase of the waves travelling at all angles θ including complex values of θ to be found. Thus for a y-directed tangential electric field in the aperture plane,

$$E(k_y) = -\frac{k}{2\pi}\int_{-\infty}^{+\infty} E_y(y,0)e^{jk_y y}\,dy \qquad (11.9)$$

and for a y-directed tangential magnetic field in the aperture plane,

$$H(k_y) = -\frac{k}{2\pi}\int_{-\infty}^{+\infty} H_y(y,0)e^{jk_y y}\,dy \qquad (11.10)$$

The integrations extend between infinite limits, but in eqn (11.9) if there is a parallel sided physical aperture cut out of an infinite perfectly conducting plane, the infinite limits can be truncated at the finite width of the aperture. But this is not so in eqn (11.10) because currents will flow on the ground plane outside the aperture, causing an additional tangential H_y field there.

Thus although two separate cases have been worked through in parallel, the above arguments show that the tangential field technique is applicable in principle to the case of a parallel sided slot aperture cut out of an infinite conducting plane only when the generating tangential field across the slot is electric and not magnetic. For this case once $E(k_y)$ is known the field anywhere in space can be found from eqn (11.5) and an illustration of this will be given in the following section.

It will be noted from eqn (11.5) that although the calculation of $E(k_y)$ depends only on the y-directed electric field in the aperture, the electric field in general contains a z-directed component normal to the aperture. This component is seen to exist even in the aperture plane, where its value is

$$E_z(y,0) = \frac{1}{k}\int_{-\infty}^{+\infty} E(k_y)_0 \frac{k_y}{\sqrt{(k^2 - k_y^2)}} e^{-jk_y y}\,dk_y$$

This may be written

$$E_z(y,0) = \frac{1}{k}\int_{-\infty}^{+\infty} E'(k_y)e^{-jk_y y}\,dk_y$$

which again is a Fourier transform relationship giving a new amplitude and phase in terms of the normal component of electric field in the aperture plane,

$$E'(k_y) = \frac{k}{2\pi}\int_{-\infty}^{+\infty} E_z(y,0)\,e^{jk_y y}\,dy$$

However, this relationship is of no practical use because $E_z(y, 0)$ will exist not only in the aperture but over the whole of the aperture plane out to infinity, and this field will in general not be known.

Returning now to the case when a tangential magnetic field exists in the aperture plane, it will be seen from the argument above that a normal magnetic field will also exist there and that this normal field will be zero outside a slot cut in the conducting plane. Hence for this case it would be preferable to work in terms of normal fields. Returning therefore to eqn (11.6) gives, for the normal magnetic field,

$$H_z(y,0) = \frac{1}{k}\int_{-\infty}^{+\infty}\left(H(k_y)\frac{k_y}{\sqrt{(k^2-k_y^2)}}\right)e^{-jk_y y}\,dk_y$$

Hence, by inversion,

$$H(k_y)\frac{k_y}{\sqrt{(k^2-k_y^2)}} = \frac{k}{2\pi}\int_{-\infty}^{+\infty} H_z(y,0)e^{+jk_y y}\,dy$$

and the total fields can thus in principle be derived from the normal component of H in the aperture of a perfectly conducting sheet.

11.2 Infinite slot of width 2*a* in planar conducting sheet

The problem of finding the fields radiated by an infinitely long slot of width $2a$ in a planar perfectly conducting sheet will now be considered. It will be assumed that an electric field E_y exists in the plane of the slot, as shown in Fig. 11.2, though this should not be taken to imply that there is no normal component of field in that plane. For analytic simplicity this field E_y will be taken to be constant. Such a distribution of field would not, however, be produced by applying equal and opposite potentials to the two halves of the conducting sheet, which would instead produce a field E_y proportional to $(1-(y/a)^2)^{-\frac{1}{2}}$.

In order to find the fields at any point in space the amplitude and phase of the angular spectrum will first be found. Thus

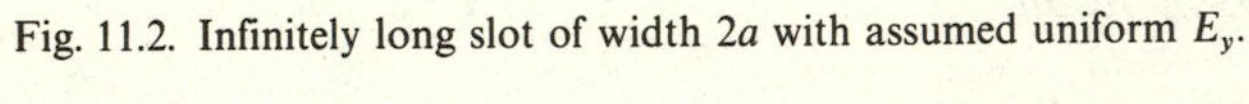

Fig. 11.2. Infinitely long slot of width $2a$ with assumed uniform E_y.

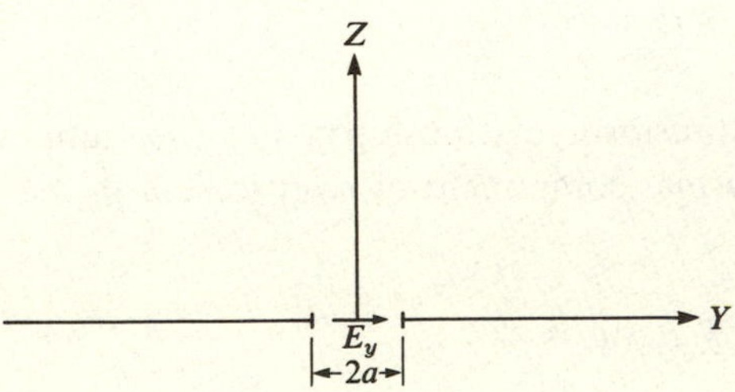

$$E(k_y) = -\frac{k}{2\pi}\int_{-\infty}^{+\infty} E_y(y,0)e^{jk_y y}\,dy$$

$$= -\frac{2E_y a}{\lambda}\frac{\sin(k_y a)}{(k_y a)} \tag{11.11}$$

since the field extends from $-a$ to $+a$ only. In the case of a very narrow slot this gives

$$E(k_y) \to -\frac{2E_y a}{\lambda}$$

which is a constant for all values of $k \sin\theta$, and hence constant for all values of θ. Thus the polar diagram of this slot is in the shape of a semi-circle since the polar diagram may be thought of as the amplitude of the single uniform plane wave, at a fixed radius large in terms of wavelengths, which travels in every direction of observation ϕ. For general slot widths $2a$ the radiation pattern given by eqn (11.11) is shown in Fig. 11.3.

To show this result analytically the field is written in the form of the integral of the spectrum as

$$\mathbf{E}(y,z) = \frac{1}{k}\int_{-\infty}^{+\infty} E(k_y)\left(-\mathbf{a}_y + \mathbf{a}_z\frac{k_y}{\sqrt{(k^2-k_y^2)}}\right)e^{-j(k_y y + k_z z)}\,dk_y \tag{11.12}$$

Changing to cylindrical coordinates (r, ϕ) where, as in Fig. 11.4(*a*),

$$y = r\sin\phi; \quad z = r\cos\phi$$

Fig. 11.3. Radiation pattern for slot of width $2a$ with uniform E_y field.

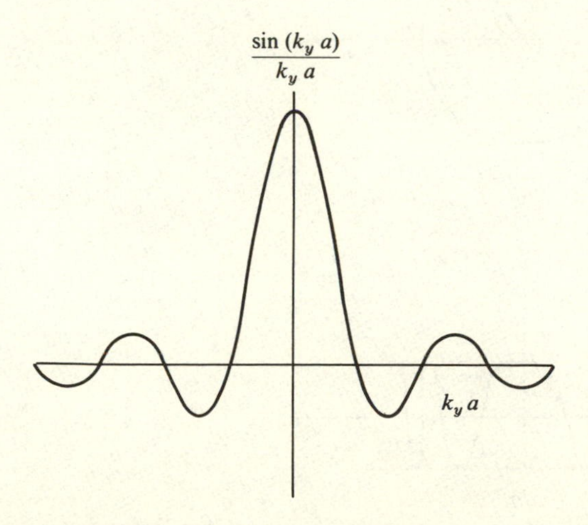

gives

$$\mathbf{E}(r, \phi) = \frac{1}{k}\int_{-\infty}^{+\infty} E(k\sin\theta)\left(-\mathbf{a}_y + \mathbf{a}_y\frac{k_y}{k_z}\right)e^{-jkr\cos(\theta-\phi)}d(k\sin\theta)$$

The exponential term in the integrand is oscillatory, as in Fig. 11.4(*b*), and at a point in the far field the contributions to the integral will come almost entirely from values of θ which are close to ϕ. Hence since $E(k\sin\theta)$ is constant for a narrow slot, the integral reduces to

$$\begin{aligned}\mathbf{E}(r, \phi) &\approx -\frac{2E_y a}{k\lambda}(-\mathbf{a}_y + \mathbf{a}_z\tan\phi)\int_{-\infty}^{+\infty} e^{-jkr(1-(\theta-\phi)^2/2)}d(k\sin\theta)\\ &\approx -\frac{2E_y a}{\lambda}(-\mathbf{a}_y + \mathbf{a}_z\tan\phi)e^{-jkr}\sqrt{\left(\frac{2\pi}{kr}\right)}e^{j(\pi/4)}\cos\phi\\ &= -\frac{2E_y a}{\sqrt{(\lambda r)}}(-\mathbf{a}_y\cos\phi + \mathbf{a}_z\sin\phi)e^{-jk(r-\lambda/8)}\end{aligned}$$

Changing to cylindrical coordinates using

$$\mathbf{a}_r = \mathbf{a}_z\cos\phi + \mathbf{a}_y\sin\phi$$
$$\mathbf{a}_\phi = -\mathbf{a}_z\sin\phi + \mathbf{a}_y\cos\phi$$

Fig. 11.4. (*a*) transformation from rectangular to cylindrical coordinates; (*b*) oscillatory nature of integrand.

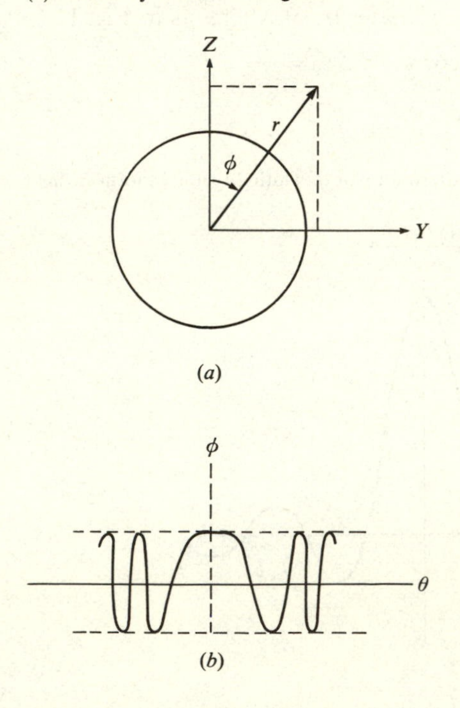

gives

$$\left.\begin{aligned} &E_r(r,\phi)=0 \\ \text{and} \\ &E_\phi(r,\phi)=\frac{2E_y a}{\sqrt{(\lambda r)}}e^{-jk(r-\lambda/8)} \end{aligned}\right\} \tag{11.13}$$

which confirms analytically the cylindrical resultant form of the spectrum of angular waves of constant magnitude.

Although the form of the polar diagram could be deduced from the physical interpretation given to the angular spectrum, the same cannot be done when it comes to deducing the size or direction of the fields in the near zone close to the slot. To examine these fields, first consider the general expression for the electric field in the case of a slot of general width, which is simpler in this case to consider than one of infinitesimal width,

$$\mathbf{E}(y,z)=\frac{1}{k}\int_{-\infty}^{+\infty}E(k_y)\left(-\mathbf{a}_y+\mathbf{a}_z\frac{k_y}{\sqrt{(k^2-k_y^2)}}\right)e^{-j(k_y y+k_z z)}\,dk_y \tag{11.14}$$

Of the two terms in this integral, the component in the y-direction at the origin can be evaluated as follows

$$E_y(0,0)=\frac{1}{k}\int_{-\infty}^{+\infty}\frac{2E_y a}{\lambda}\frac{\sin(k_y a)}{(k_y a)}\,dk_y \tag{11.15}$$

Splitting this integral into components which carry power, i.e. values of k_y between $\pm k$, and the remaining components between $\pm k$ and $\pm\infty$, gives

$$\begin{aligned} E_y(0,0)&=\frac{E_y a}{\pi}\left(\int_{-k}^{+k}\frac{\sin(k_y a)}{k_y a}+\int_{-\infty}^{-k}\frac{\sin(k_y a)}{k_y a}\right. \\ &\qquad\left.+\int_{k}^{\infty}\frac{\sin(k_y a)}{k_y a}\right)dk_y \\ &=\frac{E_y}{\pi}\left(2\int_{0}^{ka}\frac{\sin(k_y a)}{k_y a}+2\int_{ka}^{\infty}\frac{\sin(k_y a)}{k_y a}\right)d(k_y a) \\ &=\frac{2E_y}{\pi}(\mathrm{Si}(ka))+\left(\frac{\pi}{2}-\mathrm{Si}(ka)\right) \\ &=E_y \end{aligned} \tag{11.16}$$

Thus the original postulated tangential electric field in the slot has been confirmed at the origin, but it will be noted that the part of this E_y associated with the power carrying part of the spectrum is not sufficient on its own to maintain this field. Alternatively this may be expressed by saying that of the total E_y-field in the aperture the fraction which is used in supplying power to the radiation pattern is $(2/\pi)\mathrm{Si}\,(ka)$, which for a very narrow slot is approximately $4a/\lambda$. The remainder of the E_y field is associated with reactive energy, and this reactive proportion increases as the slot width decreases.

It should be noted that although E_y alone was postulated in the aperture plane that this does not mean that there is no normal electric field E_z in the same plane. Such a field does exist, as can be understood by resolving the amplitude of $E(k_y)$ into its components $E(k_y)\cos\theta$ and $E(k_y)\sin\theta$ along the y- and z-directions. However at the origin, and along the whole of the positive z-axis, the resultant travelling wave component of E_z is zero because of the cancellation of this component of field produced by waves travelling in the $\pm\theta$-directions. There does, however, exist a component of field due to values of $|k_y| > 1$, just as these components contribute to the total E_y-field in the aperture. Both these components of E_z are contained in the expression

$$E_z(y,z) = \frac{1}{k}\int_{-\infty}^{+\infty} \frac{k_y E(k_y)}{\sqrt{(k^2 - k_y^2)}} e^{-j(k_y y + k_z z)}\,dk_y \tag{11.17}$$

The travelling wave component is that part of this equation formed by substituting the limits of $\pm k$ for the infinite limits shown. This is conveniently re-arranged to avoid the pole in the integrand by reverting to θ as a variable of integration to give

$$E_{z1}(y,z) = \int_{-\pi/2}^{+\pi/2} E(\theta)\sin\theta e^{-jk(y\sin\theta + z\cos\theta)}\,d\theta \tag{11.18}$$

where

$$E(\theta) = -\frac{2E_y a}{\lambda}\frac{\sin(ka\sin\theta)}{ka\sin\theta}$$

This is readily integrable on a computer or calculator, just as the corresponding expression for $E_{y1}(y,z)$ is, viz.

$$E_{y1}(y,z) = -\int_{-\pi/2}^{+\pi/2} E(\theta)\cos\theta e^{-jk(y\sin\theta + z\cos\theta)}\,d\theta \tag{11.19}$$

which allows this tangential field to be evaluated for points of observation other than the origin.

The corresponding results for the complex range of angles $(\pm\pi/2 \pm j\infty)$ can also be evaluated numerically as follows:

$$E_{y2} = -\int_{-\pi/2-j\infty}^{-\pi/2-j0} + \int_{-\pi/2+j0}^{\pi/2+j\infty} E(\theta)\cos\theta e^{-jk(y\sin\theta + z\cos\theta)}\,d\theta$$

where $\theta = \pm\pi/2 \pm j\beta$. Consider first the positive range $[(\pi/2) + j\beta]$, for which $d\theta = jd\beta$ and

$$\sin\left(\frac{\pi}{2}+j\beta\right)=\cosh\beta$$
$$\cos\left(\frac{\pi}{2}+j\beta\right)=-j\sinh\beta$$

Then this integral becomes

$$I_1=\int_0^\infty \frac{2E_y a}{\lambda}\frac{\sin(ka\cosh\beta)}{ka\cosh\beta}\sinh\beta e^{-jky\cosh\beta}e^{-kz\sinh\beta}\,d\beta$$

Similarly the integral over the negative range $[(-\pi/2)-j\beta]$ gives

$$I_2=\int_0^\infty \frac{2E_y a}{\lambda}\frac{\sin(ka\cosh\beta)}{ka\cosh\beta}\sinh\beta e^{+jky\cosh\beta}e^{-kz\sinh\beta}\,d\beta$$

so that the field parallel to the aperture is

$$E_{y2}=\frac{4E_y a}{\lambda}\int_0^\infty \frac{\sin(ka\cosh\beta)}{ka\cosh\beta}\sinh\beta\cos(ky\cosh\beta)e^{-kz\sinh\beta}\,d\beta \tag{11.20}$$

Fig. 11.5. Radiating and evanescent electric field components E_{y1}, E_{z1} and E_{y2}, E_{z2} along z-axis of 1λ wide slot.

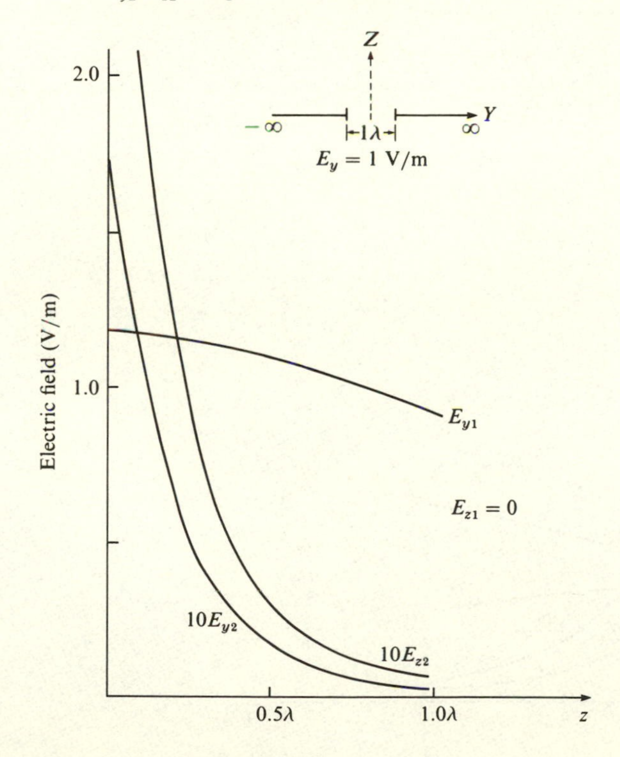

Likewise the field normal to the aperture is

$$E_{z2} = -j\frac{4E_y a}{\lambda}\int_0^\infty \frac{\sin(ka\cosh\beta)}{ka\cosh\beta}\cosh\beta\cos(ky\cosh\beta)e^{-kz\sinh\beta}\,d\beta \tag{11.21}$$

Some numerical results for the radiating components of field E_{y1}, E_{z1} as calculated from eqns (11.19) and (11.18) and for the surface wave components of field E_{y2}, E_{z2} as calculated from eqns (11.20) and (11.21), are shown in Figs. 11.5, 11.6 for a 1λ wide longitudinal slot in a perfectly conducting ground plane. The essential difference between the surface and radiating components of fields is that the surface components E_{y2}, E_{z2} effectively decay exponentially normal to the aperture plane. These fields are strongest in the middle of the slot, but even along this z-axis they have both decayed to less than 1% of the postulated tangential field within a distance of 1λ along this axis. For a slot 5λ wide the stronger, normal

Fig. 11.6. Radiating and evanescent electric field components E_{y1}, E_{z1} and E_{y2}, E_{z2} at edge of 1λ wide slot parallel to z-axis.

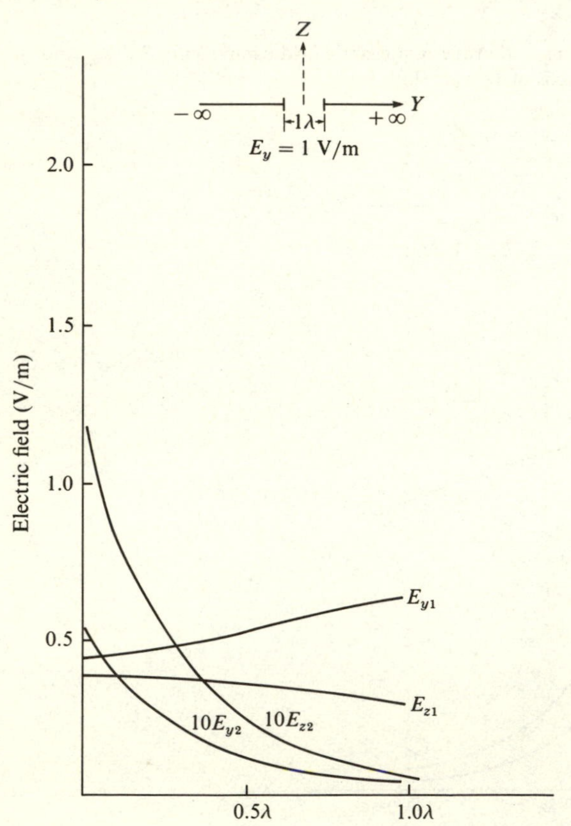

component is still only 2% of the tangential field in the aperture at the same distance. The travelling wave components behave quite differently and Fig. 11.5 shows how the tangential travelling wave component, starting at a value 18% larger than the total field in the aperture, has fallen only to 93% of this value along the z-axis. Likewise the tangential value at the edge actually rises by a factor of more than one-third at a distance of 1λ normal to the aperture plane.

With regard to the variation of fields in the direction parallel to the y-axis, Fig. 11.7 shows how these vary in the plane 0.25λ above the slot. The travelling wave component of E_y, which, from Fig. 11.5, started at the origin with a value of 1.18 has fallen to 1.16 at 0.25λ and falls from there parallel to the plane to approximately 0.43 above the edge of the slot. Likewise the E_z-component which started at zero has risen to a maximum of 0.41 at a distance of 0.4λ from the centre and then fallen to 0.38 above the edge. The E_y-field component in a plane 10λ above the slot is substantially constant over several wavelengths, as the field plot shown is in the far field of the slot, and the orthogonal E_z has substantially negligible value directly above the slot.

Some corresponding graphs for a slot of width 5λ are shown in Figs. 11.8–11.10. It will be noted that the travelling wave component of E_y is much more constant both over the aperture width, and as it travels normal

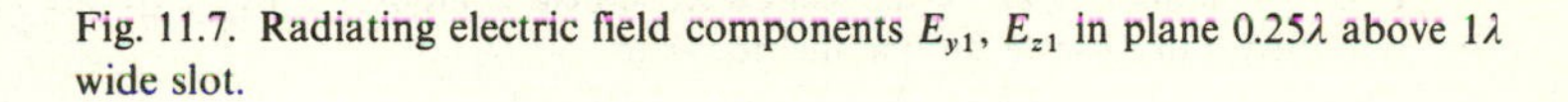
Fig. 11.7. Radiating electric field components E_{y1}, E_{z1} in plane 0.25λ above 1λ wide slot.

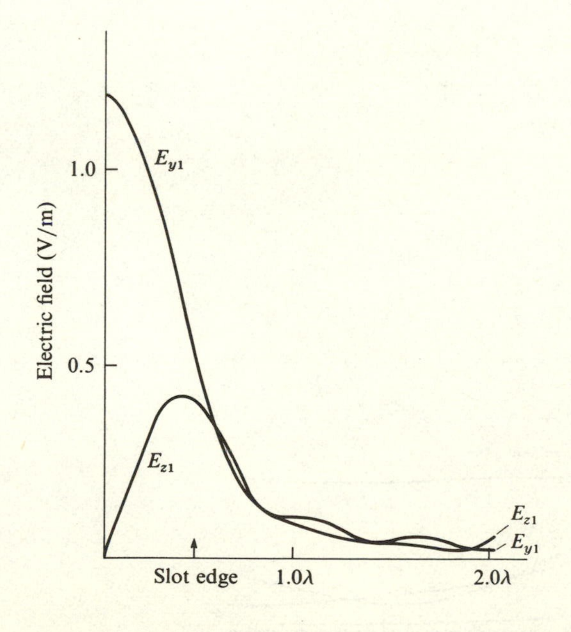

Fig. 11.8. Radiating and evanescent electric field components E_{y1}, E_{z1}, and E_{y2}, E_{z2} along z-axis of 5λ wide slot.

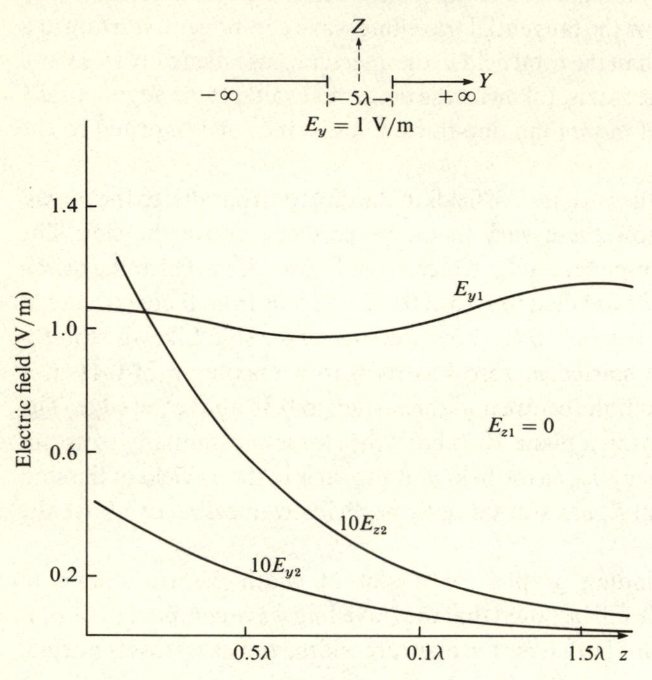

Fig. 11.9. Radiating and evanescent electric field components E_{y1}, E_{z1}, and E_{y2}, E_{z2} at edge of 5λ wide slot parallel to z-axis.

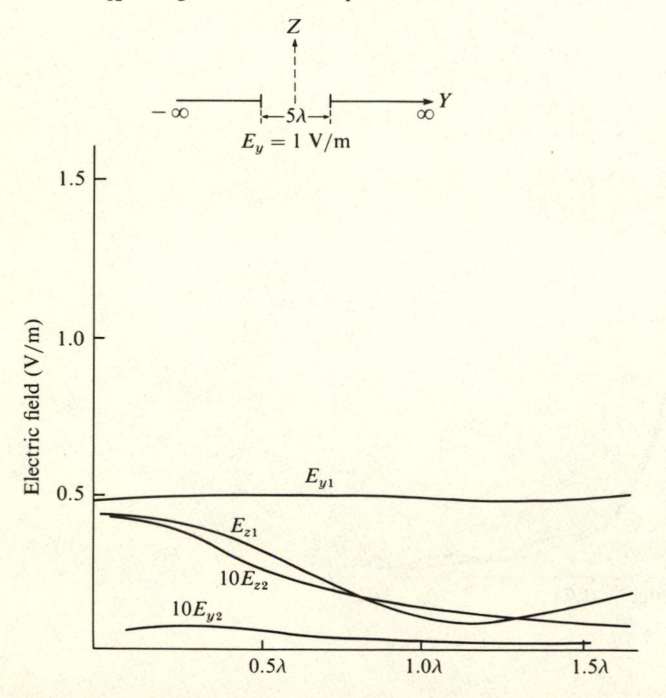

to the aperture plane. But along the line of the edge perpendicular to the aperture plane this field always falls to one-half the postulated value in the aperture.

11.3 Angular spectrum for two-dimensional aperture

With reference to Fig. 11.11 the components of electric field of a single uniform plane wave travelling in the direction (θ, ϕ) are

$$\mathbf{E} = -\mathbf{a}_x \cos\theta \cos\phi - \mathbf{a}_y \cos\theta \sin\phi + \mathbf{a}_z \sin\theta \tag{11.22}$$

and the phase variation with distance is

$$\mathbf{k}\cdot\mathbf{r} = k(x \sin\theta \cos\phi + y \sin\theta \sin\phi + z\cos\theta)$$

For a range of all possible values of (θ, ϕ) the spectrum must extend from 0 to $[(\pi/2) + j\infty]$ for θ, since the physical limits of polar angle for $z > 0$ are from 0 to $\pi/2$ only. The limits for ϕ are from 0 to 2π since there are no physical boundaries to support surface waves in this direction.

Hence at any point $P(x, y, z)$, with $z > 0$, the total field can be written in the form

$$\begin{aligned}\mathbf{E}(x, y, z) = &\int_0^{(\pi/2 + j\infty)} \int_0^{2\pi} \\ &\times E(\theta, \phi)(-\mathbf{a}_x \cos\theta\cos\phi - \mathbf{a}_y \cos\theta \sin\phi + \mathbf{a}_z \sin\theta) \\ &\times e^{-jk(x\sin\theta\cos\phi + y\sin\theta\sin\phi + z\cos\theta)}\, d\theta\, d\phi\end{aligned} \tag{11.23}$$

where $E(\theta, \phi)$ is a phasor representing the magnitude and phase of the

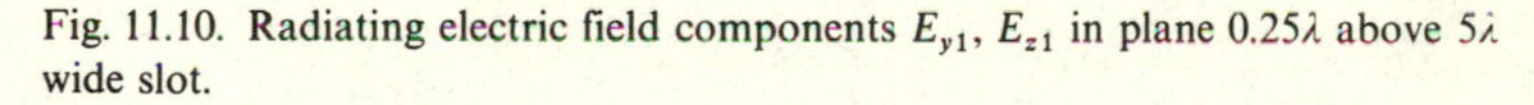

Fig. 11.10. Radiating electric field components E_{y1}, E_{z1} in plane 0.25λ above 5λ wide slot.

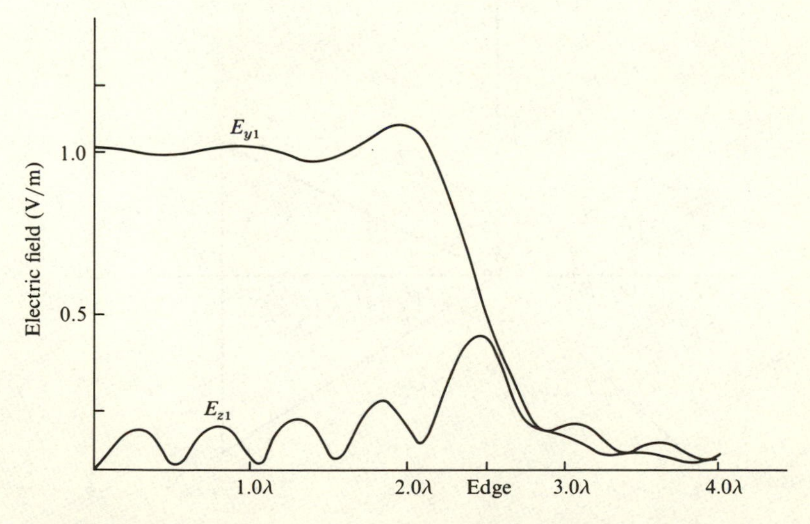

electric field of each plane wave travelling through the origin in the aperture plane at the angle (θ, ϕ). If the variable of integration were changed to $k \sin \theta$ as in the previous case, the limits of integration for θ would be 0 and $+\infty$. However, by using variables $k_x = k \sin \theta \cos \phi$ and $k_y = k \sin \theta \sin \phi$ with $\cos \phi$ and $\sin \phi$ varying between ± 1, both k_x and k_y will vary between $-\infty$ and $+\infty$, so allowing the desired Fourier transform relationship. Also in changing the variables of integration we have

$$\iint f(k_x, k_y) dk_x \, dk_y = \iint f(k_x, k_y) J \, d\theta \, d\phi$$

where J is the Jacobean defined by

$$J = \frac{\partial(k_x, k_y)}{\partial\theta \partial\phi} = \begin{vmatrix} \dfrac{\partial k_x}{\partial \theta} & \dfrac{\partial k_y}{\partial \theta} \\ \dfrac{\partial k_x}{\partial \phi} & \dfrac{\partial k_y}{\partial \phi} \end{vmatrix} = k^2 \sin \theta \cos \theta$$

The total field for $z \geqslant 0$ may therefore be written as

$$\begin{aligned} \mathbf{E}(x, y, z) = \int_{-\infty}^{+\infty} \int_{-\infty}^{+\infty} & \frac{E(k_x, k_y)}{k^2 \sin \theta \cos \theta} \\ & \times (-\mathbf{a}_x \cos \theta \cos \phi - \mathbf{a}_y \cos \theta \sin \phi + \mathbf{a}_z \sin \theta) \\ & \times e^{-j(k_x x + k_y y + k_z z)} dk_x \, dk_y \end{aligned} \tag{11.24}$$

Fig. 11.11. Rectangular components of electric field for plane wave travelling in direction (θ, ϕ).

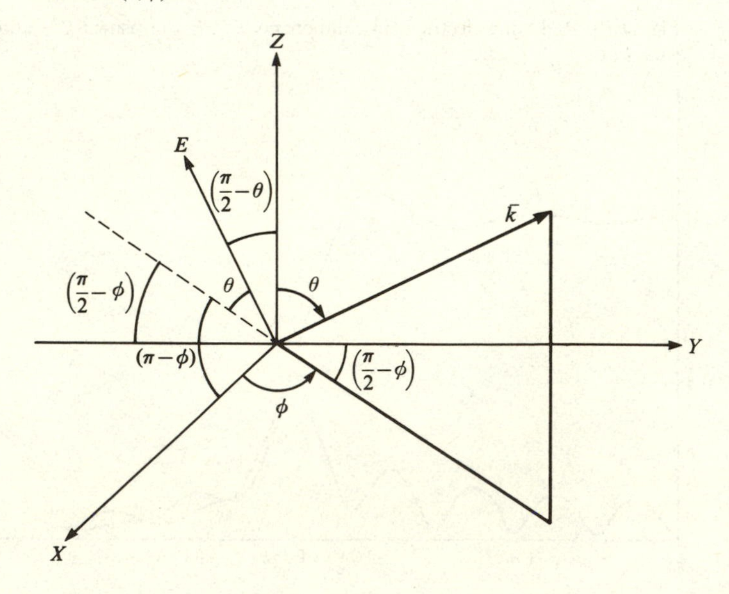

Considering now the scalar components of this expression in the aperture plane this gives, for the source fields at points $(x', y', 0)$ in this plane,

$$E_x(x', y', 0) = \frac{1}{k^2}\int_{-\infty}^{+\infty}\int_{-\infty}^{+\infty} E_{0x}(k_x, k_y)e^{-j(k_x x' + k_y y')}dk_x\,dk_y$$

$$E_y(x', y', 0) = \frac{1}{k^2}\int_{-\infty}^{+\infty}\int_{-\infty}^{+\infty} E_{0y}(k_x, k_y)e^{-j(k_x x' + k_y y')}dk_x\,dk_y$$

$$E_z(x', y', 0) = \frac{1}{k^2}\int_{-\infty}^{+\infty}\int_{-\infty}^{+\infty} E_{0z}(k_x, k_y)e^{-j(k_x x' + k_y y')}dk_x\,dk_y \tag{11.25}$$

where

$$E_{0x}(k_x, k_y) = -\frac{E(k_x, k_y)\cos\phi}{\sin\theta} \tag{11.26}$$

$$E_{0y}(k_x, k_y) = -\frac{E(k_x, k_y)\sin\phi}{\sin\theta} \tag{11.27}$$

$$E_{0z}(k_x, k_y) = \frac{E(k_x, k_y)}{\cos\theta} \tag{11.28}$$

Taking the inverse Fourier transforms of eqns (11.25) gives

$$E_{0x}(k_x, k_y) = \frac{1}{\lambda^2}\int_{-\infty}^{+\infty}\int_{-\infty}^{+\infty} E_x(x', y', 0)e^{j(k_x x' + k_y y')}dx'\,dy' \tag{11.29}$$

$$E_{0y}(k_x, k_y) = \frac{1}{\lambda^2}\int_{-\infty}^{+\infty}\int_{-\infty}^{+\infty} E_y(x', y', 0)e^{j(k_x x' + k_y y')}dx'\,dy' \tag{11.30}$$

$$E_{0z}(k_x, k_y) = \frac{1}{\lambda^2}\int_{-\infty}^{+\infty}\int_{-\infty}^{+\infty} E_z(x', y', 0)e^{j(k_x x' + k_y y')}dx'\,dy' \tag{11.31}$$

The infinite limits in eqns (11.29) and (11.30) may be replaced by the finite limits of the aperture size if the aperture is part of an infinite conducting sheet. However, this is not so for eqn (11.31) since a normal component of electric field will exist in the plane of the aperture out to infinity, and in general this will not be known. Fortunately this does not mean that it is impossible to evaluate $E_{0z}(k_x, k_y)$. Since for the travelling wave components of fields, $E(k_x, k_y)$ is orthogonal to k, this gives

$$-E(k_x, k_y)\cos\theta\cos\phi\cdot k\sin\theta\cos\phi - E(k_x, k_y)\cos\theta\sin\phi$$
$$\times\, k\sin\theta\sin\phi + E(k_x, k_y)\sin\theta\cdot k\cos\theta = 0$$

Substituting for E_{0x}, E_{0y} and E_{0z} from eqns (11.26)–(11.28) gives

$$E_{0z}(k_x, k_y) = -\frac{E_{0x}\sin\theta\cos\phi + E_{0y}\sin\theta\sin\phi}{\cos\theta}$$

Thus the total field is

$$\mathbf{E}(x,y,z)=\frac{1}{k^2}\int_{-\infty}^{+\infty}\int_{-\infty}^{+\infty}\left[\mathbf{a}_x E_{0x}(k_x,k_y)+\mathbf{a}_y E_{0y}(k_x,k_y)\right.$$
$$\left.-\mathbf{a}_z\frac{k_x E_{0x}(k_x,k_y)+k_y E_{0y}(k_x,k_y)}{k_z}\right]$$
$$\times e^{-j(k_x x+k_y y+k_z z)}\,dk_x\,dk_y \qquad (11.32)$$

which depends only on the tangential electric fields in the aperture. If one of the tangential components is zero then the total field may be expressed in terms of either the remaining tangential aperture component or the aperture normal component. But in this case E_{0z} must still be derived from E_{0x} or E_{0y} if there is a conducting ground plane present, so that the tangential component is the only useful one.

The three components of magnetic field may then be found from Maxwell's equations. Thus

$$H_x=-\frac{1}{j\omega\mu}\left(\frac{\partial E_z}{\partial y}-\frac{\partial E_y}{\partial z}\right)$$

$$\frac{\partial E_z}{\partial y}=\frac{1}{k^2}\int_{-\infty}^{+\infty}\int_{-\infty}^{+\infty}\frac{-k_x E_{0x}-k_y E_{0y}}{k_z}(-jk_y)$$
$$\times e^{-j(k_x x+k_y y+k_z z)}\,dk_x\,dk_y$$

$$\frac{\partial E_y}{\partial z}=\frac{1}{k^2}\int_{-\infty}^{+\infty}\int_{-\infty}^{+\infty}(-jk_z)E_{0y}e^{-j(k_x x+k_y y+k_z z)}\,dk_x\,dk_y$$

Hence

$$H_x=\frac{1}{\omega\mu k^2}\int_{-\infty}^{+\infty}\int_{-\infty}^{-\infty}\frac{[-k_x k_y E_{0x}-(k^2-k_x^2)E_{0y}]}{k_z}$$
$$\times e^{-j(k_x x+k_y y+k_z z)}\,dk_x\,dk_y \qquad (11.33)$$

Similarly

$$H_y=\frac{1}{\omega\mu k^2}\int_{-\infty}^{+\infty}\int_{-\infty}^{+\infty}\frac{[(k^2-k_y^2)E_{0x}+k_x k_y E_{0y}]}{k_z}$$
$$\times e^{-j(k_x x+k_y y+k_z z)}\,dk_x\,dk_y \qquad (11.34)$$

and

$$H_z=\frac{1}{\omega\mu k^2}\int_{-\infty}^{+\infty}\int_{-\infty}^{+\infty}(-k_y E_{0x}+k_x E_{0y})$$
$$\times e^{-j(k_x x+k_y y+k_z z)}\,dk_x\,dk_y \qquad (11.35)$$

11.4 Radiation field from two-dimensional aperture in conducting plane

Changing the general expression for the electric field at any point $z > 0$, into spherical coordinates gives, from eqn (11.32),

$$\mathbf{E}(r, \theta, \phi) = \frac{1}{k^2} \int_{-\infty}^{+\infty} \int_{-\infty}^{+\infty} \left[\mathbf{a}_x E_{0x} + \mathbf{a}_y E_{0y} - \mathbf{a}_z \frac{k_x E_{0x} + k_y E_{0y}}{k_z} \right] \times e^{-j(k_x r \sin\theta \cos\phi + k_y r \sin\theta \sin\phi + k_z \cos\theta)} \, dk_x \, dk_y \tag{11.36}$$

As for the slot problem previously discussed, the contribution to this integral for a far field point will come mainly from values of k_x, k_y such that $k \sin\theta' \cos\phi' \approx k \sin\theta \cos\phi$ and $k \sin\theta' \sin\phi' \approx k \sin\theta \sin\phi$. Hence let

$$\left.\begin{aligned} k_x &= k \sin(\theta + u) \cos(\phi + v) \\ k_y &= k \sin(\theta + u) \cos(\phi + v) \end{aligned}\right\} \quad \text{where } u, v \text{ are small angles}$$

Then

$$\mathbf{E}(r, \theta, \phi) = \frac{1}{k^2} \int_{-\infty}^{+\infty} \int_{-\infty}^{+\infty} \left[\mathbf{a}_x E_{0x} + \mathbf{a}_y E_{0y} - \mathbf{a}_z \frac{k_x E_{0x} + k_y E_{0y}}{k_z} \right] \times e^{-jkr(\sin\theta \cos\phi \sin(\theta+u) \cos(\phi+v) + \sin\theta \sin\phi \sin(\theta+u) \sin(\phi+v) + \cos\theta \cos(\theta+u))} \times J \, du \, dv \tag{11.37}$$

where

$$J = \frac{\partial(k_x, k_y)}{\partial(u, v)} = \begin{vmatrix} \dfrac{\partial k_x}{\partial u}, & \dfrac{\partial k_y}{\partial u} \\ \dfrac{\partial k_x}{\partial v}, & \dfrac{\partial k_y}{\partial v} \end{vmatrix} = k^2 \sin(\theta + u) \cos(\theta + u)$$

The limits for u, v are not important since although both are small the contributions from their larger values cancel, and hence the limits of $\pm\infty$ may be retained.

The trigonometric exponent of the exponential term simplifies to

$$\sin\theta \sin(\theta + u) \cos v + \cos\theta \cos(\theta + u)$$

Substituting small angle approximations for the terms involving u, v in the expanded form of this expression gives, for the exponent, $[1 - (u^2/2) - (v^2/2)\sin^2\theta]$. Hence the radiated field is

$$\mathbf{E}(r, \theta, \phi) = \int_{-\infty}^{+\infty} \int_{-\infty}^{+\infty} [\mathbf{a}_x E_{0x} \cos\theta + \mathbf{a}_y E_{0y} \cos\theta - \mathbf{a}_z (E_{0x} \sin\theta \cos\phi + E_{0y} \sin\theta \sin\phi)] \times e^{-jkr} e^{jkr(u^2/2)} e^{j(kr/2)v^2 \sin^2\theta} \sin\theta \, du \, dv \tag{11.38}$$

Now

$$\int_{-\infty}^{+\infty} e^{j(ax^2/2)}\,dx = \sqrt{\left(\frac{\pi}{a}\right)}e^{j(\pi/4)},$$

so that

$$\mathbf{E}(r,\theta,\phi) = [\mathbf{a}_x E_{0x}\cos\theta + \mathbf{a}_y E_{0y}\cos\theta - \mathbf{a}_z(E_{0x}\sin\theta\cos\phi + E_{0y}\sin\theta\sin\phi)] \times \sqrt{\left(\frac{2\pi}{kr}\right)}e^{j(\pi/4)}\sqrt{\left(\frac{2\pi}{kr\sin^2\theta}\right)}e^{j(\pi/4)}\sin\theta$$

i.e.

$$\mathbf{E}(r,\theta,\phi) = j\frac{\lambda}{r}e^{-jkr}[\mathbf{a}_x E_{0x}\cos\theta + \mathbf{a}_y E_{0y}\cos\theta - \mathbf{a}_z(E_{0x}\sin\theta\cos\phi + E_{0y}\sin\theta\sin\phi)] \tag{11.39}$$

The remarkable simplicity of this result should be noted. It says that if there is only an E_x-field in the aperture, then in the broadside direction where θ is zero the radiated field is $(\lambda/r)E_{0x}$, and similarly for an E_y-field in the aperture. E_{0x} and E_{0y} are given by the Fourier transform of the tangential field in the aperture plane. Note also that there is no E_z-component of radiation field along the z-axis, in the same way that it was absent in the slot.

The total field will now be written in terms of its spherical components using the standard transformations

$$\mathbf{a}_r = \mathbf{a}_x\sin\theta\cos\phi + \mathbf{a}_y\sin\theta\sin\phi + \mathbf{a}_z\cos\theta$$
$$\mathbf{a}_\theta = \mathbf{a}_x\cos\theta\cos\phi + \mathbf{a}_y\cos\theta\sin\phi - \mathbf{a}_z\sin\theta$$
$$\mathbf{a}_\phi = -\mathbf{a}_x\sin\phi + \mathbf{a}_y\cos\phi$$

Hence

$$E_r(r,\theta,\phi) = 0$$

$$E_\theta(r,\theta,\phi) = \frac{j}{\lambda r}e^{-jkr}\int_{-\infty}^{+\infty}\int_{-\infty}^{+\infty} \times [E_x(x',y',0)\cos\phi + E_y(x',y',0)\sin\phi] \times e^{jk(x'\sin\theta\cos\phi + y'\sin\theta\sin\phi)}\,dx'\,dy' \tag{11.40}$$

$$E_\phi(r,\theta,\phi) = \frac{-j}{\lambda r}e^{-jkr}\cos\theta\int_{-\infty}^{+\infty}\int_{-\infty}^{+\infty} \times [E_x(x',y',0)\sin\phi - E_y(x',y',0)\cos\phi] \times e^{jk(x'\sin\theta\cos\phi + y'\sin\theta\sin\phi)}\,dx'\,dy' \tag{11.41}$$

The infinite limits of integration may be replaced by the finite limits of the aperture if this is situated in a perfectly conducting plane, and the results then become identical with those previously derived by the magnetic current approach. A numerical example for a circular aperture in such a ground plane will now be worked out.

11.5 Circular aperture of radius *a* in planar conducting sheet

A circular aperture of radius a is centred at the origin in a perfectly conducting sheet lying in the xy-plane. It will be assumed that an electric field E_x exists in the plane of the aperture, as shown in Fig. 11.12, and for simplicity this field will be taken to be constant.

To find the fields at any point in space the angular spectrum resulting from this constant field will be found. Thus, from eqn (11.29),

$$E_{0x}(k_x, k_y) = \frac{1}{\lambda^2} \iint_{aperture} E_x(x', y', 0) e^{j(k_x x' + k_y y')} \, dx' \, dy'$$

$$= \frac{E_x}{\lambda^2} \iint_{aperture} e^{j(k \sin\alpha \cos\beta \cdot \rho' \cos\phi' + k \sin\alpha \sin\beta \cdot \rho' \sin\phi')} . \rho' \, d\rho' \, d\phi'$$

where (ρ', ϕ') are the polar forms of a source point (x', y'), and (α, β) are used in place of (θ, ϕ) which is used as a direction of observation later. Hence

$$E_{0x}(k_x, k_y) = \frac{E_x}{\lambda^2} \int_0^a \int_0^{2\pi} e^{jk\rho' \sin\alpha \cos(\beta - \phi')} \rho' \, d\rho' \, d\phi'$$

$$= \frac{E_x}{\lambda^2} \int_0^a 2\pi J_0(k\rho' \sin\alpha) \rho' \, d\rho'$$

$$= \frac{E_x}{\lambda^2} 2\pi a^2 \frac{J_1(ka \sin\alpha)}{ka \sin\alpha} \tag{11.42}$$

Fig. 11.12. Circular aperture of radius a in planar conducting sheet with constant electric field E_x in aperture.

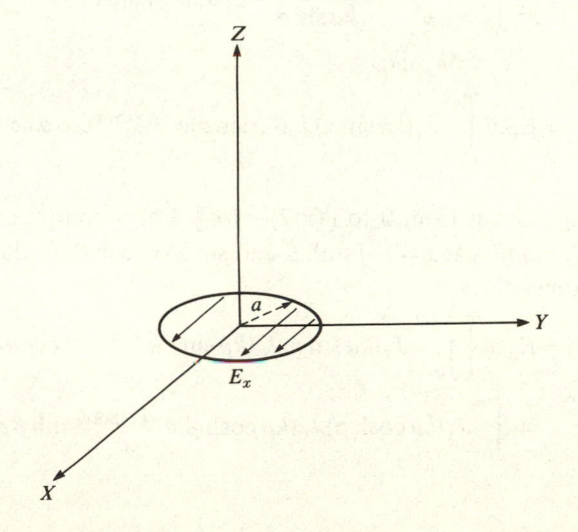

Then the x-component of the resulting field outside the aperture is, from eqn (11.32),

$$E_x(x, y, z) = \frac{1}{k^2} \int_{-\infty}^{+\infty} \int_{-\infty}^{+\infty} \frac{E_x 2\pi a^2}{\lambda^2} \frac{J_1(ka \sin \alpha)}{ka \sin \alpha} \times e^{-j(k_x x + k_y y + k_z z)} \, dk_x \, dk_y \tag{11.43}$$

To enable one integration to be performed analytically a change of variable will be carried out using

$$dk_x \, dk_y = J\begin{pmatrix} k_x k_y \\ k_{\rho'} \beta \end{pmatrix} dk_{\rho'} \, d\beta$$

where $k_x = k_{\rho'} \cos \beta$ and $k_y = k_{\rho'} \sin \beta$. Then

$$J = \begin{vmatrix} \dfrac{\partial k_x}{\partial k_{\rho'}} & \dfrac{\partial k_y}{\partial k_{\rho'}} \\ \dfrac{\partial k_x}{\partial \beta} & \dfrac{\partial k_y}{\partial \beta} \end{vmatrix} = k_{\rho'}$$

and the field E_x at any point in space with $z > 0$ becomes

$$E_x(x, y, z) = \frac{1}{k^2} \int_0^{\infty} \int_0^{2\pi} \frac{E_x 2\pi a^2}{\lambda^2} \frac{J_1(ka \sin \alpha)}{ka \sin \alpha} \times e^{-jk_{\rho'} r \sin \theta \cos(\phi - \beta)} e^{-jk_z z} k_{\rho'} \, dk_{\rho'} \, d\beta \tag{11.44}$$

Since

$$\int_0^{2\pi} e^{-jk_{\rho'} r \cdot \sin \alpha \cos(\phi - \beta)} \, d\beta = 2\pi J_0(k_{\rho'} r \sin \alpha)$$

this gives, for a point of observation (ρ, ϕ, z) in cylindrical coordinates,

$$\begin{aligned} E_x(\rho, \phi, z) &= \frac{1}{k^2} \int_0^{\infty} \frac{E_x 2\pi a^2}{\lambda^2} \frac{J_1(ka \sin \alpha)}{ka \sin \alpha} 2\pi J_0(k_{\rho'} r \sin \theta) \\ &\quad \times e^{-jk_z z} k_{\rho'} \, dk_{\rho'} \\ &= E_x ka \int_0^{\infty} J_1(ka \sin \alpha) J_0(k\rho \sin \alpha) e^{-jkz \cos \alpha} \cos \alpha \, d\alpha \end{aligned} \tag{11.45}$$

The complete range of α is from 0 to $[(\pi/2) + j\infty]$. For a complex angle $\alpha = [(\pi/2) + j\beta]$, $d\alpha = j d\beta$, $\cos \alpha = -j \sinh \beta$ and $\sin \alpha = \cosh \beta$, so that the total E_x-field becomes

$$\begin{aligned} E_x(\rho, \phi, z) = E_x ka \Bigg\{ &\int_0^{\pi/2} J_1(ka \sin \alpha) J_0(k\rho \sin \alpha) e^{-jkz \cos \alpha} \cos \alpha \, d\alpha \\ &+ \int_0^{\infty} J_1(ka \cosh \beta) J_0(k\rho \cosh \beta) e^{-kz \sinh \beta} \sinh \beta \, d\beta \Bigg\} \end{aligned} \tag{11.46}$$

Since there is no y-directed electric field in the aperture, the component $E_y(\rho, \phi, z)$ is zero, but there is a normal component E_z given by, from eqn (11.32),

$$E_z(x, y, z) = -\frac{1}{k^2}\int_{-\infty}^{+\infty}\int_{-\infty}^{+\infty}\frac{k_x}{k_z}E_{0x}e^{-j(k_x x + k_y y + k_z z)}\,dk_x\,dk_y$$

$$= -\frac{1}{k^2}\int_{-\infty}^{+\infty}\int_{-\infty}^{+\infty}\frac{k_x}{k_z}\frac{E_x 2\pi a^2}{\lambda^2}\frac{J_1(ka\sin\alpha)}{ka\sin\alpha}$$

$$\times e^{-j(k_x x + k_y y + k_z z)}\,dk_x\,dk_y \tag{11.47}$$

Changing to polar coordinates for the variables of integration gives

$$E_z(x, y, z) = -\frac{E_x 2\pi a^2}{k^2\lambda^2}\int_0^{\infty}\int_0^{2\pi}\frac{k_{\rho'}\cos\beta}{k_z}\frac{J_1(ka\sin\alpha)}{ka\sin\alpha}$$

$$\times e^{-jk_{\rho'}r\sin\theta\cos(\phi-\beta)}e^{-jk_z}k_{\rho'}\,dk_{\rho'}\,d\beta \tag{11.18}$$

$$= -\frac{E_x a}{2\pi}\int_0^{\infty}J_1(ka\sin\alpha\cdot(-)\cdot 2\pi j J_1(k_{\rho'}r\sin\theta)$$

$$\times\frac{e^{-jk_z z}}{k_z}\cos\phi k_{\rho'}\,dk_{\rho'}$$

$$= jE_x a\cos\phi\int_0^{\infty}J_1(ka\sin\alpha)J_1(k\rho\sin\alpha)$$

$$\times\frac{e^{-jk_z z}}{k_z}(k\sin\alpha)(k\cos\alpha)\,d\alpha$$

Splitting this up into components associated with real and complex values of α gives

$$E_z(\rho, \phi, z) = jE_x ka\cos\phi\left\{\int_0^{\pi/2}J_1(ka\sin\alpha)J_1(k\rho\sin\alpha)\right.$$

$$\times e^{-jkz\cos\alpha}\sin\alpha\,d\alpha + j\int_0^{\infty}J_1(ka\cosh\beta)J_1(k\rho\cosh\beta)$$

$$\left.\times e^{-kz\sinh\beta}\cosh\beta\,d\beta\right\} \tag{11.49}$$

From eqns (11.46) and (11.49) it is therefore possible to calculate the E_x- and E_z-components of electric field at any point in space, which are produced by a uniform electric field in a circular aperture of radius a. It will be noted from eqn (11.46) that the E_x-component is independent of the angular position ϕ of the point of observation, and is therefore easier to illustrate graphically than the E_z-component. But to enable a numerical comparison of these field values an angle of 45° will be chosen for ϕ and the fields for other values of ϕ will be related to those at 45° by the equation

$$E_z(\rho, \phi, z) = \sqrt{2}\cos\phi\, E_z\left(\rho, \frac{\pi}{4}, z\right)$$

Consider first a small circular aperture of diameter 0.1λ, over which the total tangential electric field is specified as having unit magnitude in the x-direction in the plane of the aperture. For such a small antenna it would be expected that the component of E_x associated with the complex range of angles, and giving rise to surface wave fields, would be a large part of the total E_x. This is confirmed by the results derived from eqn (11.46) which are shown in Fig. 11.13. It is in fact possible to derive analytically the proportion of the applied E_x at the origin which is associated with such fields as follows. From eqn (11.46) the total E_x-field at the origin is given by

$$\begin{aligned} E_x(0,0,0) &= E_x ka\left\{\int_0^{\pi/2} J_1(ka\sin\alpha)\cos\alpha\, d\alpha \right. \\ &\qquad \left. + \int_0^{\infty} J_1(ka\cosh\beta)\sinh\beta\, d\beta\right\} \\ &= E_x\left[\int_0^{ka} J_1(u)\,du + \int_{ka}^{\infty} J_1(v)\,dv\right] \end{aligned} \tag{11.50}$$

where $u = \sin\alpha$, $v = \cosh\beta$

Fig. 11.13. Radiating and evanescent electric field components in aperture plane of circular aperture of diameter 0.1λ.

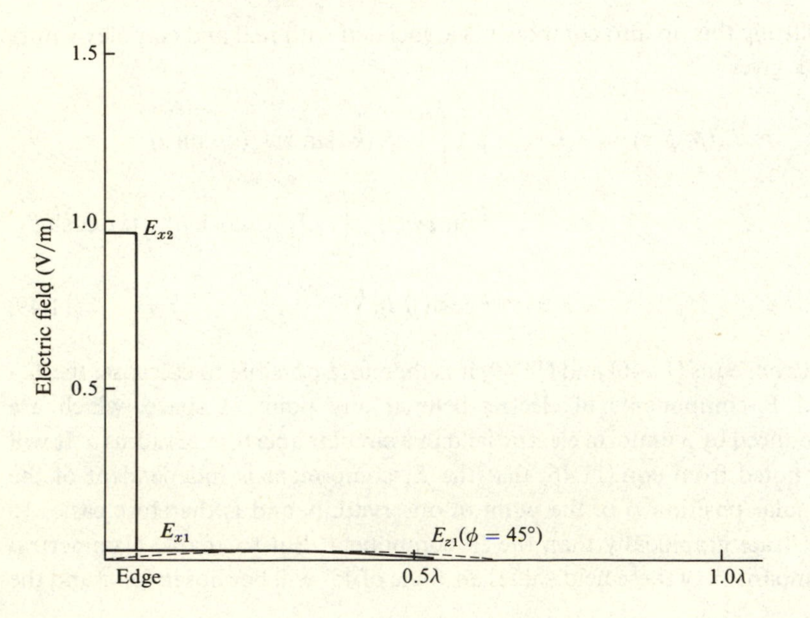

i.e.

$$E_x(0,0,0) = E_x\{[1 - J_0(ka)] + J_0(ka)\}$$

Thus the radiated field at the origin uses up a fraction $[1 - J_0(ka)]$ of the applied E_x, and the remainder is associated with the stored energy of the aperture. For the case of the 0.1λ diameter aperture these figures are approximately 3% and 97%, and the graph shows that this ratio remains constant over the full aperture. Outside the aperture, but still in its plane, the total E_x must of course be zero because of the presence of the perfectly conducting sheet, but it is forced to have zero value as a result of these radiating and evanescent components having equal and antiphase value. The radiating component is shown in the diagram out to a radius of 1λ.

The same general equation for E_x can be used to plot the variation of its two components along the axis normal to the plane of the aperture, and this variation is shown in Fig. 11.14 out to a distance of 1λ from the aperture plane. The rapid decay of the evanescent field is illustrated by its having fallen to 3% of its initial value in a distance of approximately 0.2λ.

But in addition to these x-directed electric fields there also exists a z-directed field across the aperture. This can be thought of as being caused by the charge distribution which produces E_x in the first place. Positive charges will exist round the edge of the aperture for $x > 0$, and likewise negative charges for $x < 0$. Consequently there are no charges along the y-

Fig. 11.14. Radiating and evanescent electric field components along z-axis normal to plane of circular aperture of diameter 0.1λ.

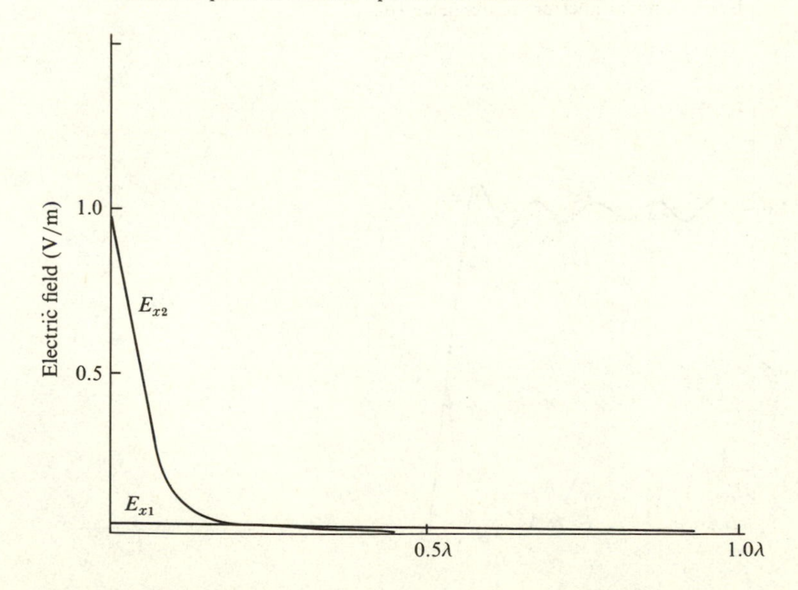

axis and the yz-plane will be a neutral plane with only a E_x-directed field across it, as is confirmed by the equation for E_z with ϕ equal to $\pi/2$. Since the z-axis lies in this plane E_z is zero along it, but in the plane of the aperture for $\rho \neq 0$ a small value of E_z does exist, as shown by the curve of Fig. 11.13.

Consider next a more representative antenna aperture of 10λ diameter, with the same postulated constant E_x-field of unit magnitude in its plane. The proportion of this which is associated with the radiation field at the origin has now risen from 3% to 90%, as shown in Fig. 11.15. As the observation point moves in the plane of the aperture, in any angular direction out to the edge, this proportion oscillates about its full value of unity until close to the edge before falling to half this value at the edge itself. Since the total field is constant in the aperture this means that at the edge-point half the field there is associated with evanescent waves. Both the radiation and evanescent E_x-fields in the plane of the aperture are shown in the diagram. Outside the aperture the two fields are equal in magnitude and opposite in phase, so that their resultant satisfies the boundary condition in the conducting plane. The variation of the normal component of electric field in the aperture is also shown in the diagram, starting from zero value on the z-axis.

Along the z-axis the radiation component of the E_x-electric field varies between a very low value and almost twice its value in the aperture, which occurs at a distance of $[(D^2/4\lambda) - \lambda/4]$. Thereafter, for still greater distances

Fig. 11.15. Radiating and evanescent electric field components in aperture plane of circular aperture of diameter 10λ.

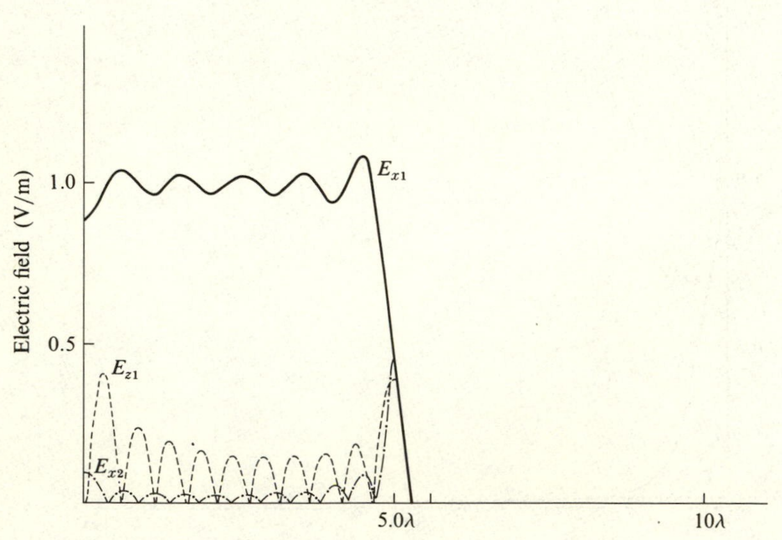

normal to the aperture, the field progressively decays towards the $1/r$ variation with distance characteristic of the radiation field. This variation is shown in Fig. 11.16. Unlike this result the variation of field from the edge parallel to the z-axis is shown for a distance of 5λ in Fig. 11.17. In this case the field remains substantially constant in magnitude, so that the radiation may be thought of as travelling in a parallel beam of diameter equal to that of the aperture before spreading out to form its radiation pattern.

Fig. 11.16. Radiating electric field component E_{x1} along z-axis normal to circular aperture of diameter 10λ.

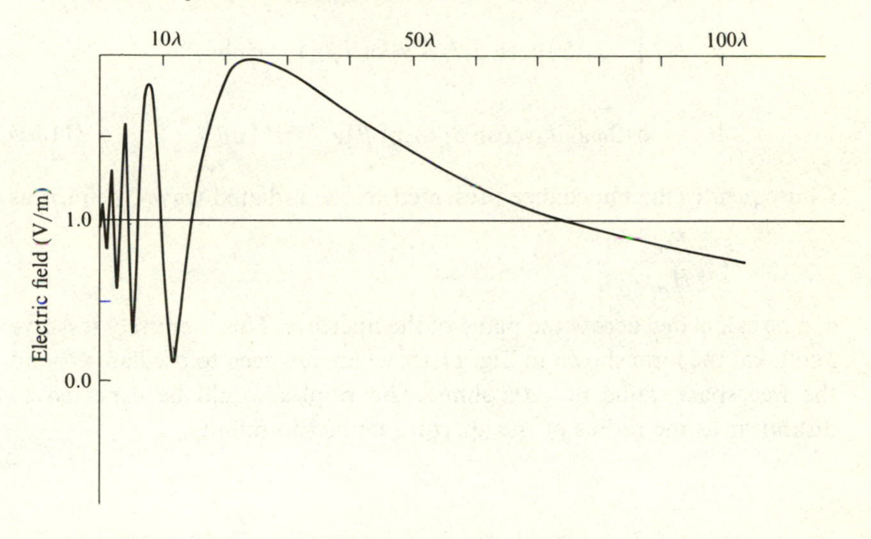

Fig. 11.17. Radiating electric field components E_{x1}, E_{z1} at edge of 10λ diameter circular aperture parallel to z-axis.

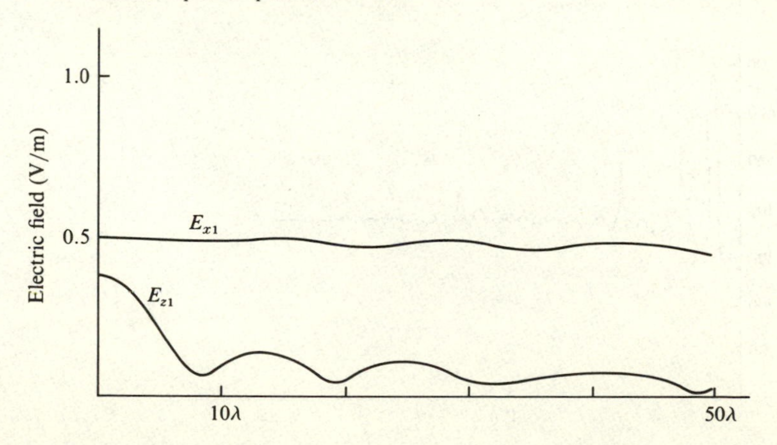

Although the earlier analysis of the field anywhere in space was carried out only for the electric fields, the use of Maxwell's equations at the observation point clearly provides all the magnetic field components in terms of the derivatives of these. The result of performing these operations gives for the magnetic component H_y, in terms of the postulated E_x in the aperture,

$$\begin{aligned} H_y = \frac{E_x ka}{2Z_0} \Bigg\{ \int_0^{\pi/2} & J_1(ka \sin\alpha)[J_0(k\rho \sin\alpha)(1+\cos^2\alpha) \\ & - \cos 2\phi J_2(k\rho \sin\alpha) \sin^2\alpha] e^{-jkz\cos\alpha}\, d\alpha \\ & + j \int_0^{\infty} J_1(ka \cosh\beta) J_0(k\rho\cosh\beta)(1-\sinh^2\beta) \\ & - \cos 2\phi J_2(k\rho \cosh\beta)\cosh^2\beta] e^{-kz\sinh\beta}\, d\beta \Bigg\} \end{aligned} \tag{11.51}$$

Consequently the impedance presented to the radiated waves, defined as

$$Z = \frac{E_{xr}}{H_{yr}}$$

can be calculated across the plane of the aperture. This is entirely resistive and takes the form shown in Fig. 11.18, where it is seen to oscillate around the free space value of 120π ohms. The ripples would be expected to disappear as the radius of the aperture tended to infinity.

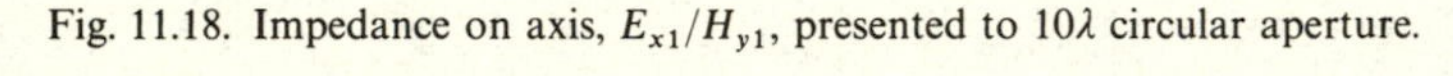
Fig. 11.18. Impedance on axis, E_{x1}/H_{y1}, presented to 10λ circular aperture.

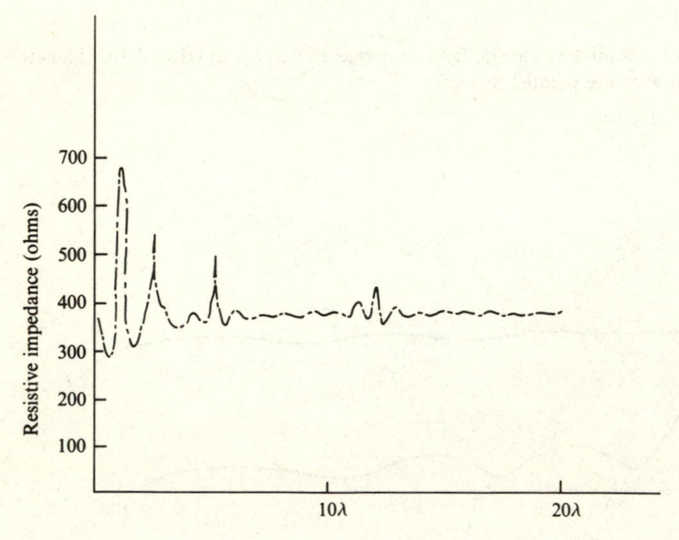

11.6 Radiation fields of uniformly excited circular aperture

It has previously been shown for a circular aperture of radius a in an infinite perfectly conducting screen that the radiation field is given by

$$E_\theta = \frac{j}{\lambda r} e^{-jkr} \iint_{aperture} (E_x \cos\phi + E_y \sin\phi) \times e^{jk\rho' \sin\theta \cos(\phi - \phi')} \rho' \, d\rho' \, d\phi' \tag{11.52}$$

and

$$E_\phi = -\frac{j}{\lambda r} e^{-jkr} \cos\theta \iint_{aperture} (E_x \sin\phi - E_y \cos\phi) \times e^{jk\rho' \sin\theta \cos(\phi - \phi')} \rho' \, d\rho' \, d\phi' \tag{11.53}$$

Hence when the excitation consists of a uniform E_x-field only,

$$E_\theta = \frac{jE_x}{\lambda} \frac{e^{-jkr}}{r} \cos\phi \int_0^a \int_0^{2\pi} e^{jk\rho' \sin\theta \cos(\phi - \phi')} \rho' \, d\rho' \, d\phi'$$

$$= \frac{jkE_x e^{-jkr} \cos\phi}{r} \int_0^a J_0(k\rho' \sin\theta) \rho' \, d\rho'$$

Let $v = k\rho' \sin\theta$; $dv = k \sin\theta \, d\rho'$, so that

$$E_\theta = \frac{jE_x e^{-jkr} \cos\phi}{k \sin^2\theta} \int_0^{ka \sin\theta} J_0(v) v \, dv$$

$$= \frac{jE_x}{\lambda r} e^{-jkr} \cos\phi \, 2\pi a^2 \frac{J_1(ka \sin\theta)}{ka \sin\theta} \tag{11.54}$$

For a y-directed electric field $E_x \cos\phi$ should be replaced by $E_y \sin\phi$. Similarly

$$E_\phi = -\frac{jE_x}{\lambda r} e^{-jkr} \cos\theta \sin\phi \, 2\pi a^2 \frac{J_1(ka \sin\theta)}{ka \sin\theta} \tag{11.55}$$

and for a y-directed field replace $E_x \cos\theta \sin\phi$ by $-E_y \cos\theta \cos\phi$. Normally it is sufficient to consider radiation patterns in two perpendicular planes, the E-plane, corresponding in this case to the xz-plane, which is parallel to the electric field in the aperture; and the H-plane, which for the E_x-field is the yz-plane

11.6.1 *E-plane pattern: $\phi = 0°$ for E_x-field and $\phi = \pi/2$ for E_y-field*

The expressions for E_θ, E_ϕ simplify for points of observation in the E-plane to give

$$E_\theta = \frac{jE}{\lambda r} e^{-jkr} 2\pi a^2 \frac{J_1(ka \sin\theta)}{ka \sin\theta};$$

$$E_\phi = 0 \quad \text{for both} \quad E_x \quad \text{and} \quad E_y \quad \text{aperture fields.} \tag{11.56}$$

The half-power beamwidth for E_θ, corresponding to the angles at which the field has fallen to $1/\sqrt{2}$ of its maximum value, occur at $\sin\theta = 0.51\lambda/2a$, and the first sidelobes have an amplitude -17.5 dB below the main lobe. These sidelobes form complete circles round the main lobe, when viewed along the negative z-axis.

11.6.2 *H-plane pattern: $\phi = \pi/2$ for E_x-field and $\phi = \pi$ for E_y-field*

For points of observation in the H-plane the field expressions become

$$E_\theta = 0; \quad E_\phi = -\frac{jE}{\lambda r} e^{-jkr} \cos\theta 2\pi a^2 \frac{J_1(ka\sin\theta)}{ka\sin\theta} \tag{11.57}$$

The E_ϕ-pattern in this plane is thus simply $\cos\theta$ times the E_θ-pattern in the xz-plane. For large apertures, corresponding to large values of ka, the pattern is determined mainly by the $[J_1(ka\sin\theta)/ka\sin\theta]$ term, so that the main difference between the patterns lies in the outer sidelobes, as shown in Fig. 11.19 for a 5λ diameter aperture. Note that the patterns are known only for $\theta \leqslant \pi/2$.

11.7 Power gain of uniformly excited circular aperture

Power gain has previously been defined as

$$G = \frac{\text{maximum radiation intensity}}{\text{average radiation intensity}} = \frac{\Phi_{max}}{\Phi_{av}}$$

Fig. 11.19. Radiation pattern for 5λ diameter circular aperture.

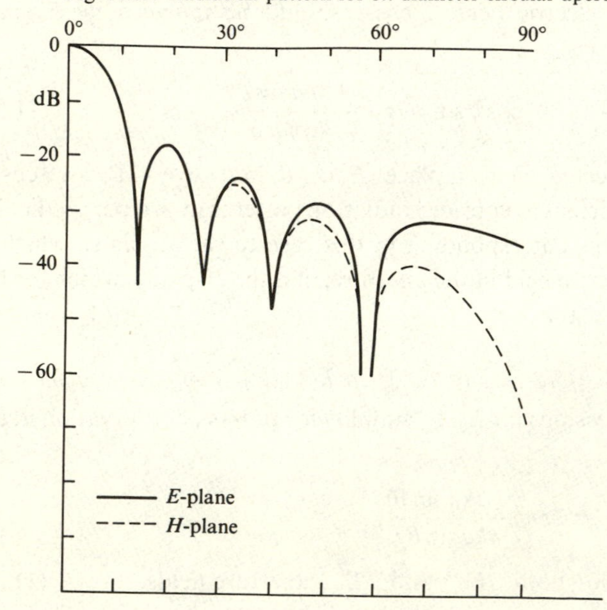

The maximum radiation intensity occurs in the forward direction along the z-axis where θ is zero, and either E_θ or E_ϕ may be used to calculate the maximum field, since they are antiparallel vectors in this direction. Then

$$\Phi_{max} = \frac{E_x^2}{\lambda^2 r^2} \frac{4\pi^2 a^4}{4 \cdot 2Z_0}$$

since

$$\underset{x \to 0}{J_1(x)} \approx \frac{x}{2}$$

The average radiation intensity is most simply approximated by assuming that the power density supplied to the aperture is that of a uniform plane wave, so that

$$\Phi_{av} = \frac{E_x^2 \pi a^2}{2Z_0 \cdot 4\pi r^2}$$

Hence the power gain of this uniformly excited circular aperture is

$$G = \frac{4\pi}{\lambda^2} \pi a^2 = \frac{4\pi}{\lambda^2} (\text{area of aperture}) \tag{11.58}$$

This result of the power gain being directly proportional to the physical area of the aperture has been found to be so useful that it is applied to all aperture antennas through a defining equation

$$G = \frac{4\pi}{\lambda^2} (\text{effective area of aperture}) = \frac{4\pi}{\lambda^2} A_{eff}$$

where in general $A_{eff} < A$. The ratio of the two areas is then defined as the aperture efficiency of the antenna

$$\eta = \frac{A_{eff}}{A} \tag{11.59}$$

11.8 Radiation fields of uniformly excited rectangular aperture

For a rectangular aperture of dimension a, b in the x, y-directions, centred on the origin and surrounded in the xy-plane by a perfectly conducting sheet, it has previously been shown that the radiation fields produced by a tangential electric field in the aperture are given by

$$E_\theta = \frac{j}{\lambda} \frac{e^{-jkr}}{r} \iint_{aperture} (E_x \cos\phi + E_y \sin\phi) \times e^{jk(x' \sin\theta \cos\phi + y' \sin\theta \sin\phi)} \, dx' \, dy' \tag{11.60}$$

$$E_\phi = -\frac{j}{\lambda} \frac{e^{-jkr}}{r} \cos\theta \iint_{aperture} (E_x \sin\phi - E_y \cos\phi) \times e^{jk(x' \sin\theta \cos\phi + y' \sin\theta \sin\phi)} \, dx' \, dy' \tag{11.61}$$

Hence when the excitation consists of a uniform field in the x-direction only,

$$E_\theta = \frac{jE_x e^{-jkr}}{\lambda r}\cos\phi \int_{-a/2}^{+a/2}\int_{-b/2}^{+b/2} e^{jk(x'\sin\theta\cos\phi + y'\sin\theta\sin\phi)}\,dx'\,dy'$$

$$= \frac{jE_x e^{-jkr}}{\lambda r}\cos\phi \left[\frac{e^{jkx'\sin\theta\cos\phi}}{jk\sin\theta\cos\phi}\right]_{-a/2}^{+a/2}$$

$$\times \left[\frac{e^{jky'\sin\theta\sin\phi}}{jk\sin\theta\sin\phi}\right]_{-b/2}^{+b/2}$$

$$= \frac{jE_x ab e^{-jkr}}{\lambda r}\cos\phi \frac{\sin\left(\dfrac{ka}{2}\sin\theta\cos\phi\right)}{\dfrac{ka}{2}\sin\theta\cos\phi}$$

$$\times \frac{\sin\left(\dfrac{kb}{2}\sin\theta\sin\phi\right)}{\dfrac{kb}{2}\sin\theta\sin\phi} \tag{11.62}$$

Likewise when the aperture field is y directed, $E_x\cos\phi$ should be replaced by $E_y\sin\phi$. Similarly the E_ϕ field for an x-directed uniform aperture excitation is

$$E_\phi = -\frac{jE_x ab e^{-jkr}}{\lambda r}\cos\theta\sin\phi \frac{\sin\left(\dfrac{ka}{2}\sin\theta\cos\phi\right)}{\dfrac{ka}{2}\sin\theta\cos\phi}$$

$$\times \frac{\sin\left(\dfrac{kb}{2}\sin\theta\sin\phi\right)}{\dfrac{kb}{2}\sin\theta\sin\phi} \tag{11.63}$$

and for a y-direction $E_x\cos\theta\sin\phi$ should be replaced by $-E_y\cos\theta\cos\phi$.

11.8.1 *E-plane pattern: $\phi = 0°$ for E_x-field and $\phi = \pi/2$ for E_y-field*

The expressions for E_θ, E_ϕ given above simplify for points of observation in the E-plane to give

$$\left.\begin{aligned} E_\theta &= \frac{jEab}{\lambda r} e^{-jkr} \frac{\sin\left(\dfrac{ka}{2}\sin\theta\right)}{\dfrac{ka}{2}\sin\theta} \\ E_\phi &= 0 \end{aligned}\right\} \tag{11.64}$$

for both E_x- and E_y-aperture fields. It will be noted that the shape of the E_θ pattern depends exclusively on the electrical width of the aperture in the plane of the pattern, although the strength of its field is proportional to the area of the aperture. The half-power or 3dB beamwidth occurs when $\sin\theta = 0.44\lambda/a$, so that the angle between the half-power points is calculated from twice this ratio. If this angle is small so that $\sin\theta \approx \theta$ this means that the angle between half-power points is $0.88\lambda/a$, i.e. the beamwidth is inversely proportional to the width of the aperture in the xz-plane. The first sidelobe level is -13.2dB down on the main lobe which occurs along the z-axis in the $\theta = 0°$ direction.

11.8.2 *H-plane pattern: $\phi = \pi/2$ for E_x-field and $\phi = \pi$ for E_y-field*

The general expressions for E_θ, E_ϕ simplify in this case to

$$\left.\begin{aligned} E_\theta &= 0 \\ E_\phi &= -\frac{jEabe^{-jkr}}{\lambda r}\cos\theta\,\frac{\sin\left(\dfrac{kb}{2}\sin\theta\right)}{\dfrac{kb}{2}\sin\theta} \end{aligned}\right\} \quad (11.65)$$

for both E_x- and E_y-aperture fields. It will be noted that the pattern width depends on the width of the aperture in the plane of the pattern, and for large apertures this depends mainly on its electrical width kb, except for the outer sidelobes where the $\cos\theta$ factor becomes important. For such large apertures the beam width between half-power points is approximately $0.88\lambda/b$ radians or $50.4\lambda/b$ degrees.

11.9 Power gain of uniformly excited rectangular aperture

Maximum field occurs in the broadside direction along the z-axis, and the value of this field, from eqn (11.64), is

$$|E_{\theta max}| = \frac{Eab}{\lambda r}$$

Hence the maximum radiation intensity is

$$\Phi_{max} = \frac{E^2a^2b^2}{\lambda^2 r^2 \cdot 2Z_0}$$

Assuming as for the circular aperture that the total power delivered to it is equal to the power density of a normally incident uniform plane wave multiplied by the area of the aperture gives, for the average radiation intensity,

$$\Phi_{av} = \frac{E^2ab}{2Z_0 \cdot 4\pi r^2}$$

Hence the power gain is

$$G = \frac{4\pi ab}{\lambda^2} = \frac{4\pi}{\lambda^2}(\text{area of aperture}) \tag{11.66}$$

giving an aperture efficiency of 100% for the uniformly illuminated case, as before.

11.10 Measured radiation patterns

For a given tangential electric field distribution in the aperture of an antenna eqns (11.60) and (11.61) enable the orthogonal components $E_\theta(\theta, \phi)$ and $E_\phi(\theta, \phi)$ to be calculated provided the integrations can be carried out. If, however, these distributions are not known with sufficient accuracy, or resort is made to measurement, the problem is faced that a completely three-dimensional measurement system is generally not available. While a complete revolution of an antenna in the horizontal plane in azimuth is possible the presence of the ground prevents a corresponding 360° rotation is elevation. This problem may be overcome as follows.

Consider first a planar aperture antenna with a vertical tangential electric field E_x which transmits a signal to a receiving vertically polarised antenna at the same height above the ground. This height is sufficient to allow ground reflection effects to be ignored. Then as the planar antenna is rotated in azimuth the signal picked up by the receiving antenna represents the H-plane pattern of the transmitting antenna. Now let the transmitting antenna be rotated through 90° so that its electric field E_y is horizontal, and let the receiving antenna be similarly rotated by 90°. As the transmitting antenna is rotated in azimuth again the E-plane pattern is obtained. Hence for these two principal planes the absence of a rotating facility in elevation has been overcome.

Next consider the same planar aperture antenna rotated by an intermediate angle so that its tangential electric field in the aperture makes an angle ϕ with the vertical. This gives a vertical component $E \cos \phi$ and a horizontal component $E \sin \phi$, so that the receiving antenna will have to be rotated by ϕ also to pick up maximum signal. The signal picked up is therefore the sum of the signals produced by each of these components, and the pattern traced out is referred to as the copolar plot. Its magnitude in spherical components is

$$E_c = E_\theta(\theta, \phi) \cos \phi - E_\phi(\theta, \phi) \sin \phi$$

when the transmitted field is polarised in the x-direction and

$$E_c = E_\theta(\theta, \phi) \sin \phi + E_\phi(\theta, \phi) \cos \theta$$

when it is polarised in the y-direction.

Likewise when the receiving antenna is rotated by $\pi/2$ so that zero signal is received in the forward direction $\theta = 0°$, the resultant cross-polar signal picked up is

$$E_x = E_\theta(\theta, \phi) \sin \phi + E_\phi(\theta, \phi) \cos \phi$$

for a x-directed transmitting field and

$$E_x = E_\theta(\theta, \phi) \cos \phi - E_\phi(\theta, \phi) \sin \phi$$

when it is polarised in the y-direction.

The importance attached to reducing cross-polarised fields arises because if two orthogonally polarised sources, aligned towards each other, launch only copolarised fields, then their fields will be orthogonal everywhere in space and can thus allow doubling of the frequency channel use. In the case of patterns which can be described by the equation

$$R = f_1(\theta) \cos \phi \bar{a}_\theta - f_2(\theta) \sin \phi \bar{a}_\phi$$

where $f_1(\theta)$, $f_2(\theta)$ are the E, H-plane patterns respectively, the equation for E can be rewritten as

$$E = (f_1(\theta) \cos^2 \phi + f_2(\theta) \sin^2 \phi)(\bar{a}_\theta \cos \phi - \bar{a}_\phi \sin \phi) + \frac{\sin^2 \phi}{2}(f_1(\theta) - f_2(\theta))(\bar{a}_\theta \sin \phi + \bar{a}_\phi \cos \phi)$$

It is thus seen that where the E- and H-plane patterns are identical there is zero cross-polarisation, and that in general the cross-polarisation pattern is maximum in the $\phi = \pm 45°$ planes.

Further reading

H.G. Booker and P.C. Clemmow: 'The concept of an angular spectrum of plane waves and its relation to that of polar diagram and aperture distribution': *JIEE*, **97**, 1950, 1–11.

J. Brown: 'A theoretical analysis of some errors in aerial measurements': *Proc. IEE,* Part C, Sept. 1958, pp. 343–51.

P.K. Chum: Aperture radiation: PhD thesis, University of Birmingham, 1977.

P.C. Clemmow: *The Plane Wave Spectrum Representation of Electromagnetic Fields*: Pergamon Press, 1966.

R.E. Collin and F.J. Zucker: *Antenna Theory*: McGraw-Hill, 1969.

D.R. Rhodes: *Synthesis of Planar Antenna Sources*: Clarendon Press, 1974.

12

+−+

Waveguide radiators

The radiation considered in the preceding chapter was from two-dimensional apertures, which took no account of how the tangential fields in the aperture were formed, or how, having been formed, the effect of the aperture boundaries enabled the fields under consideration to be considered terminated at the aperture. In this chapter this matter will be considered further.

It will be recalled that the radiation outside a closed surface can always be found provided the tangential components of both electric and magnetic fields over the surface are known. Consequently in Fig. 12.1, representing a waveguide with an aperture at one end, a knowledge of both tangential **E** and **H** over the aperture, together with a knowledge of tangential **H** over the outside walls of the guide, would be sufficient to determine such fields. The

Fig. 12.1. Cylindrical waveguide radiator.

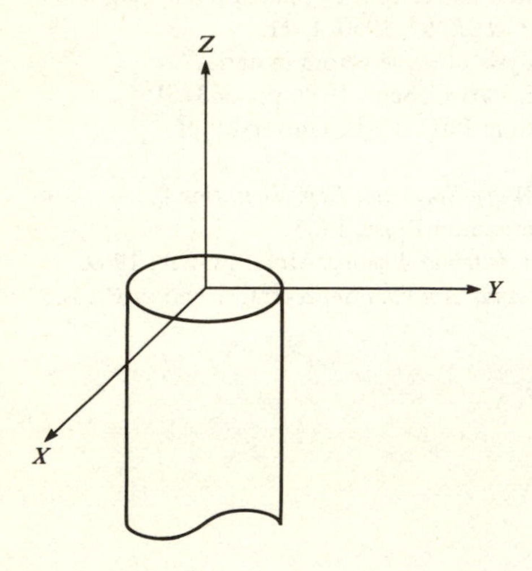

major difficulty in using this technique is that the current on the outside walls of the waveguide is in general not known and it will depend to some extent on the dimensions of the aperture. Consequently the approximation is usually made in practice that the total radiation takes place from the aperture alone.

Alternatively a closed surface covering both the inside and outside walls of the guide can be used. The currents on the outside surface are again not known, but the currents on the inside are, to an adequate degree of approximation. Note that the result for a finite length of guide using this approach depends on the length of the guide, though the analytical result is based on a semi-infinite guide. Numerical results for the two methods differ slightly, but are substantially the same over the main beam of radiation.

As a third method, since the radiation from the current on the outside walls is being ignored, the aperture plane may be extended to $\pm\infty$ by means of a perfectly conducting surface, and only the tangential electric field in the aperture considered. Again this will produce slightly different results from the other methods, but if, for a circular aperture, the radius is greater than one wavelength, the patterns will agree very closely with those of the first method for the whole of forward half-space. In backward half-space there is strictly no radiation at all produced by this technique, since the postulated perfect conductor outside the aperture implies that no field can penetrate this plane.

Although the radiation from the currents on the outside walls of the waveguide is being ignored, it is of interest to draw attention to the fact that there are two separate causes for the existence of such currents. Firstly, backward radiation from the aperture and, secondly, currents flowing on the inside walls of the guide turn back along the outside walls when they reach the aperture end. Although no analytic solution for this is available, analogy with the case of the half-plane can be drawn. Here two separate cases must be distinguished, one in which current is flowing parallel to the edge and the other in which it is perpendicular to it. For the parallel flow case the current falls to a small percentage of its value at the edge within a distance of one wavelength from the edge. But for the case of current flow perpendicular to the edge the current on the outside walls is likely to flow for several wavelengths before its magnitude becomes negligible.

12.1 Radiation pattern of rectangular waveguide carrying TE_{01} mode

The rectangular waveguide is aligned along the z-axis with its aperture positioned symmetrically about the origin in the xy-plane, as shown in Fig. 12.2. For an x-directed electric field in the aperture, the tangential field components required for finding the approximate radiation pattern are

$$\left.\begin{aligned} \mathbf{E} &= \mathbf{a}_x A \cos\frac{\pi y}{b} \\ \mathbf{H} &= \mathbf{a}_y \frac{\beta}{\omega\mu} A \cos\frac{\pi y}{b} \end{aligned}\right\} \begin{aligned} -\frac{a}{2} \leqslant x \leqslant \frac{a}{2} \\ -\frac{b}{2} \leqslant y \leqslant \frac{b}{2} \end{aligned} \qquad (12.1)$$

From eqn (10.41) with r_0 being replaced by r, the θ- and ϕ-components of radiated field are given by, when E_y, H_x are zero.

$$\left.\begin{aligned} E_\theta &= \frac{j}{2\lambda}\frac{e^{-jkr}}{r}\iint (E_x \cos\phi + Z_0 H_y \cos\theta\cos\phi) \\ &\quad \times e^{jk(x'\sin\theta\cos\phi + y'\sin\theta\sin\phi)}\,dx'\,dy' \\ E_\phi &= \frac{-j}{2\lambda}\frac{e^{-jkr}}{r}\iint (E_x \cos\theta\sin\phi + Z_0 H_y \sin\phi) \\ &\quad \times e^{jk(x'\sin\theta\cos\phi + y'\sin\theta\sin\phi)}\,dx'\,dy' \end{aligned}\right\} \qquad (12.2)$$

Substituting for the tangential components of fields and using primed notation to indicate the variables of integration gives

$$\left.\begin{aligned} E_\theta &= \frac{j}{2\lambda}\frac{e^{-jkr}}{r} A \int_{-a/2}^{a/2}\int_{-b/2}^{b/2} \cos\phi\cos\frac{\pi y'}{b}\left(1 + \frac{\beta}{k}\cos\theta\right) \\ &\quad \times e^{jk(x'\sin\theta\cos\phi + y'\sin\theta\sin\phi)}\,dx'\,dy' \\ E_\phi &= \frac{-j}{2\lambda}\frac{e^{-jkr}}{r} A \int_{-a/2}^{a/2}\int_{-b/2}^{b/2} \sin\phi\cos\frac{y'}{b}\left(\cos\theta + \frac{\beta}{k}\right) \\ &\quad \times e^{jk(x'\sin\theta\cos\phi + y'\sin\theta\sin\phi)}\,dx'\,dy' \end{aligned}\right\} \qquad (12.3)$$

Fig. 12.2. Rectangular waveguide radiator.

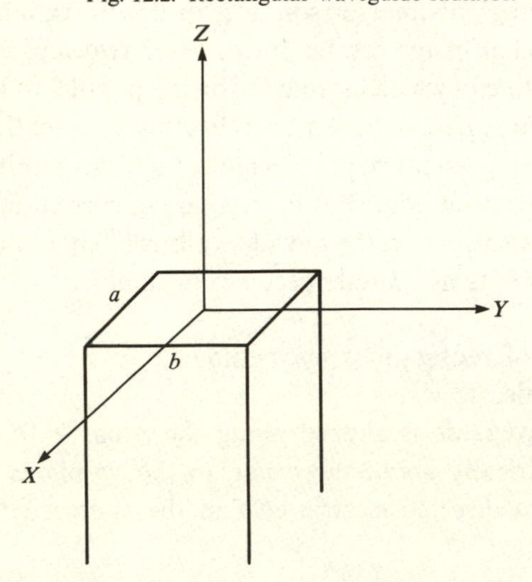

Apart from different constants the two integrands are identical. Denoting them by I gives

$$I = \int_{-a/2}^{a/2} \int_{-b/2}^{b/2} \cos \frac{\pi y'}{b} e^{j(kx' \sin\theta\cos\phi + y' \sin\theta\sin\phi)} \, dx' \, dy'$$

$$= \frac{a \sin\left(\frac{ka}{2} \sin\theta\cos\phi\right)}{\frac{ka}{2} \sin\theta\cos\phi} \frac{1}{2} \int_{-b/2}^{b/2} [e^{j(\pi y'/b + ky' \sin\theta\sin\phi)}$$

$$+ e^{-j(\pi y'/b - ky' \sin\theta\sin\phi)}] \, dy'$$

$$= \frac{a \sin\left(\frac{ka}{2} \sin\theta\cos\phi\right)}{\frac{ka}{2} \sin\theta\cos\phi} \cdot \frac{2\pi}{b} \cdot \frac{\cos\left(\frac{kb}{2} \sin\theta\sin\phi\right)}{\left(\frac{\pi^2}{b^2} - k^2 \sin^2\theta\sin^2\phi\right)} \tag{12.4}$$

The E-plane pattern corresponding to $\phi = 0°$ thus gives

$$E_\theta(\phi = 0°) = \frac{jab}{\pi\lambda} \frac{e^{-jkr}}{r} A\left(1 + \frac{\beta}{k} \cos\theta\right) \frac{\sin\left(\frac{ka}{2} \sin\theta\right)}{\frac{ka}{2} \sin\theta} \tag{12.5}$$

and the H-plane pattern corresponding to $\phi = \pi/2$ gives

$$E_\phi(\phi = 90°) = -\frac{j\pi ab e^{-jkr}}{\lambda r} A\left(\cos\theta + \frac{\beta}{k}\right) \frac{\cos\left(\frac{kb}{2} \sin\theta\right)}{(\pi^2 - k^2 b^2 \sin^2\theta)} \tag{12.6}$$

The E-plane pattern has its first nulls at angle θ given by

$$\left.\begin{aligned} &\frac{ka}{2} \sin\theta = \pi \\ \text{or} \quad & \\ &\theta = \sin^{-1} \frac{\lambda}{a} \end{aligned}\right\} \tag{12.7}$$

Likewise the H-plane pattern has its first zeros at

$$\theta = \sin^{-1} \frac{3\lambda}{2b} \tag{12.8}$$

corresponding to $(kb/2) \sin\theta$ having a value of $3\pi/2$, rather than $\pi/2$ when the denominator also becomes zero. The first sidelobes are at a level of -13.2 dB referred to the main lobe in the E-plane pattern, and -23 dB in the H-plane pattern, because of the amplitude taper in the aperture in the y-direction in that plane.

To find the power gain of this rectangular waveguide it is assumed that all the power incident on the aperture is radiated, this power being given by

$$W = \tfrac{1}{2}\,\mathrm{Re}\iint E_x H_y^*\,dx'\,dy'$$

$$= \frac{1}{2}\frac{\beta}{\omega\mu}A^2 \int_{-a/2}^{a/2}\int_{-b/2}^{b/2} \cos^2\frac{\pi y}{b}\,dx'\,dy' \tag{12.9}$$

$$= \frac{\beta ab A^2}{4\omega\mu} \tag{12.10}$$

The average power density is thus

$$\Phi_{av} = \frac{\beta ab A^2}{16\omega\mu\pi r^2} \tag{12.11}$$

Since the maximum radiated field occurs in the forward direction along the positive z-axis and is given by $E_\theta(\theta = 0°) = -E_\phi(\theta = 0°)$, the maximum power density is

$$\Phi_{max} = \frac{|E_\theta(\theta = 0°)|^2}{2Z_0}$$

$$= \left(1 + \frac{\beta}{k}\right)^2 \frac{a^2 b^2 A^2}{\pi^2 \lambda^2 \gamma^2}\frac{1}{2Z_0} \tag{12.12}$$

Hence the power gain is

$$G = \frac{\left(1 + \dfrac{\beta}{k}\right)^2 ab\omega\mu}{15\beta\pi^2\lambda^2} \tag{12.13}$$

which for a dominant mode waveguide is typically 4–5 dB.

12.2 *H*-plane sectoral horn

In order to make the radiation from a rectangular waveguide more directional it is necessary to increase its radiating aperture. This may be achieved by flaring the waveguide out in the H-plane, i.e. broadening the y-dimensions of the waveguide to form an H-plane sectoral horn. Alternatively it may be flared out orthogonally to form a E-plane sectoral horn, or in both orthogonal planes to form a pyramidal horn. An analysis of the H-plane horn will be carried out and some results for the other two cases will be stated for comparison.

Consider the H-plane horn shown in Fig. 12.3. The tapered sides of the horn are suggestive of a linearly polarised cylindrical wave spreading out from a fictitious radiating line source at the intersection of the two tapered sides. Such a cylindrical wave will produce a phase variation along the aperture of the horn with the phase at the centre leading the phase at a point y from the centre and $\sqrt{(\rho^2 + y^2)}$ from the line source, by

$$\text{phase lead at centre} = k[\sqrt{(\rho^2 + y^2)} - \rho] \tag{12.14}$$

Taking a phase reference at this centre point gives a phase lag elsewhere, so that the aperture field is given by

$$E_x = A \cos \frac{\pi y}{b} e^{-jk[\sqrt{(\rho^2 + y^2)} - \rho]} \tag{12.15}$$

When the length ρ of the horn centre line is $\gg y$, the exponent of the exponential can be approximated by the use of the binomial theorem to give an aperture electric field.

$$E_x \simeq A \cos \frac{\pi y}{b} e^{-j(ky^2/2\rho)} \tag{12.16}$$

It will be observed from the analysis of the radiation from the rectangular waveguide that the radiation from the tangential magnetic field in the aperture can be obtained from that of the electric field by multiplying its E_θ-pattern by $[1 + (\beta/k)\cos\theta]$ and its E_ϕ-pattern by $[\cos\theta + (\beta/k)]$. Consequently we shall apply this result to the horn and calculate initially only the

Fig. 12.3. H-plane horn radiator.

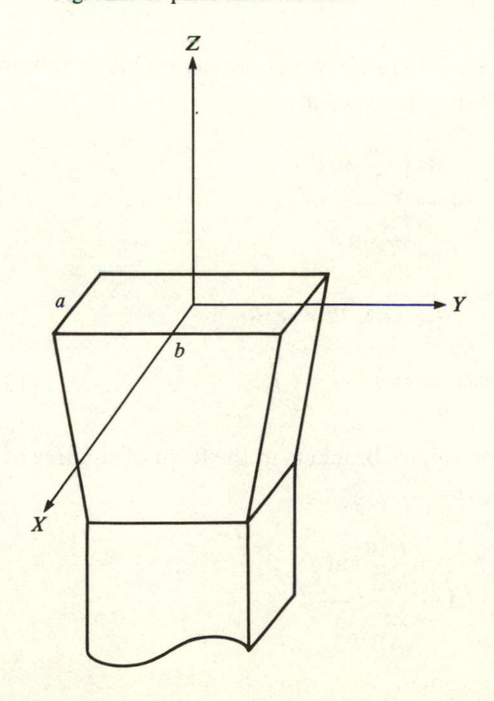

radiation from the electric field in the aperture. This is given by

$$\left.\begin{aligned} E_\theta &= \frac{j}{2\lambda}\frac{e^{-jkr}}{r}A\int_{-a/2}^{a/2}\int_{-b/2}^{b/2}\cos\phi\cos\frac{\pi y'}{b}e^{-j(k(y')^2/2\rho)} \\ &\quad\times e^{jk(x'\sin\theta\cos\phi+y'\sin\theta\sin\phi)}\,dx'\,dy' \\ \text{and} \\ E_\phi &= \frac{-j}{2\lambda}\frac{e^{-jkr}}{r}A\int_{-a/2}^{a/2}\int_{-b/2}^{b/2}\sin\phi\cos\frac{\pi y'}{b}e^{-j(k(y')^2/2\rho)} \\ &\quad\times e^{jk(x'\sin\theta\cos\phi+y'\sin\theta\sin\phi)}\,dx'\,dy' \end{aligned}\right\} \tag{12.17}$$

In the E-plane corresponding to $\phi = 0°$ this gives

$$E_\theta(\phi=0°) = \frac{ja}{2\lambda}\frac{e^{-jkr}}{r}A\frac{\sin\left(\frac{ka}{2}\sin\theta\right)}{\frac{ka}{2}\sin\theta} \times\frac{1}{2}\int_{-b/2}^{b/2}\{e^{j[\pi y'/b-k(y')^2/2\rho]}+e^{-j[\pi y'/b+k(y')^2/2\rho]}\}\,dy'$$

It is desired to express this integrand in the form of the standard integral

$$\int_0^z e^{-j(\pi/2)t^2}\,dt = C(z) - jS(z)$$

where $C(z)$ and $S(z)$ are Fresnel integrals which are available in tabulated form. To do this $E_\theta(\phi = 0°)$ may be written

$$E_\theta(\phi=0°) = \frac{ja}{2\lambda}\frac{e^{-jkr}}{r}A\frac{\sin\left(\frac{ka}{2}\sin\theta\right)}{\frac{ka}{2}\sin\theta} \times\frac{1}{2}\int_{-b/2}^{b/2}\{e^{-j(\pi/2)[2(y')^2/\lambda\rho-2y'/b]} + e^{-j(\pi/2)[2(y')^2/\lambda\rho+2y'/b]}\}\,dy' \tag{12.18}$$

Now express the terms in the square brackets in the form of squares of the variable of integration to give

$$E_\theta(\phi=0°) = \frac{ja}{2\lambda}\frac{e^{-jkr}}{r}A\frac{\sin\left(\frac{ka}{2}\sin\theta\right)}{\frac{ka}{2}\sin\theta}$$

$$\times \frac{1}{2}\int_{-b/2}^{b/2} \{e^{-j(\pi/2)[(\sqrt{2}y'/\sqrt{(\lambda\rho)} - \sqrt{(\lambda\rho)}/\sqrt{2}b)^2 - \lambda\rho/2b^2]}$$

$$+ e^{-j(\pi/2)[(\sqrt{2}y'/\sqrt{(\lambda\rho)} + \sqrt{(\lambda\rho)}/\sqrt{2}b)^2 - \lambda\rho/2b^2]}\}\, dy'$$

Writing $t_{\frac{1}{2}} = [\sqrt{2}y'/(\lambda\rho)] \mp [\sqrt{(\lambda\rho)}/\sqrt{2}b]$, so that $dt_{\frac{1}{2}} = \sqrt{(2/\lambda\rho)}dy'$, gives

$$E_\theta(\phi = 0°) = \frac{jAa}{2\lambda}\frac{e^{-jkr}}{r} e^{j(\pi\lambda\rho/4b^2)} \frac{\sin\left(\frac{ka}{2}\sin\theta\right)}{\frac{ka}{2}\sin\theta} \frac{1}{2}\frac{\sqrt{(\lambda\rho)}}{\sqrt{2}}$$

$$\times \left\{ \int e^{-j(\pi/2)t_1^2}\, dt_1 + \int e^{-j(\pi/2)t_2^2}\, dt_2 \right\} \tag{12.19}$$

The limits on t_1 are $\{\mp[b/\sqrt{(2\lambda\rho)}] - [\sqrt{(\lambda\rho)}/\sqrt{2}b]\}$ and on t_2, $\{\mp[b/\sqrt{(2\lambda\rho)}] + [\sqrt{(\lambda\rho)}/\sqrt{2}b]$, but since the Fresnel integrals must have lower limits of zero, the ranges of integration for the first integral may be subdivided into $\{-[b/\sqrt{(2\lambda\rho)}] - [\sqrt{(\lambda\rho)}/\sqrt{2}b]\}$ to 0 and 0 to $-\sqrt{(\lambda\rho)}/\sqrt{2}b$ followed by $-\sqrt{(\lambda\rho)}/\sqrt{2}b$ to 0 and 0 to $\{+[b/\sqrt{(2\lambda\rho)}] - [\sqrt{(\lambda\rho)}/\sqrt{2}b]\}$. The middle integrals cancel leaving for the resultant E_θ,

$$E_\theta(\phi = 0°) = \frac{jAa}{2\lambda}\frac{e^{-jkr}}{r} e^{j(\pi/4)(\lambda\rho/b^2)} \frac{\sin\left(\frac{ka}{2}\sin\theta\right)}{\frac{ka}{2}\sin\theta} \frac{1}{2}\frac{\sqrt{(\lambda\rho)}}{\sqrt{2}}$$

$$\times \left\{ C\left[\frac{b}{\sqrt{(2\lambda\rho)}} - \frac{\sqrt{(\lambda\rho)}}{\sqrt{2}b}\right] - jS\left[\frac{b}{\sqrt{(2\lambda\rho)}} - \frac{\sqrt{(\lambda\rho)}}{\sqrt{2}b}\right] \right.$$

$$- C\left[-\frac{b}{\sqrt{(2\lambda\rho)}} - \frac{\sqrt{(\lambda\rho)}}{\sqrt{2}b}\right] + jS\left[-\frac{b}{\sqrt{(2\lambda\rho)}} - \frac{\sqrt{(\lambda\rho)}}{\sqrt{2}b}\right]$$

$$+ C\left[\frac{b}{\sqrt{(2\lambda\rho)}} + \frac{\sqrt{(\lambda\rho)}}{\sqrt{2}b}\right] - jS\left[\frac{b}{\sqrt{(2\lambda\rho)}} + \frac{\sqrt{(\lambda\rho)}}{\sqrt{2}b}\right]$$

$$\left. - C\left[-\frac{b}{\sqrt{(2\lambda\rho)}} + \frac{\sqrt{(\lambda\rho)}}{\sqrt{2}b}\right] + jS\left[-\frac{b}{\sqrt{(2\lambda\rho)}} + \frac{\sqrt{(\lambda\rho)}}{\sqrt{2}b}\right] \right\}$$

but since $C(-z) = -C(z)$ and $S(-z) = -S(z)$ this gives

$$E_\theta(\phi = 0°) = \frac{jAa}{2\lambda}\frac{e^{-jkr}}{r} e^{j(\pi/4)(\lambda\rho/b^2)} \frac{\sin\left(\frac{ka}{2}\sin\theta\right)}{\frac{ka}{2}\sin\theta} \sqrt{\left(\frac{\lambda\rho}{2}\right)}$$

$$\times \{C(u) - C(v) - jS(u) + jS(v)\} \tag{12.20}$$

where

$$u=\left(\frac{b}{\sqrt{(2\lambda\rho)}}+\frac{\sqrt{(\lambda\rho)}}{\sqrt{2b}}\right),\quad v=\left(\frac{\sqrt{(\lambda\rho)}}{\sqrt{2b}}-\frac{b}{\sqrt{(2\lambda\rho)}}\right) \tag{12.21}$$

Superimposing the component of radiation field in the same direction due to the tangential magnetic field in the aperture, gives a total E_θ in the E-plane of

$$E_\theta(\phi=0^\circ)=\frac{jAa}{2\lambda}\frac{e^{-jkr}}{r}e^{j(\pi/4)(\lambda\rho/b^2)}\left(1+\frac{\beta}{k}\cos\theta\right)$$
$$\times\frac{\sin\left(\frac{ka}{2}\sin\theta\right)}{\frac{ka}{2}\sin\theta}\sqrt{\left(\frac{\lambda\rho}{2}\right)}\{C(u)-C(v)-jS(u)+jS(v)\} \tag{12.22}$$

Likewise for the H-plane pattern for $\phi=90^\circ$

$$E_\phi(\phi=90^\circ)=-\frac{jA}{2\lambda}\frac{e^{-jkr}}{r}\left(\cos\theta+\frac{\beta}{k}\right)$$
$$\times\int_{-a/2}^{a/2}\int_{-b/2}^{b/2}\cos\frac{\pi y'}{b}e^{-j(k(y')^2/2\rho)}e^{jky'\sin\theta}\,dx'\,dy'$$
$$=-\frac{jAa}{4\lambda}\frac{e^{-jkr}}{r}\left(\cos\theta+\frac{\beta}{k}\right)\int_{-b/2}^{b/2}e^{j[\pi y'/b-k(y')^2/2\rho+ky'\sin\theta]}$$
$$+e^{-j[\pi y'/b+k(y')^2/2\rho-ky'\sin\theta]}\cdot dy'$$

This may be expressed in the form of Fresnel integrals and the result may be shown to be

$$E_\phi(\phi=90^\circ)=-j\frac{Aa}{4\lambda}\frac{e^{-jkr}}{r}\left(\cos\theta+\frac{\beta}{k}\right)\sqrt{\left(\frac{\lambda\rho}{2}\right)}$$
$$\times\{e^{j(\pi/4)\lambda\rho(1/b+2\sin\theta/\lambda)^2}[C(u_1)-C(v_1)-jS(u_1)$$
$$+jS(v_1)]+e^{j(\pi/4)\lambda\rho(1/b-2\sin\theta/\lambda)^2}$$
$$\times[C(u_2)-C(v_2)-jS(u_2)+jS(v_2)]\} \tag{12.23}$$

where

$$\left.\begin{aligned}
u_1&=\left[\frac{b}{\sqrt{(2\lambda\rho)}}+\sqrt{\left(\frac{\lambda\rho}{2}\right)}\left(\frac{1}{b}+\frac{2\sin\theta}{\lambda}\right)\right];\\
v_1&=\left[\sqrt{\left(\frac{\lambda\rho}{2}\right)}\left(\frac{1}{b}+\frac{2\sin\theta}{\lambda}\right)-\frac{b}{\sqrt{(2\lambda\rho)}}\right]\\
u_2&=\left[\frac{b}{\sqrt{(2\lambda\rho)}}+\sqrt{\left(\frac{\lambda\rho}{2}\right)}\left(\frac{1}{b}-\frac{2\sin\theta}{\lambda}\right)\right];\\
v_2&=\left[\sqrt{\left(\frac{\lambda\rho}{2}\right)}\left(\frac{1}{b}-\frac{2\sin\theta}{\lambda}\right)-\frac{b}{\sqrt{(2\lambda\rho)}}\right]
\end{aligned}\right\} \tag{12.24}$$

In the forward direction $u_1 = u_2$ and $v_1 = v_2$, and the expression for $E_\phi(\phi = 90°, \theta = 0°)$ reduces to the negative of $E_\theta(\phi = 0°, \theta = 0°)$ as it must do.

To find the power gain of this H-plane sectoral horn the power delivered to the horn will be taken equal to the power delivered to the waveguide feeding it, and this has been earlier shown to be given by

$$W = \frac{\beta ab A^2}{4\omega\mu}$$

Hence the average power density of the horn is

$$\Phi_{av} = \frac{\beta ab A^2}{16\omega\mu\pi r^2} \tag{12.25}$$

The maximum radiated field occurs in the forward direction and corresponds to $E_\theta(\phi = 0°, \theta = 0°)$ or $E_\phi(\phi = 90°, \theta = 0°)$. In this direction

$$\Phi_{max} = \frac{A^2 a^2 \lambda\rho\left(1 + \frac{\beta}{k}\right)^2 \{[C(u) - C(v)]^2 + [S(u) - S(v)]^2\}}{16\lambda^2 r^2 Z_0} \tag{12.26}$$

The power gain is thus

$$G_H = \frac{\pi a\rho\left(1 + \frac{\beta}{k}\right)^2 \omega\mu}{\lambda b\beta Z_0}\{[C(u) - C(v)]^2 + [S(u) - S(v)]^2\} \tag{12.27}$$

As the aperture width b increases for a given horn length ρ the gain increases to a maximum, and then decreases because of the antiphase contribution starting from the edges.

12.3 Power gain of E-plane sectoral horn

For a tangential electric field in the aperture given by

$$E_x = A\cos\frac{\pi y}{b}e^{-j(kx^2/2\rho)}$$

which represents a sectoral horn flared in the orthogonal E-direction, it may similarly be shown that the power gain is

$$G_E = \frac{16\rho b\left(1 + \frac{\beta}{k}\right)^2}{\pi\lambda a}\left[C^2\left(\frac{a}{\sqrt{(2\lambda\rho)}}\right) + S^2\left(\frac{a}{\sqrt{(2\lambda\rho)}}\right)\right] \tag{12.28}$$

12.4 Power gain of pyramidal horn

When the horn is flared out in both its H- and E-planes giving a tangential electric field in the aperture

$$E_x = A\cos\frac{\pi y}{b}e^{-jk(x^2/2\rho_1 + y^2/2\rho_2)}$$

the gain may be shown to be

$$G_p = \frac{\pi}{32}\left(\frac{\lambda}{a}G_H\right)\left(\frac{\lambda}{b}G_E\right) \tag{12.29}$$

12.5 Cylindrical waveguide: derivation of wave equation

Starting from Maxwell's equations in cylindrical coordinates,

$$\left.\begin{aligned} j\omega\varepsilon E_\rho &= \frac{1}{\rho}\frac{\partial H_z}{\partial \phi} - \frac{\partial H_\phi}{\partial z} && \text{(a)} \\ j\omega\varepsilon E_\phi &= \frac{\partial H_\rho}{\partial z} - \frac{\partial H_z}{\partial \rho} && \text{(b)} \\ j\omega\varepsilon E_z &= \frac{1}{\rho}\frac{\partial(\rho H_\phi)}{\partial \rho} - \frac{1}{\rho}\frac{\partial H_\rho}{\partial \phi} && \text{(c)} \end{aligned}\right\} \tag{12.30}$$

$$\left.\begin{aligned} -j\omega\mu H_\rho &= \frac{1}{\rho}\frac{\partial E_z}{\partial \phi} - \frac{\partial E_\phi}{\partial z} && \text{(a)} \\ -j\omega\mu H_\phi &= \frac{\partial E_\rho}{\partial z} - \frac{\partial E_z}{\partial \rho} && \text{(b)} \\ -j\omega\mu H_z &= \frac{1}{\rho}\frac{\partial(\rho E_\phi)}{\partial \rho} - \frac{1}{\rho}\frac{\partial E_\rho}{\partial \phi} && \text{(c)} \end{aligned}\right\} \tag{12.31}$$

and assuming a propagating wave $e^{-\gamma z}$, so that $\partial/\partial z = -\gamma$, gives, for equations (12.30(b)) and (12.31(a)),

$$-\gamma H_\rho = j\omega\varepsilon E_\phi + \frac{\partial H_z}{\partial \rho}$$

$$-j\omega\mu H_\rho = \frac{1}{\rho}\frac{\partial E_z}{\partial \phi} + \gamma E_\phi$$

Eliminating H_ρ from these equations gives E_ϕ in terms of the longitudinal fields E_z and H_z,

$$E_\phi = \frac{1}{(\gamma^2 + k^2)}\left[j\omega\mu\frac{\partial H_z}{\partial \rho} - \frac{\gamma}{\rho}\frac{\partial E_z}{\partial \phi}\right]$$

Similarly, from equations (12.30(a)) and (12.31(b)),

$$\left.\begin{aligned} j\omega\varepsilon E_\rho &= \frac{1}{\rho}\frac{\partial H_z}{\partial \phi} + \gamma H_\phi \\ -j\omega\mu H_\phi &= -\gamma E_\rho - \frac{\partial E_z}{\partial \rho} \end{aligned}\right\} \tag{12.32}$$

Eliminating E_ρ from these equations gives H_ϕ in terms of E_z and H_z,

$$H_\phi = -\frac{1}{(\gamma^2 + k^2)}\left[j\omega\varepsilon\frac{\partial E_z}{\partial \rho} + \frac{\gamma}{\rho}\frac{\partial H_z}{\partial \phi}\right]$$

Then from eqns (12.30(b)) and (12.31(a)) H_ρ can be expressed in terms of the same parameters

$$H_\rho = \frac{1}{(\gamma^2 + k^2)}\left[\frac{j\omega\varepsilon}{\rho}\frac{\partial E_z}{\partial \phi} - \gamma\frac{\partial H_z}{\partial \rho}\right]$$

and from eqns (12.30(a)) and (12.31(b)) E_ρ likewise, giving

$$E_\rho = -\frac{1}{(\gamma^2 + k^2)}\left[\frac{j\omega\mu}{\rho}\frac{\partial H_z}{\partial \phi} + \gamma\frac{\partial E_z}{\partial \rho}\right]$$

Now substitute for E_ϕ and E_ρ in eqn (12.31(c)) to obtain an equation in H_z, thus

$$\begin{aligned}
-j\omega\mu\rho H_z &= E_\phi + \rho\frac{\partial E_\phi}{\partial \rho} - \frac{\partial E_\rho}{\partial \phi} \\
&= \frac{1}{(\gamma^2 + k^2)}\left[j\omega\mu\frac{\partial H_z}{\partial \rho} - \frac{\gamma}{\rho}\frac{\partial E_z}{\partial \phi} + j\omega\mu\rho\frac{\partial^2 H_z}{\partial \rho^2}\right. \\
&\quad \left. - \gamma\frac{\partial^2 E_z}{\partial \rho \partial \phi} + \frac{\gamma}{\rho}\frac{\partial E_z}{\partial \phi} + \frac{j\omega\mu}{\rho}\frac{\partial^2 H_z}{\partial \phi^2} + \gamma\frac{\partial^2 E_z}{\partial \rho \partial \phi}\right] \\
&= \frac{j\omega\mu}{(\gamma^2 + k^2)}\left[\frac{\partial H_z}{\partial \rho} + \rho\frac{\partial^2 H_z}{\partial \rho^2} + \frac{1}{\rho}\frac{\partial^2 H_z}{\partial \phi^2}\right]
\end{aligned}$$

Hence the wave equation in H_z is

$$\frac{\partial^2 H_z}{\partial \rho^2} + \frac{1}{\rho}\frac{\partial H_z}{\partial \rho} + \frac{1}{\rho^2}\frac{\partial^2 H_z}{\partial \phi^2} + (\gamma^2 + k^2)H_z = 0 \tag{12.33}$$

12.6 Cylindrical waveguide: solution of wave equation for TE modes

For TE modes the wave equation in H_z is

$$\frac{\partial^2 H_z}{\partial \rho^2} + \frac{1}{\rho}\frac{\partial H_z}{\partial \rho} + \frac{1}{\rho^2}\frac{\partial^2 H_z}{\partial \phi^2} + (\gamma^2 + k^2)H_z = 0 \tag{12.34}$$

Assuming a solution of product form,

$$H_z = R(\rho)P(\phi)e^{-\gamma z}$$

gives for the wave equation after dividing by $R(\rho)P(\phi)$

$$\frac{1}{R}\frac{\partial^2 R}{\partial \rho^2} + \frac{1}{\rho R}\frac{\partial R}{\partial \rho} + \frac{1}{\rho^2 P}\frac{\partial^2 P}{\partial \phi^2} + (y^2 + k^2) = 0$$

or

$$\frac{\rho^2}{R}\frac{\partial^2 R}{\partial \rho^2} + \frac{\rho}{R}\frac{\partial R}{\partial \rho} + (\gamma^2 + k^2)\rho^2 = -\frac{1}{P}\frac{\partial^2 P}{\partial \phi^2}$$

Since both sides of this equation are to be equal for all values of ρ and ϕ they must both be constants. Hence

$$\frac{\partial^2 P}{\partial \phi^2} = -n^2 P \tag{12.35}$$

and

$$\frac{\partial^2 R}{\partial \rho^2} + \frac{1}{\rho}\frac{\partial R}{\partial \rho} + \left(\gamma^2 + k^2 - \frac{n^2}{\rho^2}\right) R = 0 \tag{12.36}$$

Eqn (12.35) gives a solution

$$P = A_n \cos n\phi + B_n \sin n\phi$$

and eqn (12.36),

$$R = C_n J_n(h\rho) + D_n N_n(h\rho)$$

where

$$h^2 = \gamma^2 + k^2$$

and J_n, N_n are Bessel functions of the first and second kind of order n. Since, however, the axis corresponding to $\rho = 0$ is included in the solution, this means that the constant D_n must always be zero. Likewise the reference angle for ϕ can be rotated to make B_n zero, so that the final solution for H_z is

$$H_z = A_n J_n(h\rho) \cos n\phi \tag{12.37}$$

Since for TE modes E_z is zero, the remaining fields become

$$H_\rho = -\frac{\gamma}{h} A_n J_n'(h\rho) \cos n\phi$$

$$H_\phi = \frac{n\gamma}{h^2 \rho} A_n J_n(h\rho) \sin n\phi$$

$$E_\rho = \frac{j\omega\mu n}{h^2 \rho} A_n J_n(h\rho) \sin n\phi$$

$$E_\phi = \frac{j\omega\mu}{h} A_n J_n'(h\rho) \cos n\phi$$

The boundary condition requires that tangential E must vanish on the perfectly conducting walls at radius a. Hence

$$E_\phi(r = a) = 0, \quad \text{or} \quad J_n'(ha) = 0$$

For the TE_{11} mode, this means that ha can only have the value 1.841, and hence $h = 1.841/a$. Thus the fields for this mode are, denoting 1.841 by δ, and γ by $j\beta$ for lossless propagation,

$$H_z = A J_1\left(\frac{\delta}{a}\rho\right) \cos \phi$$

$$H_\rho = \frac{j\beta a}{\delta} A J_1'\left(\frac{\delta}{a}\rho\right) \cos \phi$$

$$\left.\begin{aligned} H_\phi &= \frac{j\beta a^2}{\delta^2 \rho} A J_1\left(\frac{\delta}{a}\rho\right)\sin\phi \\ E_z &= 0 \\ E_\rho &= \frac{j\omega\mu a^2}{\delta^2 \rho} A J_1\left(\frac{\delta}{a}\rho\right)\sin\phi \\ E_\phi &= \frac{j\omega\mu a}{\delta} A J_1'\left(\frac{\delta}{a}\rho\right)\cos\phi \end{aligned}\right\} \quad (12.38)$$

It will be noted that the ratio of E_ρ/H_ϕ, which may be defined as the guide impedance is $\omega\mu/\beta$ for this TE mode.

12.7 Tangential electric and magnetic fields in aperture of cylindrical waveguide carrying TE_{11} mode

It can be seen from Fig. 12.4 that the rectangular and cylindrical components of the tangential electric field in the aperture of the cylindrical waveguide are related by the equations

$$\left.\begin{aligned} E_x &= E_\rho \cos\phi - E_\phi \sin\phi \\ E_y &= E_\rho \sin\phi + E_\phi \cos\phi \end{aligned}\right\} \quad (12.39)$$

Substituting in eqn (12.38) gives

$$E_x = \frac{j\omega\mu a^2}{\delta^2 \rho} A J_1\left(\frac{\delta}{a}\rho\right)\sin\phi\cos\phi - \frac{j\omega\mu a}{\delta} A J_1'\left(\frac{\delta}{a}\rho\right)\cos\phi\sin\phi$$

Fig. 12.4. Cylindrical to rectangular coordinate conversion.

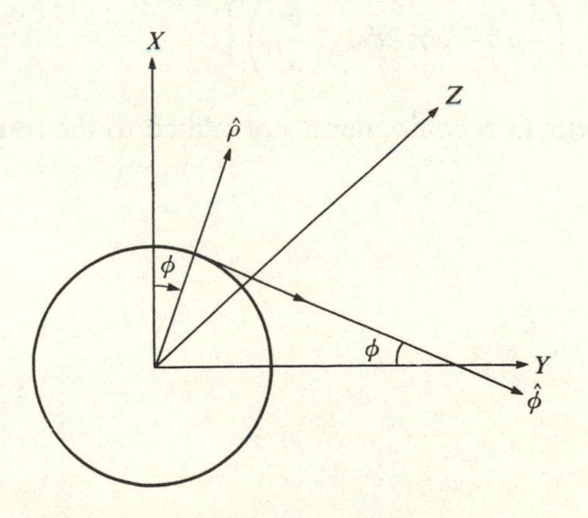

$E_x = E_\rho \cos\phi - E_\phi \sin\phi$
$E_y = E_\rho \sin\phi + E_\phi \cos\phi$

$$= \frac{j\omega\mu a}{\delta} A \sin\phi \cos\phi \left[\frac{J_1\left(\frac{\delta}{a}\rho\right)}{\left(\frac{\delta}{a}\rho\right)} - J'_1\left(\frac{\delta}{a}\rho\right) \right]$$

$$= \frac{j\omega\mu a}{2\delta} A \sin 2\phi J_2\left(\frac{\delta}{a}\rho\right) \tag{12.40}$$

Similarly

$$E_y = \frac{j\omega\mu a^2}{\delta^2 \rho} A J_1\left(\frac{\delta}{a}\rho\right) \sin^2\phi + \frac{j\omega\mu a}{\delta} A J'_1\left(\frac{\delta}{a}\rho\right) \cos^2\phi$$

$$= \frac{j\omega\mu a}{\delta} A \left[\frac{J_1\left(\frac{\delta}{a}\rho\right)}{\left(\frac{\delta}{a}\rho\right)} \left(\frac{1}{2} - \frac{\cos 2\phi}{2}\right) + J'_1\left(\frac{\delta}{a}\rho\right)\left(\frac{1}{2} + \frac{\cos 2\phi}{2}\right) \right]$$

$$= \frac{j\omega\mu a}{2\delta} A \left\{ \left[\frac{J_1\left(\frac{\delta}{a}\rho\right)}{\left(\frac{\delta}{a}\rho\right)} + J'_1\left(\frac{\delta}{a}\rho\right) \right] - \cos 2\phi \left[\frac{J_1\left(\frac{\delta}{a}\rho\right)}{\left(\frac{\delta}{a}\rho\right)} - J'_1\left(\frac{\delta}{a}\rho\right) \right] \right\}$$

$$= \frac{j\omega\mu a}{2\delta} A \left[J_0\left(\frac{\delta}{a}\rho\right) - \cos 2\phi J_2\left(\frac{\delta}{a}\rho\right) \right] \tag{12.41}$$

The transverse magnetic field components are related to the transverse electric by

$$\mathbf{H}_t = \frac{\beta}{\omega\mu} (\hat{\mathbf{z}} \times \mathbf{E}_t)$$

Hence

$$H_x = -\frac{\beta}{\omega\mu} E_y \tag{12.42}$$

and

$$H_y = \frac{\beta}{\omega\mu} E_x \tag{12.43}$$

12.8 Radiation pattern of cylindrical waveguide carrying TE_{11} mode

It has previously been shown that the radiation pattern of a circular aperture antenna with tangential electric and magnetic fields in the aperture can be expressed as

$$\left.\begin{aligned} E_\theta &= \frac{j}{2\lambda}\frac{e^{-jkr}}{r}\iint [E_x\cos\phi + E_y\sin\phi \\ &\quad - Z_0H_x\cos\theta\sin\phi + Z_0H_y\cos\theta\cos\phi] \\ &\quad \times e^{jk\rho'\sin\theta\cos(\phi-\phi')}\rho'\,d\rho'\,d\phi' \\ E_\phi &= \frac{j}{2\lambda}\frac{e^{-jkr}}{r}\iint [-E_x\cos\theta\sin\phi \\ &\quad + E_y\cos\theta\cos\phi - Z_0H_x\cos\phi - Z_0H_y\sin\phi] \\ &\quad \times e^{jk\rho'\sin\theta\cos(\phi-\phi')}\rho'\,d\rho'\,d\phi' \end{aligned}\right\} \tag{12.44}$$

Substituting for the components H_x, H_y from eqns (12.42) and (12.43) gives

$$E_\theta = \frac{j}{2\lambda}\frac{e^{-jkr}}{r}\iint\left(1 + \frac{\beta_{11}}{k}\cos\theta\right)(E_x\cos\phi + E_y\sin\phi)$$
$$\times e^{jk\rho'\sin\theta\cos(\phi-\phi)}\rho'\,d\rho'\,d\phi'$$

$$E_\phi = \frac{j}{2\lambda}\frac{e^{-jkr}}{r}\iint\left(\cos\theta + \frac{\beta_{11}}{k}\right)(-E_x\sin\phi + E_y\cos\phi)$$
$$\times e^{jk\rho'\sin\theta\cos(\phi-\phi)}\rho'\,d\rho'\,d\phi'$$

The components E_x, E_y will now be expressed in terms of the form they take in the aperture of the cylindrical waveguide, so that

$$\begin{aligned} E_\theta &= \frac{j}{2\lambda}\frac{e^{-jkr}}{r}\left(1 + \frac{\beta_{11}}{k}\cos\theta\right)\int_0^a\int_0^{2\pi}\frac{j\omega\mu a}{2\delta}A \\ &\quad \times\left[\sin 2\phi'\cos\phi J_2\left(\frac{\delta}{a}\rho'\right) + \sin\phi J_0\left(\frac{\delta}{a}\rho'\right)\right. \\ &\quad \left. - \cos 2\phi'\sin\phi J_2\left(\frac{\delta}{a}\rho'\right)\right] \\ &\quad \times e^{jk\rho'\sin\theta\cos(\phi-\phi')}\rho'\,d\rho'\,d\phi' \end{aligned} \tag{12.45}$$

where the primed coordinates refer to the variables of integration. Similarly

$$\begin{aligned} E_\phi &= \frac{j}{2\lambda}\frac{e^{-jkr}}{r}\left(\cos\theta + \frac{\beta_{11}}{k}\right)\int_0^a\int_0^{2\pi}\frac{j\omega\mu a}{2\delta}A \\ &\quad \times\left[-\sin 2\phi'\sin\phi J_2\left(\frac{\delta}{a}\rho'\right) + \cos\phi J_0\left(\frac{\delta}{a}\rho'\right)\right. \end{aligned}$$

$$-\cos 2\phi' \cos\phi J_2\left(\frac{\delta}{a}\rho'\right)\bigg]$$

$$\times e^{jk\rho'\sin\theta\cos(\phi-\phi')}\rho'\,d\rho'\,d\phi' \tag{12.46}$$

Using the integrals

$$\int_0^{2\pi} e^{ju\cos(\phi-\phi')}\cos n\phi'\,d\phi' = 2\pi j^n \cos n\phi J_n(u) \tag{12.47}$$

and

$$\int_0^{2\pi} e^{ju\cos(\phi-\phi')}\sin n\phi'\,d\phi' = 2\pi j^n \sin n\phi J_n(u) \tag{12.48}$$

gives

$$E_\theta = -\frac{\omega\mu a A}{4\delta\lambda}\frac{e^{-jkr}}{r}\left(1+\frac{\beta_{11}}{k}\cos\theta\right)\bigg[-2\pi\sin 2\phi\cos\phi$$

$$\times\int_0^a J_2(k\rho'\sin\theta)J_2\left(\frac{\delta}{a}\rho'\right)$$

$$+2\pi\sin\phi\int_0^a J_0(k\rho'\sin\theta)J_0\left(\frac{\delta}{a}\rho'\right)+2\pi\cos 2\phi\sin\phi$$

$$\times\int_0^a J_2(k\rho'\sin\theta)J_2\left(\frac{\delta}{a}\rho'\right)\bigg]\rho'\,d\rho'$$

and

$$E_\phi = \frac{\omega\mu a A}{4\delta\lambda}\frac{e^{-jkr}}{r}\left(\cos\theta+\frac{\beta_{11}}{k}\right)\bigg[-2\pi\sin 2\phi\sin\phi$$

$$\times\int_0^a J_2(k\rho'\sin\theta)J_2\left(\frac{\delta}{a}\rho'\right)-2\pi\cos\phi$$

$$\times\int_0^a J_0(k\rho'\sin\theta)J_0\left(\frac{\delta}{a}\rho'\right)-2\pi\cos 2\phi\cos\phi$$

$$\times\int_0^a J_2(k\rho'\sin\theta)J_2\left(\frac{\delta}{a}\rho'\right)\bigg]\rho'\,d\rho'$$

Collecting terms gives the results for E_θ, E_ϕ in the form of two single integrals,

$$E_\theta = -\frac{\omega\mu a\pi A}{2\lambda\delta}\frac{e^{-jkr}}{r}\left(1+\frac{\beta_{11}}{k}\cos\theta\right)\sin\phi$$

$$\times\int_0^a\bigg[J_0(k\rho'\sin\theta)J_0\left(\frac{\delta}{a}\rho'\right)$$

$$-J_2(k\rho'\sin\theta)J_2\left(\frac{\delta}{a}\rho'\right)\bigg]\rho'\,d\rho' \tag{12.49}$$

and

$$E_\phi = -\frac{\omega\mu a\pi A}{2\lambda\delta}\frac{e^{-jkr}}{r}\left(\cos\theta + \frac{\beta_{11}}{k}\right)\cos\phi$$
$$\times \int_0^a \left[J_0(k\rho'\sin\theta) J_0\left(\frac{\delta}{a}\rho'\right) + J_2(k\rho'\sin\theta) J_2\left(\frac{\delta}{a}\rho'\right)\right]\rho'\,d\rho' \qquad (12.50)$$

Although it would be straightforward to evaluate these integrals numerically by computer they can be evaluated analytically as follows. Consider the Lommel result:

$$\int_0^a J_n(\alpha x)J_n(\beta x)x\,dx = \frac{a}{\alpha^2-\beta^2}\left\{J_n(\alpha a)\left[\frac{d}{dx}J_n(\beta x)\right]_{x=a} - J_n(\beta a)\left[\frac{d}{dx}J_n(\alpha x)\right]_{x=a}\right\}$$
$$= \frac{a}{\alpha^2-\beta^2}\{\beta J_n(\alpha a)J_n'(\beta a) - \alpha J_n(\beta a)J_n'(\alpha a)\} \qquad (12.51)$$

where $J_n'(x)$ is the derivative with respect to the argument of $J_n(x)$.

Then

$$\int_0^a J_n(k\rho'\sin\theta)J_n\left(\frac{\delta}{a}\rho'\right)\rho'\,d\rho' = \frac{a}{\left(k^2\sin^2\theta - \dfrac{\delta^2}{a^2}\right)}\left\{\frac{\delta}{a}J_n(ka\sin\theta)J_n'(\delta) - k\sin\theta J_n(\delta)J_n'(ka\sin\theta)\right\}$$

Writing $u^2 = k^2a^2\sin^2\theta$, enables this integral to be written as

$$I = \frac{a^2}{u^2-\delta}[\delta J_n(u)J_n'(\delta) - uJ_n(\delta)J_n'(u)]$$

so that E_θ may be written

$$E_\theta = -\frac{\omega\mu a\pi A}{2\lambda\delta}\frac{e^{-jkr}}{r}\left(1 + \frac{\beta_{11}}{k}\cos\theta\right)\sin\phi\frac{a^2}{u^2-\delta^2}$$
$$\times\{\delta[J_0(u)J_0'(\delta) - J_2(u)J_2'(\delta)] - u[J_0(\delta)J_0'(u) - J_2(\delta)J_2'(u)]\} \qquad (12.52)$$

and

$$E_\phi = -\frac{\omega\mu a\pi A}{2\lambda\delta}\frac{e^{-jkr}}{r}\left(\cos\theta + \frac{\beta_{11}}{k}\right)\cos\phi\frac{a^2}{u^2-\delta^2}$$
$$\times\{\delta[J_0(u)J'_0(\delta) + J_2(u)J'_2(\delta)]$$
$$-u[J_0(\delta)J'_0(u) + J_2(\delta)J'_2(u)]\} \qquad (12.53)$$

Now J_2 and J'_2 can be written as

$$J_2(x) = \frac{2J_1(x)}{x} - J_0(x)$$

$$J'_2(x) = \frac{2J_0(x)}{x} + \left(1 - \frac{4}{x^2}\right)J_1(x)$$

and $J'_0 = -J_1$, so that the expression in curly brackets for E_θ can be written as

$$\Bigg\{-\delta J_0(u)J_1(\delta) - \delta\left[\frac{2J_1(u)}{u} - J_0(u)\right]$$
$$\times\left[\frac{2J_0(\delta)}{\delta} + \left(1 - \frac{4}{\delta^2}\right)J_1(\delta)\right]$$
$$+ uJ_0(\delta)J_1(u) + u\left[\frac{2J_1(\delta)}{\delta} - J_0(\delta)\right]$$
$$\times\left[\frac{2J_0(u)}{u} + \left(1 - \frac{4}{u^2}\right)J_1(u)\right]\Bigg\}$$

This readily simplifies to

$$\frac{2(u^2-\delta^2)}{u\delta}J_1(\delta)J_1(u)$$

so that E_θ becomes

$$E_\theta = -\frac{\omega\mu a^3\pi J_1(\delta)A}{\lambda\delta^2}\frac{e^{-jkr}}{r}\left(1 + \frac{\beta_{11}}{k}\cos\theta\right)\sin\phi\frac{J_1(ka\sin\theta)}{ka\sin\theta} \qquad (12.54)$$

and likewise E_ϕ can be simplified to

$$E_\phi = -\frac{\omega\mu a^3\pi J_1(\delta)A}{\lambda\delta^2}\frac{e^{-jkr}}{r}\left(\cos\theta + \frac{\beta_{11}}{k}\right)\cos\phi$$
$$\times\frac{\delta^2 J'_1(ka\sin\theta)}{(\delta^2 - k^2a^2\sin^2\theta)} \qquad (12.55)$$

It will be noted that the ϕ dependence in the expressions for both E_θ and E_ϕ is particularly simple, and the expression for the total E-field may be

Fig. 12.5. Radiation patterns of open-ended TE_{11} mode circular waveguide: (*a*) $b = 0.383\lambda$ ($\beta/k = 0.644$); (*b*) $b = 1.056\lambda$ ($\beta/k = 0.961$).

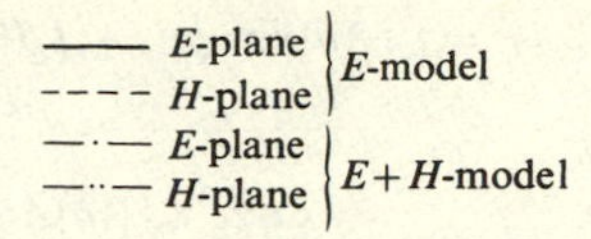

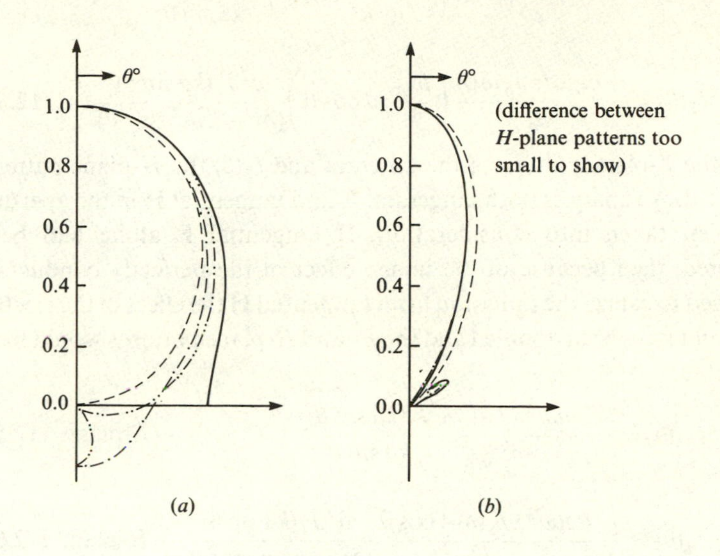

Fig. 12.6. *H*-plane patterns of 10 mm radius cylindrical waveguide at 10.7 GHz: experimental and theoretical.

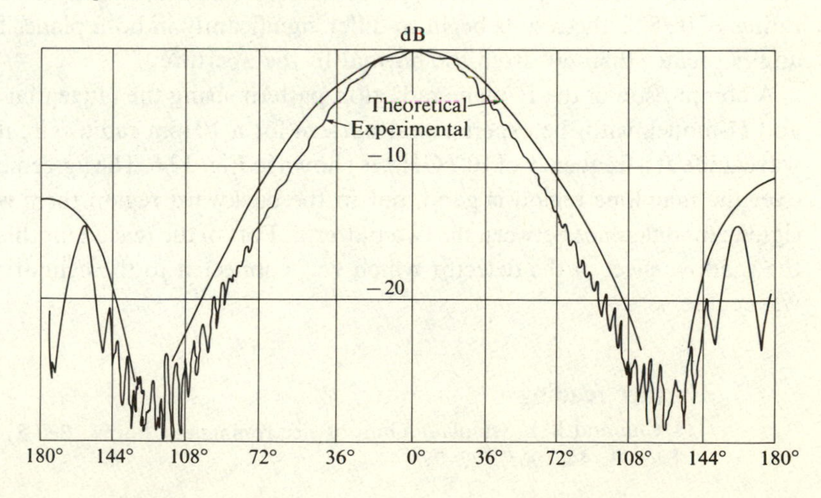

written

$$\bar{E} = [f_1(\theta)\sin\phi\bar{a}_\theta + f_2(\theta)\cos\phi\bar{a}_\phi]\frac{e^{-jkr}}{r} \tag{12.56}$$

where

$$f_1(\theta) = -\frac{\omega\mu a^3\pi J_1(\delta)A}{\lambda\delta^2}\left(1 + \frac{\beta_{11}}{k}\cos\theta\right)\frac{J_1(ka\sin\theta)}{ka\sin\theta} \tag{12.57}$$

and

$$f_2(\theta) = -\frac{\omega\mu a^3\pi J_1(\delta)A}{\lambda\delta^2}\left(\frac{\beta_{11}}{k} + \cos\theta\right)\frac{\delta^2 J'_1(ka\sin\theta)}{(\delta^2 - k^2a^2\sin^2\theta)} \tag{12.58}$$

$f_1(\theta)$ is the E-plane pattern of the antenna and $f_2(\theta)$ the H-plane pattern.

In the above analysis both tangential **E** and tangential **H** in the aperture have been taken into consideration. If tangential **E** alone had been considered, then because of the image effect of the perfectly conducting plane used to cancel the radiation from tangential **H** the effect of the electric field would have been doubled and the E- and H-plane patterns would have become

$$f_3(\theta) = -\frac{2\omega\mu a^3\pi J_1(\delta)A}{\lambda\delta^2}\frac{J_1(ka\sin\theta)}{ka\sin\theta} \quad \cdots E\text{-plane} \tag{12.59}$$

and

$$f_4(\theta) = -\frac{2\omega\mu a^3\pi J_1(\delta)A\cos\theta}{\lambda\delta^2}\frac{\delta^2 J'_1(ka\sin\theta)}{(\delta^2 - k^2a^2\sin^2\theta)} \cdots H\text{-plane} \tag{12.60}$$

Two numerical examples illustrating the differences between the combined tangential **E**- and **H**-model on the one hand, and the **E**-model with an image plane on the other, are shown in Fig. 12.5. For a waveguide radius of 1.056λ, the two methods differ by less than approximately 1% over the main lobes of the E- and H-plane patterns. On the other hand, for the smaller radius of 0.383λ, the results begin to differ significantly in both planes for angles greater than 50° from the normal to the aperture.

A comparison of the H-plane radiation pattern, using the tangential **E**- and **H**-model, with the experimental pattern for a 10 mm radius circular waveguide at a frequency of 10.7 GHz, is shown in Fig. 12.6. The agreement over the mainlobe region is good, but in the backward region there is a significant difference between the two patterns. Part of the reason for this is the shadow effect of the detector which was connected to the cylindrical waveguide.

Further reading

T.S. Chu and R.A. Semplak: 'Gain of electromagnetic horns': *Bell Syst. Tech. J.*, **44**, 1965, 527–37.

R.E. Collin and F.J. Zucker: *Antenna Theory*: McGraw-Hill, 1969.
R.S. Elliott: *Antenna Theory and Design*: Prentice-Hall, Englewood Cliffs, New Jersey, 1981.
H. Jasik (Ed.): *Antenna Engineering Handbook*: McGraw-Hill, New York, 1961.
D.R. Rhodes: 'An experimental investigation of the radiation patterns of electromagnetic horn antennas': *Proc. IRE*, **36**, 1948, 1101–5.
A.W. Rudge, K. Milne, A.D. Olver, P. Knight (Ed.): *The Handbook of Antenna Design*: Peter Peregrinus, 1982.
S. Silver: *Microwave Antenna Theory and Design*: McGraw-Hill, 1949.
W.M. Truman and C.A. Balanis: 'Optimum design of horn feeds for reflector antennas': *IEEE Trans. Antennas and Propagation*, **AP-22**, 1974, 585–6.

13

+−+

Paraboloidal reflector

A paraboloid is the locus of a point which moves in space so that its distance from a fixed point is equal to its distance from a fixed plane. The geometrical properties of the paraboloid follow in general from those of its generating parabola.

13.1 Equations of parabola

Consider the equation $r = f\sec^2(\theta/2)$ in polar coordinates, centred on the origin O in Fig. 13.1. The (x, z)-coordinates of points on the curve represented by this equation are given by

$$x = r\sin\theta = f\sec^2\frac{\theta}{2}\sin\theta = 2f\tan\frac{\theta}{2}$$

$$z = r\cos\theta = f\sec^2\frac{\theta}{2}\cos\theta$$

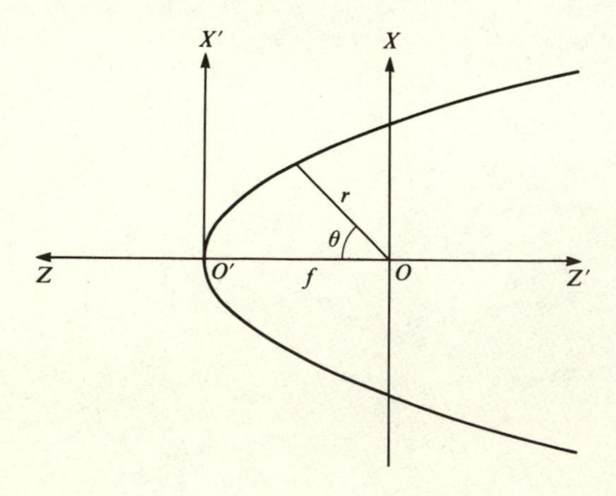

Fig. 13.1. Equations of parabola in polar and rectangular forms.

Let the origin now be altered to the point O', so that with respect to this new origin the (x', z')-coordinates of points on the curve are given by

$$x' = x = 2f\tan\frac{\theta}{2}$$

$$z' = (f - z) = f\left(1 - 2 + \sec^2\frac{\theta}{2}\right) = f\tan^2\frac{\theta}{2}$$

Hence the equation of the curve in the new coordinate system is

$$(x')^2 = 4f(z')$$

which is the well known form of the equation for a parabola.

13.1.1 *Equation of paraboloid*

Referring to the spherical coordinates centred at O in Fig. 13.2 the equation is again given by

$$r = f\sec^2\frac{\theta}{2} \qquad (13.1)$$

The symbol ρ is sometimes used in place of r.

13.1.2 *Element of area for paraboloid*

Referring to Fig. 13.2 an element of area dS on the paraboloid is equal to the product of the element of length ds on the generating parabola multiplied by the orthogonal element of length $r\sin\theta d\phi$, i.e.

$$dS = r\sin\theta d\phi ds$$

Since, however,

$$ds^2 = dx^2 + dz^2$$

Fig. 13.2. Element of arc length on paraboloid.

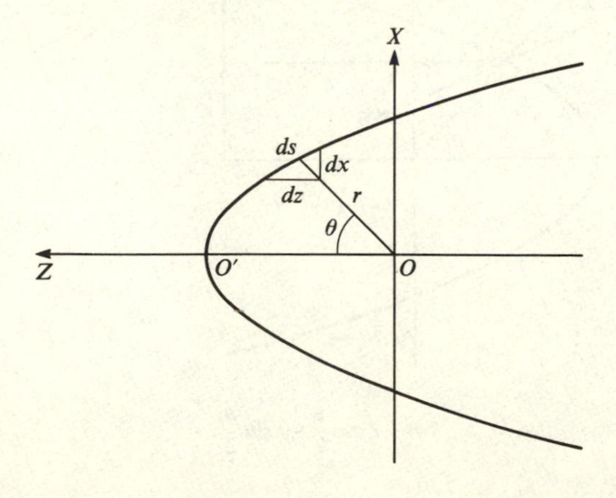

where

$$x = 2f\tan\frac{\theta}{2} \quad \text{so that} \quad \frac{dx}{d\theta} = f\sec^2\frac{\theta}{2}$$

and

$$z = 2f - f\sec^2\frac{\theta}{2} \quad \text{so that} \quad \frac{dz}{d\theta} = -f\sec^2\frac{\theta}{2}\tan\frac{\theta}{2}$$

it follows that

$$\frac{ds}{d\theta} = \left[\left(\frac{dx}{d\theta}\right)^2 + \left(\frac{dz}{d\theta}\right)^2\right]^{\frac{1}{2}} = \left[f^2\sec^4\frac{\theta}{2} + f^2\sec^4\frac{\theta}{2}\tan^2\frac{\theta}{2}\right]^{\frac{1}{2}}$$

$$= f\sec^3\frac{\theta}{2} = r\sec\frac{\theta}{2}$$

Hence the element of area dS on the paraboloid is given by

$$dS = r\sin\theta d\phi \cdot r\sec\frac{\theta}{2}d\theta = r^2\sin\theta\sec\frac{\theta}{2}d\theta d\phi$$

$$= f^2\sin\theta\sec^5\frac{\theta}{2}d\theta d\phi \tag{13.2}$$

13.1.3 *Equation of inward normal to paraboloid*

The inward unit normal $\hat{\mathbf{n}}$ at a point on the surface is seen from Fig. 13.3 to be given by

$$\hat{\mathbf{n}} = -\hat{\mathbf{r}}\cos\frac{\theta}{2} + \hat{\boldsymbol{\theta}}\sin\frac{\theta}{2} \tag{13.3}$$

Fig. 13.3. Unit inward normal to paraboloid.

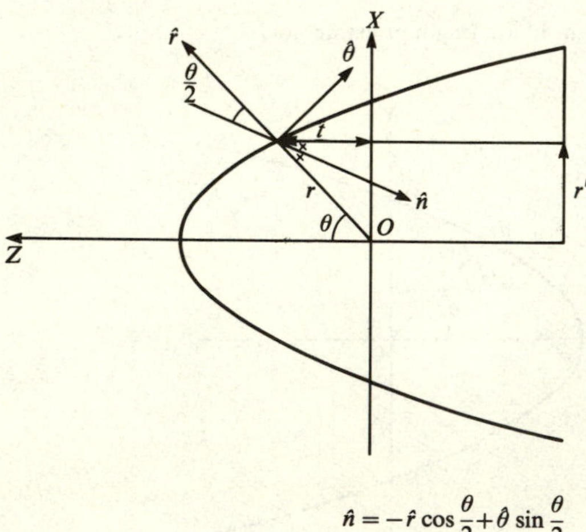

where $\hat{\mathbf{r}}$, $\hat{\boldsymbol{\theta}}$ are unit vectors in the r- and θ-directories, given by

$$\hat{\mathbf{r}} = \hat{\mathbf{x}} \sin\theta \cos\phi + \hat{\mathbf{y}} \sin\theta \sin\phi + \hat{\mathbf{z}} \cos\theta$$
$$\hat{\boldsymbol{\theta}} = \hat{\mathbf{x}} \cos\theta \cos\phi + \hat{\mathbf{y}} \cos\theta \sin\phi - \hat{\mathbf{z}} \sin\theta$$

Substituting these equations into eqn (13.3) gives

$$\hat{\mathbf{n}} = -\hat{\mathbf{x}} \sin\frac{\theta}{2} \cos\phi - \hat{\mathbf{y}} \sin\frac{\theta}{2} \sin\phi - \hat{\mathbf{z}} \cos\frac{\theta}{2} \tag{13.4}$$

13.1.4 *Additional geometrical properties of paraboloid*

(i) *Distance from surface point (r, θ, ϕ) on paraboloid to any point (x, y, z) in space*

Let u represent the distance between the surface point (r, θ, ϕ) and the observation point (x, y, z). Then

$$u^2 = (r \sin\theta \cos\phi - x)^2 + (r \sin\theta \sin\phi - y)^2 + (r \cos\theta - z)^2 \tag{13.5}$$

If the observation point is defined in cylindrical coordinates (ρ, α, z) then

$$\begin{aligned} u^2 &= (r \sin\theta \cos\phi - \rho \cos\alpha)^2 + (r \sin\theta \sin\phi - \rho \sin\alpha)^2 \\ &\quad + (r \cos\theta - z)^2 \\ &= r^2 + z^2 + \rho^2 - 2\rho r \sin\theta \cos(\phi - \alpha) - 2rz \cos\theta \end{aligned} \tag{13.6}$$

(ii) *Perpendicular distance from surface point (r, θ, ϕ) to focal plane*

Denoting this perpendicular distance by t gives

$$t = r\cos\theta = 2f - r \tag{13.7}$$

(iii) *Relation between point in aperture plane at radius r' from centre and angle θ at point of reflection*

Referring to Fig. 13.3:

$$\sin\theta = \frac{r'}{r} = \frac{r'}{f\left(1 + \tan^2\dfrac{\theta}{2}\right)} = \frac{2\tan\dfrac{\theta}{2}}{\left(1 + \tan^2\dfrac{\theta}{2}\right)}$$

Hence

$$\sin\theta = \frac{4fr'}{4f^2 + (r')^2} \tag{13.8}$$

13.2 Paraboloid excited by normally incident uniform plane wave

A uniform plane wave with its magnetic vector in the x-direction is assumed to travel parallel to the z-axis towards the paraboloid, as shown in Fig. 13.4. Currents are thereby induced in the paraboloid, which will set up fields everywhere in space. In particular we are interested in fields close to the geometric focus, which is expected to be a region where there is a high concentration of the power contained in that portion of the wave which falls within the aperture of the paraboloid.

If the currents over the whole of the paraboloidal surface, both front and back, were known, then an application of the equivalence principle would enable the fields everywhere in space to be found exactly. The difficulty arises because the currents on the back surface are not known and the currents on the front surface are only known approximately. The currents on the front surface would be known exactly if the reflector consisted of an infinite planar sheet which was perfectly conducting. In such a case they would be given by

$$\mathbf{J} = 2(\mathbf{n} \times \mathbf{H}) \tag{13.9}$$

The two respects in which a paraboloid differs from such a surface are that, firstly, the paraboloid is finite in extent and, secondly, it has a finite radius of curvature. To separate these two effects it is necessary to consider, most simply, the problem of the conducting half-plane, to find out how far away from the edge the current distribution differs significantly from that given by eqn (13.9). The answer to this question depends on whether the incident electromagnetic wave has its electric field or its magnetic field parallel to the edge. If it has neither parallel then the answer can be obtained by resolving the wave into two components, one with the electric field parallel and the other with the magnetic component parallel.

With the electric field parallel to the bounding edge the solution to the half-plane problem shows that for an observation point further than $\lambda/2$ from the edge the current density differs insignificantly from that of the infinite plane case. When the incident magnetic field is parallel to the edge the effect extends appreciably further. The reason for this is simply that **H**

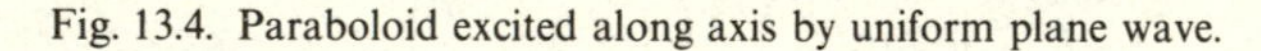

Fig. 13.4. Paraboloid excited along axis by uniform plane wave.

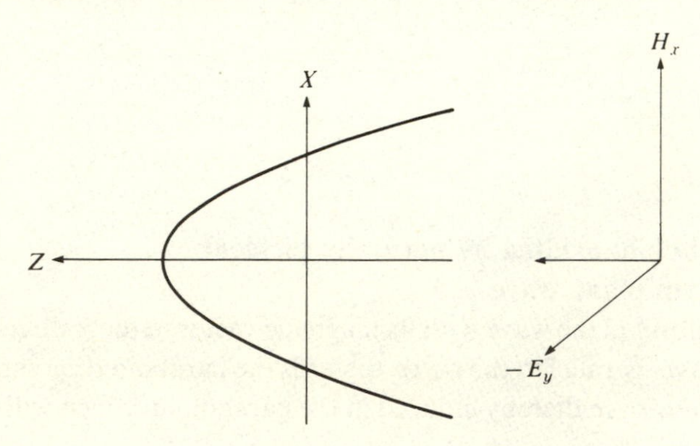

parallel to the edge implies a current density **J** perpendicular to the edge. When this current density reaches the edge its direction of current flow immediately reverses, but the current now flows on the back face of the plane. But the value at the edge is exactly one-half of that for the infinite plane, so that there is a reflected wave on the front face edge as well as an incident wave. The result is that at a distance of 1λ from the edge the current still differs by approximately 10% from that of the infinite plane, and even at a distance of 100λ the effect is of the order of 1%. Thus the effect is more far ranging than for E parallel to the edge, though its magnitude is not excessively large even at the short distance of 1λ from it. The approximation will therefore be made that for paraboloids greater than 10λ in diameter the effect of the edge will be ignored.

With regard to the effect of curvature it is helpful to consider the case of an infinitely long parabolic cylinder, for which computed results which have taken the effect of coupling between its parallel generating strips into account are available. They also take into account the effect of the edge and confirm the results stated above for the case of E parallel to the edge, which is the operating condition for such a cylinder, that at distances more that $\lambda/2$ the effect is negligible. For the case of a parabolic cylinder with a ratio of focal length/aperture height of 0.4, and an aperture of 5.0λ the computed results for current have shown a current distribution not departing more than 10% from the $2(\mathbf{n} \times \mathbf{H})$ result, except within the edge region. Hence for paraboloids of the same diameter and f/D ratio, it appears reasonable to assume that $\mathbf{J} = 2(\mathbf{n} \times \mathbf{H})$ is a reasonable approximation.

But the agreement with the $\mathbf{J} = 2(\mathbf{n} \times \mathbf{H})$ approximation is even greater than this. Using this approximation for parabolic cylinders with an aperture of 5.0λ as before, but decreasing the radii of curvature down to an f/D of 0.1, gave results for the focal region fields which agreed within 5% with those of $\mathbf{J} = 2(\mathbf{n} \times \mathbf{H})$. Hence again there is evidence that this approximation is a good one.

13.3 Analysis of fields produced by paraboloid irradiated by normally incident plane wave

It has previously been postulated that the wave travelling along the normal to the paraboloid in the positive z-direction has a magnetic field given by

$$\mathbf{H} = \mathbf{a}_x H_x^i e^{-jkz}$$

The unit normal to the surface of the reflector has been shown to be

$$\mathbf{n} = -\mathbf{a}_x \sin\frac{\theta}{2}\cos\phi - \mathbf{a}_y \sin\frac{\theta}{2}\sin\phi - \mathbf{a}_z \cos\frac{\theta}{2}$$

so that the current density **J** is given by its scalar components

$$J_x = 2(n_y H_z - n_z H_y) = 0$$

$$J_y = 2(n_z H_x - n_x H_z) = -2\cos\frac{\theta}{2} H_x^i e^{-jkt}$$

where t is the distance between the focal plane and the point on the surface where J is being considered, and

$$J_z = 2(n_x H_y - n_y H_x) = 2\sin\frac{\theta}{2}\sin\phi H_x^i e^{-jkt}$$

Due to these current densities the vector potential at a point $P\begin{pmatrix} x, y, z \\ \rho, \alpha, z \end{pmatrix}$ distant u from this source has components

$$\left.\begin{aligned} dA_x &= 0 \\ dA_y &= -\frac{\mu H_x^i}{2\pi u}\cos\frac{\theta}{2} e^{-jk(t+u)} dS \\ dA_z &= \frac{\mu H_x^i}{2\pi u}\sin\frac{\theta}{2}\sin\phi e^{-jk(t+u)} dS \end{aligned}\right\} \tag{13.10}$$

where

$$\begin{aligned} u &= [(x_s - x)^2 + (y_s - y)^2 + (z_s - z)^2]^{\frac{1}{2}} \\ &= [r^2 + z^2 + \rho^2 - 2\rho r\sin\theta\cos(\phi - \alpha) - 2rz\cos\theta]^{\frac{1}{2}} \end{aligned}$$

$\begin{pmatrix} x_s, y_s, z_s \\ r, \theta, \phi \end{pmatrix}$ are the coordinates of the source current density and

$$dS = r^2 \sec\frac{\theta}{2}\sin\theta d\theta d\phi$$

Then using

$$\mathbf{H} = \frac{1}{\mu}\nabla \times \mathbf{A}$$

gives

$$\begin{aligned} dH_x &= \frac{1}{\mu}\left(\frac{\partial A_z}{\partial y} - \frac{\partial A_y}{\partial z}\right) = \frac{H_x^i}{2\pi} e^{-jkt} dS \\ &\quad \times \left\{\sin\frac{\theta}{2}\sin\phi\frac{\partial}{\partial y}\left(\frac{e^{-jku}}{u}\right) + \cos\frac{\theta}{2}\frac{\partial}{\partial z}\left(\frac{e^{-jku}}{u}\right)\right\} \\ &= \frac{H_x^i}{2\pi} e^{-jkt} dS \left\{\sin\frac{\theta}{2}\sin\phi\left[\frac{e^{-jku}}{u}(-jk) - \frac{e^{-jku}}{u^2}\right]\right. \\ &\quad \left. \times \frac{\partial u}{\partial y} + \cos\frac{\theta}{2}\left[\frac{e^{-jku}}{u}(-jk) - \frac{e^{-jku}}{u^2}\right]\frac{\partial u}{\partial z}\right\} \end{aligned}$$

$$= \frac{H_x^i}{2\pi} e^{-jkt} \frac{e^{-jku}}{u} \left(-jk - \frac{1}{u} \right) dS$$

$$\times \left\{ \sin\frac{\theta}{2} \sin\phi \left(-\frac{(y_s - y)}{u} \right) - \cos\frac{\theta}{2} \frac{(z_s - z)}{u} \right\}$$

$$= \frac{H_x^i}{2\pi} e^{-jkt} \frac{e^{-jku}}{u^2} \left(-jk - \frac{1}{u} \right) \left\{ (\rho \sin\alpha - r \sin\theta \sin\phi) \right.$$

$$\left. \times \sin\frac{\theta}{2} \sin\phi + (z - r\cos\theta) \cos\frac{\theta}{2} \right\} dS$$

13.4 Scattered magnetic fields on axis

Considering first the case when P is on the z-axis, so that u is independent of ϕ, and integrating over the surface gives

$$H_x = \frac{H_x^i}{2\pi} \int_0^{\theta_0} \int_0^{2\pi} \frac{e^{-jk(t+\mu)}}{u^2} \left(-jk - \frac{1}{u} \right)$$

$$\times \left\{ (\rho \sin\alpha - r\sin\theta\sin\phi) \sin\frac{\theta}{2} \sin\phi \right.$$

$$\left. + (z - r\cos\theta)\cos\frac{\theta}{2} \right\} r^2 \sec\frac{\theta}{2} \sin\theta \, d\theta \, d\phi$$

$$= \frac{H_x^i}{2} \int_0^{\theta_0} \frac{e^{-jk(t+u)}}{u^2} \left(-jk - \frac{1}{u} \right)$$

$$\times \left\{ -\pi r \sin\frac{\theta}{2} \sin\theta + 2\pi (z - r\cos\theta) \cos\frac{\theta}{2} \right\} r^2 \sec\frac{\theta}{2} \sin\theta \, d\theta$$

$$= \frac{H_x^i}{2} \int_0^{\theta_0} \frac{e^{-jk(t+u)}}{u^2} \left(-jk - \frac{1}{u} \right)$$

$$\times \left\{ 2\cos\frac{\theta}{2} \left[-r\sin^2\frac{\theta}{2} + z - r\cos\theta \right] \right\} r^2 \sec\frac{\theta}{2} \sin\theta \, d\theta$$

$$= H_x^i \int_0^{\theta_0} \frac{e^{-jk(t+u)}}{u^2} \left(-jk - \frac{1}{u} \right) r^2 \sin\theta \left[z - r\cos^2\frac{\theta}{2} \right] d\theta$$

$$= H_x^i (f - z) \int_0^{\theta_0} \frac{r^2}{u^2} \left(jk + \frac{1}{u} \right) e^{-jk(t+u)} \sin\theta \, d\theta \qquad (13.11)$$

The integral on θ has, in general, to be performed numerically but this is a straightforward operation. Similarly it may be shown that for all points of observation on the axis

$$H_y = H_z = 0$$

13.5 Scattered magnetic field at focus

If the point of observation is restricted to the focus, however, then it is possible to obtain an analytic expression for the magnetic field H_x there. To show this we note that at the focus the following simplifications occur:

$$u = r; \quad (t+u) = r(\cos\theta + 1) = 2r\cos^2\frac{\theta}{2} = 2f; \quad z = 0$$

Then

$$H_{x_{focus}} = H_x^i f \int_0^{\theta_0} \left(\frac{1}{f\sec^2\frac{\theta}{2}} + jk \right) e^{-j2kf} \sin\theta \, d\theta$$

$$= H_x^i f e^{-jk2f} \int_0^{\theta_0} \left(\frac{\sin\theta}{2f} + \frac{\sin 2\theta}{4f} + jk\sin\theta \right) d\theta \qquad (13.12)$$

since

$$2\sin\theta + \sin 2\theta = \frac{4\sin\theta}{\sec^2\frac{\theta}{2}}$$

The integrals in eqn (13.12) give, for the magnetic field at the focus,

$$H_{x_{focus}} = H_x^i e^{-j2kf} \left[\frac{\sin^2\theta_0}{4} + \sin^2\frac{\theta_0}{2} + j2kf\sin^2\frac{\theta_0}{2} \right] \qquad (13.13)$$

and it will be recalled that the only approximation involved in producing this result is the approximation that the current density is given by $2(\mathbf{n} \times \mathbf{H}^i)$.

The term in square brackets in eqn (13.13) may be thought of as the magnification produced by the paraboloid. Since the first term cannot exceed 0.25 and the second term unity, the major contribution is provided by the third term whose magnification is directly proportional to frequency. Thus at high frequencies,

$$|\text{magnification}| = 2kf\sin^2\frac{\theta_0}{2} = \frac{\pi}{2}\frac{D}{\lambda}\sin\theta_0 \qquad (13.14)$$

which for a focal plane paraboloid for which θ_0 equals $\pi/2$ and f/D equals 0.25, gives

$$|\text{magnification}| = \frac{\pi}{2}\frac{D}{\lambda}$$

For a 100λ diameter paraboloid therefore, one effect of the focussing action of the paraboloid is to concentrate the incoming field by a factor of 157 times at its geometric focus. It will also be noted from eqn (13.13) that the phase of the field there is $\pi/2$ ahead of the phase which would be calculated from the geometric path length phase of e^{-jk2f}. This effect is sometimes referred to in optics as a phase jump, but since there is no discontinuity of phase there, it is better to avoid this term.

13.6 Scattered electric fields on axis including focus

Turning now to the electric field produced by the currents flowing on the paraboloid, this can be obtained, once the magnetic field is known, from

$$j\omega\varepsilon\mathbf{E} = \nabla \times \mathbf{H}$$

Since the incoming magnetic field was x-directed, the incoming electric field is along the negative y-direction, and hence the re-radiated electric field is expected to be predominantly along the positive y-axis. In general

$$j\omega\varepsilon E_y = \left(\frac{\partial H_x}{\partial z} - \frac{\partial H_z}{\partial x}\right)$$

but it must be remembered that although H_z itself is zero on the z-axis $\partial H_z/\partial x$ is not necessarily so. Consequently there is a greater amount of algebra to carry out for the evaluation of the electric field components. Because this is straightforward, but lengthy, the details of the working are not given here, but only the results on axis. These are

$$\left.\begin{aligned} E_y = \frac{jE_y^i}{2k}\int_0^{\theta_0} \frac{r^2}{u^2}&\left\{\left(j\frac{3k}{u^2} + \frac{3}{u^3} - \frac{k^2}{u}\right)\right. \\ &\times\left[rz(1+3\cos\theta) - 2r^2\cos^2\frac{\theta}{2} - 2z^2\right] + \left.\frac{4}{u} + j4k\right\} \\ &\times e^{-jk(t+u)}\sin\theta\, d\theta \end{aligned}\right\} \tag{13.15}$$

$$E_x = E_z = 0$$

At the focus E_y simplifies to the analytic result

$$\begin{aligned} E_{y_{focus}} = -\frac{jE_y^i}{2k}e^{-j2kf}&\left\{\frac{jk}{2}\left(4\sin^2\frac{\theta_0}{2} - 3\sin^2\theta_0\right)\right. \\ &+\frac{1}{8f}[2(8k^2f^2-1) - \cos\theta_0(1+16k^2f^2) \\ &\left.+ 2\cos 2\theta_0 + \cos 3\theta_0]\right\} \end{aligned} \tag{13.16}$$

which for high frequencies gives a magnification of the incoming electric field,

$$\begin{aligned} |\text{magnification}| &= \frac{1}{16kf}(16k^2f^2 - 16k^2f^2\cos\theta_0) \\ &= 2kf\sin^2\frac{\theta_0}{2} \end{aligned}$$

which is the same as for the magnetic field. It will also be noted that for these high frequencies the ratio of electric to magnetic field at the focus is exactly Z_0.

13.7 Asymmetry of scattered fields on axis about focus

It has previously been shown that for any point on the axis the scattered magnetic field is given in the form of the integral

$$H_x = H_x^i(f-z)\int_0^{\theta_0}\frac{r^2}{u^2}\left(jk+\frac{1}{u}\right)e^{-jk(t+u)}\sin\theta\,d\theta$$

Since the origin of coordinates is at the focus the multiplying factor $(f-z)$ suggests that the field may be asymmetric about the focus. To confirm this, consider the high frequency case when $k \gg 1/u$, with

$$u = [z^2 + r^2 - 2rz\cos\theta]^{\frac{1}{2}}$$

$$\approx r\left(1-\frac{z}{r}\cos\theta\right) \quad \text{for} \quad z^2 \ll r^2$$

i.e. consideration is being restricted to points near the focus. Then for this case,

$$(t+u) \approx r\cos\theta + r\left(1-\frac{z}{r}\cos\theta\right)$$

$$= (2f - z\cos\theta)$$

and the scattered magnetic field becomes

$$H_x \approx H_x^i(f-z)e^{-j2kf}\int_0^{\theta_0}\frac{jk\sin\theta}{\left(1-\frac{z}{r}\cos\theta\right)^2}e^{jkz\cos\theta}\,d\theta$$

$$\approx H_x^i(f-z)e^{-j2kf}jk\int_0^{\theta_0}\sin\theta e^{jkz\cos\theta}\,d\theta \quad \text{since} \quad \frac{z}{r} \ll 1$$

This integral is readily evaluated to give

$$H_x = j2kH_x^i\sin^2\frac{\theta_0}{2}\left[(f-z)\frac{\sin\left(kz\sin^2\frac{\theta_0}{2}\right)}{kz\sin^2\frac{\theta_0}{2}}\right] \times e^{-jk(2f - z\cos^2(\theta_0/2))} \tag{13.17}$$

It is clear that the amplitude factor in the square brackets is not symmetric about the focus but falls away more slowly on the side further from the vertex. This is particularly the case for values of f/D greater than 0.35.

13.8 Numerical values of scattered fields on axis for normally irradiated paraboloid

Calculations of the fields on the axis of a small commercial paraboloid of diameter 31.95 cm have been carried out for frequencies of 5, 10 and 20 GHz, corresponding to aperture diameters between 5.33 and 21.3 wavelengths, when it is irradiated by a normally incident plane wave. A range of values of f/D for the paraboloid between 0.1 and 1.0 has been considered, and in Fig. 13.5 the electric field E_y is shown for the 10 GHz case for four different values of f/D within this range, assuming unit incident field. It is seen that maximum field exists for the case of a f/D of 0.25, and for this case alone this maximum field exists at the geometric focus. For all other cases the maximum field is displaced from the geometric focus, though for small values of f/D this displacement is small.

Fig. 13.6 shows how the displacement varies with both frequency and f/D ratio. The first thing to notice in this figure is that as the frequency increases the displacement decreases, and indeed for optical frequencies the position of maximum received field always occurs at the geometric focus, no

Fig. 13.5. Axial electric field for paraboloid excited along axis by uniform plane wave.

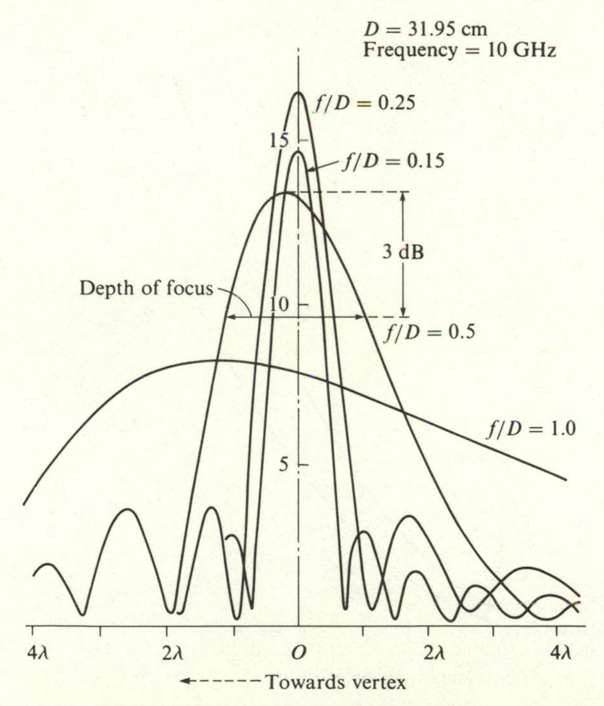

matter what the f/D ratio is. But for the microwave frequencies considered in this example the displacement may be several wavelengths for large values of f/D, though in such cases the field variation along the axis is a very flat one. Consequently, in this example, the field strength at the geometric focus is never more than 0.5 dB less than it is at the maximum field position, even when this separation is several wavelengths.

Defining the depth of focus along the z-axis as the distance between the two positions where the field has fallen to $1/\sqrt{2}$ of its peak value gives the graphs shown in Fig. 13.7. As the frequency increases so the depth of focus increases in terms of wavelengths, except for the case of a f/D of 0.25 where the depth of focus is independent of frequency over the range plotted. This type of variation is of interest for optical frequencies in pointing the advantage of a large f/D, i.e. a long focal length, in ease of positioning for finding the focus.

Fig. 13.8 shows how the maximum field value depends on the f/D ratio for different frequencies. Although the field at the geometric focus has previously been calculated analytically, this is not the same as the maximum field value shown here, except for the f/D of 0.25 when the two

Fig. 13.6. Displacement of true focus from geometrical focus.

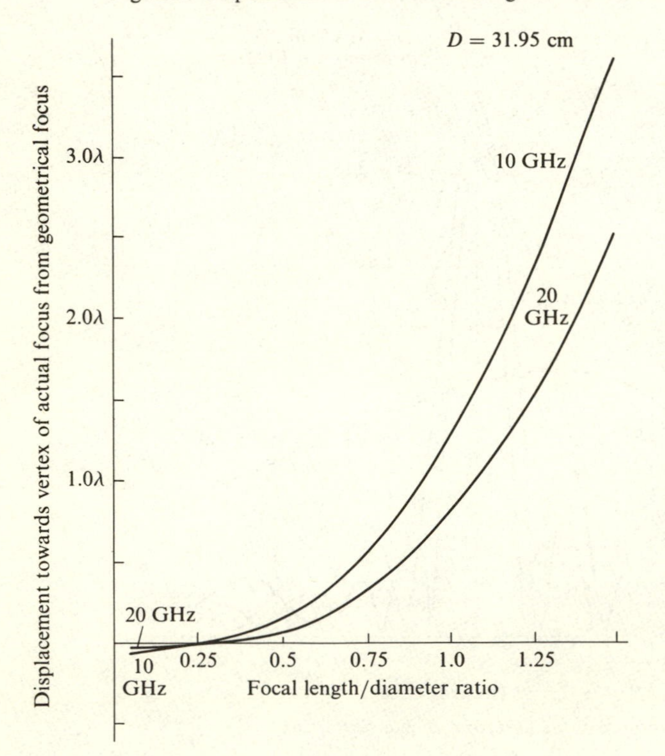

positions coincide. For this particular value of f/D the maximum field has a magnitude of $(\pi/2)(D/\lambda)$ times the incident field.

Experimental measurements of the electric field along the axis of a paraboloid irradiated by a normally incident uniform plane wave have been carried out by Landry and Chasse. For their paraboloid with a diameter of 69.1λ and f/D of 0.353, calculated results from eqn (13.15) are compared with their measurements in Fig. 13.9 and it can be seen that the agreement is excellent both with regard to the magnitude and position of each sidelobe.

13.9 Focal plane magnetic fields for paraboloid with normally incident plane wave

To examine the scattered field distribution in a plane orthogonal to the z-axis it is natural to consider first a plane though the maximum field point on this axis. To do this exactly requires a double numerical integration if a general point of observation in this plane is being considered. However, if attention is restricted to a region of the order of one

Fig. 13.7. Axial depth of focus vs focal length to diameter ratio.

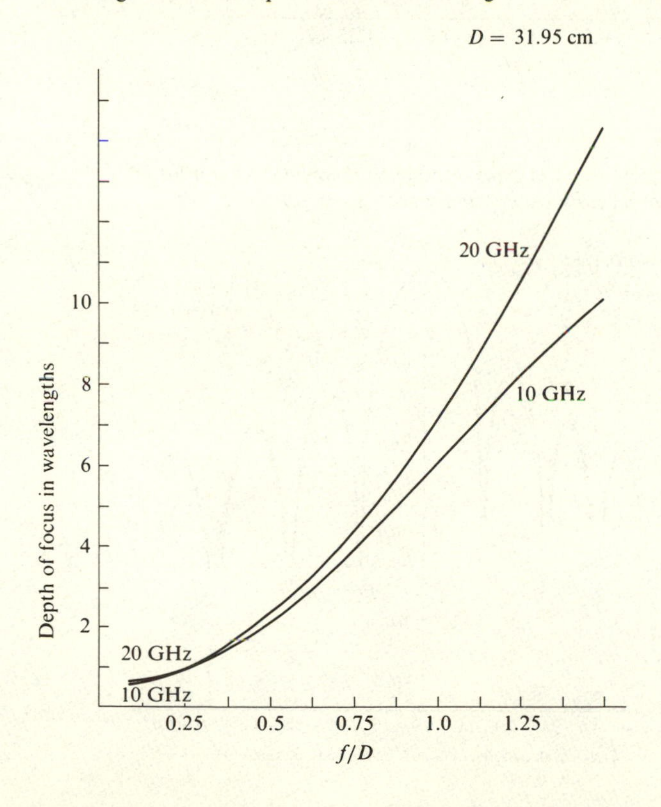

Fig. 13.8. Maximum field at true focus of paraboloid.

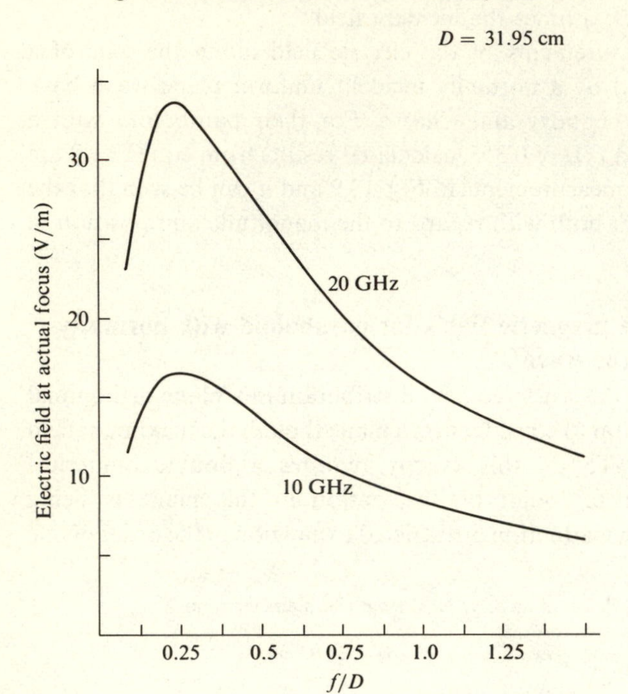

Fig. 13.9. Comparison of experimental and theoretical axial fields of paraboloidal reflector. (After Landry and Chasse, 1971.)

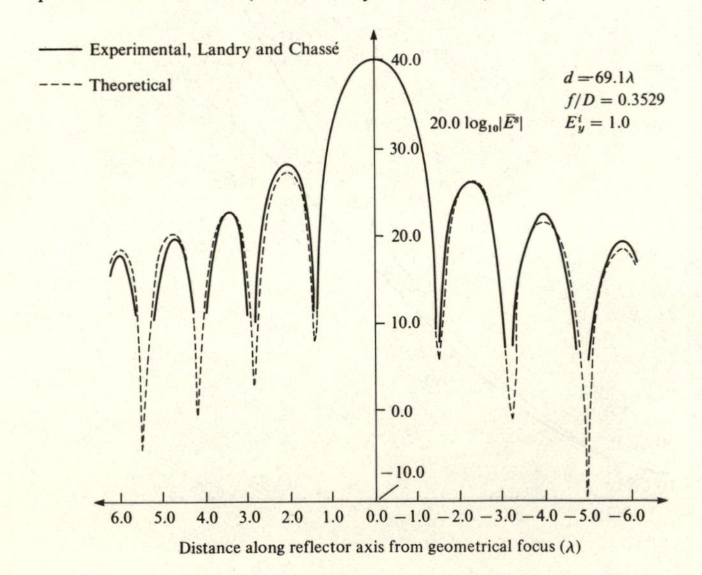

wavelength radius from the axis, it is found that the transverse field variation in this plane is virtually the same as the transverse variation in the plane of the geometric focus. Although the field on axis is, in general, slightly larger in the plane of the true focus it becomes slightly less beyond a radius of the order of half a wavelength. Consequently the fields in the plane of the geometric focus only will be calculated since the analysis and computation for this case are much simpler.

It will be recalled that the general expressions for the vector potentials due to an element of current associated with an incremental area dS of the paraboloid, when it is irradiated along its axis with an H_x field are,

$$dA_x = 0$$

$$dA_y = -\frac{\mu H_x^i}{2\pi u}\cos\frac{\theta}{2}e^{-jk(t+u)}dS$$

$$dA_z = \frac{\mu H_x^i}{2\pi u}\sin\frac{\theta}{2}\sin\phi e^{-jk(t+u)}dS$$

where

$$u = [r^2 + z^2 + \rho^2 - 2\rho r\sin\theta\cos(\phi - \alpha) - 2rz\cos\theta]^{\frac{1}{2}}$$

$$t = 2f - r$$

and

$$dS = r^2\sin\theta\sec\frac{\theta}{2}d\theta d\phi$$

The approximation will now be made that, for points in the focal plane with $\rho \ll r$,

$$u \approx r\left[1 - \frac{2\rho}{r}\sin\theta\cos(\phi - \alpha)\right]^{\frac{1}{2}} \approx r - \rho\sin\theta\cos(\phi - \alpha)$$

Then the components of vector potential will be written

$$dA_y \approx -\frac{\mu H_x^i}{2\pi r}\cos\frac{\theta}{2}e^{-jk[2f - \rho\sin\theta\cos(\phi-\alpha)]}dS$$

$$dA_z \approx \frac{\mu H_x^i}{2\pi r}\sin\frac{\theta}{2}\sin\phi e^{-jk[2f - \rho\sin\theta\cos(\phi-\alpha)]}dS$$

Then since

$$dH_x = \frac{1}{\mu}\left(\frac{\partial A_z}{\partial y} - \frac{\partial A_y}{\partial z}\right)$$

this gives

$$H_x = \frac{H_x^i}{2\pi}\int_0^{\theta_0}\int_0^{2\pi}\frac{e^{-jk(t+u)}}{r^2}\left(-jk - \frac{1}{r}\right)$$

$$\times \left\{ (\rho \sin\alpha - r\sin\theta\sin\phi)\sin\frac{\theta}{2}\sin\phi - r\cos\theta\cos\frac{\theta}{2} \right\}$$

$$\times r^2 \sin\theta \sec\frac{\theta}{2}\, d\theta\, d\phi$$

$$= \frac{H_x^i}{2\pi} e^{-jk2f}\left(-jk - \frac{1}{r}\right)\int_0^{\theta_0}\int_0^{2\pi}$$

$$\times \left\{ (\rho \sin\alpha - r\sin\theta\sin\phi)\sin\frac{\theta}{2}\sin\phi - r\cos\theta\cos\frac{\theta}{2} \right\}$$

$$\times \sin\theta \sec\frac{\theta}{2} e^{jk\rho\sin\theta\cos(\phi-\alpha)}\, d\theta\, d\phi$$

Using the standard integrals

$$\int_0^{2\pi} e^{ju\cos(\phi-\phi')} \begin{matrix}\sin\\ \cos\end{matrix} n\phi'\, d\phi' = 2\pi j^n \begin{matrix}\sin\\ \cos\end{matrix} n\phi J_n(u)$$

gives

$$H_x = \frac{H_x^i}{2\pi} e^{-jk2f}\left(-jk - \frac{1}{r}\right)$$

$$\times \int_0^{\theta_0} \left\{ j\rho \sin^2\alpha \sin\theta \tan\frac{\theta}{2} J_1(k\rho\sin\theta) \right.$$

$$- \frac{r}{2}\sin^2\theta \tan\frac{\theta}{2} J_0(k\rho\sin\theta) - \frac{r}{2}\sin^2\theta\tan\frac{\theta}{2}$$

$$\left. \times \cos 2\alpha J_2(k\rho\sin\theta) - r\cos\theta\sin\theta J_0(k\rho\sin\theta) \right\} 2\pi\, d\theta$$

$$= H_x^i e^{-jk2f}\left(-jk - \frac{1}{r}\right)\int_0^{\theta_0}\left[-f\sin\theta J_0(k\rho\sin\theta) \right.$$

$$+ j\rho\sin^2\alpha\sin\theta\tan\frac{\theta}{2} J_1(k\sin\theta)$$

$$\left. - f\sin\theta\tan^2\frac{\theta}{2} J_2(k\rho\sin\theta)\cos 2\alpha \right] d\theta \tag{13.18}$$

Since normally $\rho \ll f$ and $k \gg 1/r$ this may be simplified to

$$H_x = jkfH_x^i e^{-jk2f}\int_0^{\theta_0}\left[\sin\theta J_0(k\rho\sin\theta) \right.$$

$$\left. + \sin\theta\tan^2\frac{\theta}{2} J_2(k\rho\sin\theta)\cos 2\alpha \right] d\theta \tag{13.19}$$

Similarly, since

$$dH_y = \frac{1}{\mu}\left(\frac{\partial A_x}{\partial z} - \frac{\partial A_z}{\partial x}\right) = -\frac{1}{\mu}\frac{\partial A_z}{\partial x}$$

this gives

$$dH_y = -\frac{H_x^i}{2\pi}\sin\frac{\theta}{2}\sin\phi e^{-jkt}dS\frac{\partial}{\partial x}\left(\frac{e^{-jku}}{u}\right)$$
$$= -\frac{H_x^i}{2\pi}\sin\frac{\theta}{2}\sin\phi e^{-jkt}dS\frac{e^{-jku}}{u}\left(-jk-\frac{1}{u}\right)\left(\frac{x-x_s}{u}\right)$$
$$= -\frac{H_x^i}{2\pi}\sin\frac{\theta}{2}\sin\phi\frac{e^{-jk(t+u)}}{u^2}\left(-jk-\frac{1}{u}\right)$$
$$\times(\rho\cos\alpha - r\sin\theta\cos\phi)dS$$

Integrating over the paraboloidal surface gives, with $u \approx r$ in the amplitude term,

$$H_y = -\frac{H_x^i}{2\pi}\int_0^{\theta_0}\int_0^{2\pi}\frac{e^{-jk(t+u)}}{r^2}\left(-jk-\frac{1}{r}\right)$$
$$\times\left\{(\rho\cos\alpha - r\sin\theta\cos\phi)\sin\frac{\theta}{2}\sin\phi r^2\sin\theta\sec\frac{\theta}{2}d\theta\,d\phi\right.$$
$$= -\frac{H_x^i}{2\pi}e^{-jk2f}\left(-jk-\frac{1}{r}\right)\int_0^{\theta_0}$$
$$\times\left[j\rho\sin\alpha\cos\alpha\sin\theta\tan\frac{\theta}{2}J_1(k\rho\sin\theta)\right.$$
$$\left.+\frac{r}{2}\sin^2\theta\tan\frac{\theta}{2}\sin 2\alpha J_2(k\rho\sin\theta)\right]2\pi\,d\theta \tag{13.20}$$

For $\rho \ll f$ and $k \gg 1/r$ this simplifies to

$$H_y = jkfH_x^ie^{-jk2f}\int_0^{\theta_0}\sin\theta\tan^2\frac{\theta}{2}J_2(k\rho\sin\theta)\sin 2\alpha\,d\theta \tag{13.21}$$

Likewise since

$$dH_z = \frac{1}{\mu}\left(\frac{\partial A_y}{\partial x}-\frac{\partial A_x}{\partial y}\right) = \frac{1}{\mu}\frac{\partial A_y}{\partial x}$$

this gives

$$dH_z = -\frac{H_x^i}{2\pi}\cos\frac{\theta}{2}\frac{e^{-jk(t+u)}}{u^2}\left(-jk-\frac{1}{u}\right)$$
$$\times(\rho\cos\alpha - r\sin\theta\cos\phi)r^2\sin\theta\sec\frac{\theta}{2}d\theta d\phi$$

Integrating over θ and ϕ, with $u \approx r$ in the amplitude term, gives

$$H_z = -\frac{H_x^i}{2\pi}e^{-jk2f}\left(-jk-\frac{1}{r}\right)\int_0^{\theta_0}\int_0^{2\pi}$$
$$\times(\rho\cos\alpha - r\sin\theta\cos\phi)\sin\theta e^{jk\rho\sin\theta\cos(\phi-\alpha)}\,d\theta\,d\phi$$

$$= -\frac{H_x^i}{2\pi} e^{-jk2f}\left(-jk - \frac{1}{r}\right)\int_0^{\theta_0} [j\rho\cos^2\alpha\sin\theta J_1(k\rho\sin\theta)$$

$$-jr\sin^2\theta J_1(k\rho\sin\theta)\cos\alpha]2\pi\, d\theta \tag{13.22}$$

For $\rho \ll f$ and $k \gg 1/r$ this simplifies to

$$H_z = 4kfH_x^i e^{-jk2f}\int_0^{\theta_0} \sin^2\frac{\theta}{2} J_1(k\rho\sin\theta)\cos\alpha\, d\theta \tag{13.23}$$

The expressions for the three scalar components of **H** can be integrated in closed form provided the angle θ_0 is small, so that the trigonometric terms can be approximated by the first term in their power series representation. Thus for such cases, writing $k\rho\theta$ as x,

$$H_x \approx jkfH_x^i e^{-j2kf}\left\{\int \frac{x}{k^2\rho^2} J_0(x)\, dx + \cos 2\alpha \int \frac{x^3}{4k^4\rho^4} J_2(x)\, dx\right\}$$

$$= jkfH_x^i e^{-j2kf}\left\{\theta_0^2 \frac{J_1(k\rho\theta_0)}{k\rho\theta_0} + \frac{\theta_0^4}{4}\frac{J_3(k\rho\theta_0)}{k\rho\theta_0}\cos 2\alpha\right\}$$

$$H_y \approx jkfH_x^i e^{-jk2f}\left\{\frac{\theta_0^4}{4}\frac{J_3(k\rho\theta_0)}{k\rho\theta_0}\sin 2\alpha\right\}$$

$$H_z \approx 4kfH_x^i e^{-jk2f}\cos\alpha\left\{\int \frac{x^2}{4k^3\rho^3} J_1(x)\, dx\right\}$$

$$= kfH_x^i e^{-jk2f}\theta_0^3 \frac{J_2(k\rho\theta_0)}{k\rho\theta_0}\cdot\cos\alpha$$

However, these analytic results apply only when $f/D > 2.5$ approximately, and consequently are not applicable to normal paraboloid antennas which rarely have an $f/D > 1.0$. Consequently the magnetic field components in the focal plane have to be obtained by numerical integration of eqns (13.19), (13.21) and (13.23).

These integrals have been evaluated for a 31.95 cm diameter paraboloid at a frequency of 10 GHz for an f/D ratio of 0.25, and the results for H_x, H_y and H_z are shown in Fig. 13.10 in the form of contour plots. A reference value of unity has been taken for the amplitude of H_x at the geometrical focus and a phase reference of 0° at the same point. The contour plots for H_x are markedly elliptical and sidelobes occur along the y-axis with an amplitude of 35% of the main peak at a radius of approximately 0.7λ from the focus. The intervening minima have amplitudes less than 0.1 along this axis but less than 0.05 along the orthogonal x-axis.

The cross-polar H_y plot has a maximum amplitude of 0.184 at 45° to the x-axis and at a radius of 0.5λ from the focus, and is zero along both the x- and y-axis. It should be noted that a loop placed at the cross-polar peak with its plane parallel to the xz-plane would give an output voltage 18.4% of that from the same loop at the focus placed parallel to the yz-plane. Similarly a loop placed in the xy-plane at approximately 0.3λ from the focus along the x-axis would give an output voltage 64.5% of that from the main H_x-field.

Although the numerical results illustrated are for a small paraboloid of diameter 10.65λ it has been found that for aperture diameters greater than approximately 5.0λ contour plots are independent of aperture diameter when normalised to unit value at the focus. They do depend markedly, however, on f/D ratio. But before considering this dependence an examination will be made of how the electric field varies in the focal plane of the paraboloid.

Fig. 13.10. Contours of scattered magnetic field in geometrical focal plane of paraboloid: $f/D = 0.25$; (*a*) H_x^S; (*b*) H_y^S; (*c*) H_z^S.

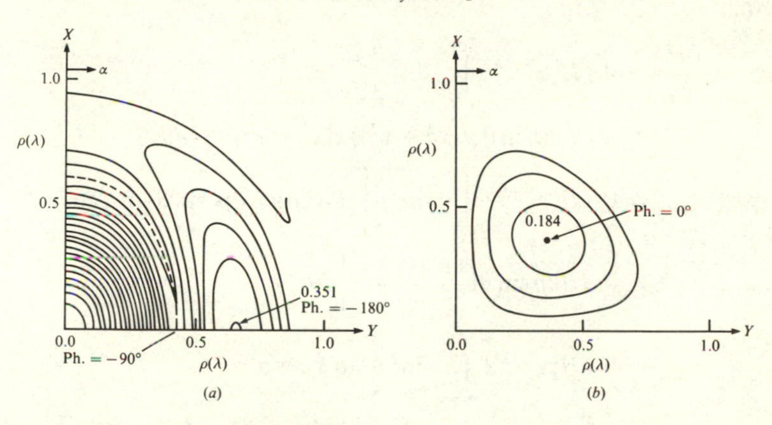

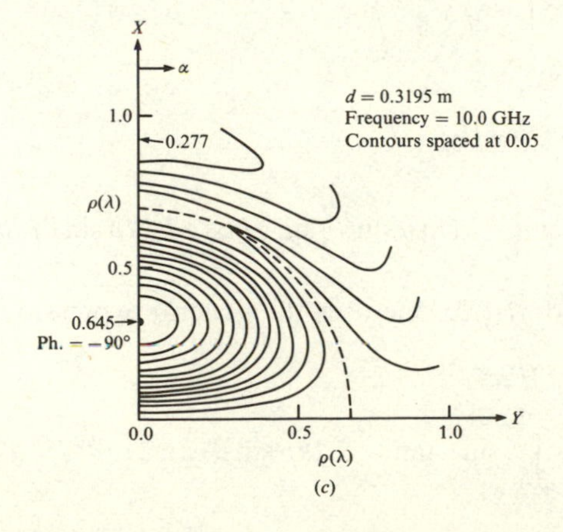

13.10 Focal plane electric fields for paraboloid with normally incident plane wave

From Maxwell's equations,

$$j\omega\varepsilon E_x = \frac{\partial H_z}{\partial y} - \frac{\partial H_y}{\partial z}$$

now

$$\frac{\partial H_z}{\partial y} = \frac{\partial}{\partial y}\left\{4kfH_x^i e^{-j2kf}\int_0^{\theta_0}\sin^2\frac{\theta}{2}J_1(k\rho\sin\theta)\cos\alpha\,d\theta\right\}$$

where

$$\rho = [x^2+y^2]^{\frac{1}{2}} \quad \text{and} \quad \cos\alpha = \frac{x}{[x^2+y^2]^{\frac{1}{2}}}$$

From Dwight,

$$\frac{d}{dc}\int_p^q f(x,c)\,dx = \int_p^q \frac{\partial}{\partial c}f(x,c)\,dx + f(q,c)\frac{dq}{dc} - f(p,c)\frac{dp}{dc}$$

so that

$$\frac{\partial H_z}{\partial y} = 4kfH_x^i e^{-j2kf}\int_0^{\theta_0}\sin^2\frac{\theta}{2}\frac{\partial}{\partial y}$$

$$\times[J_1(k\sin\theta\sqrt{(x^2+y^2)})x(x^2+y^2)^{-\frac{1}{2}}]\,d\theta$$

$$= 4kfH_x^i e^{-j2kf}\int_0^{\theta_0}\sin^2\frac{\theta}{2}\left[J_1'(k\rho\sin\theta)k\sin\theta\sin\alpha\cos\alpha\right.$$

$$\left. - J_1(k\rho\sin\theta)\frac{\sin\alpha\cos\alpha}{\rho}\right]d\theta$$

$$= 4kfH_x^i e^{-j2kf}\int_0^{\theta_0}\sin^2\frac{\theta}{2}\sin\alpha\cos\alpha$$

$$\times\left[k\sin\theta J_0(k\rho\sin\theta) - \frac{J_1(k\rho\sin\theta)}{\rho} - \frac{J_1(k\rho\sin\theta)}{\rho}\right]d\theta$$

i.e.

$$\frac{\partial H_z}{\partial y} = -4kfH_x^i e^{-j2kf}$$

$$\times\int_0^{\theta_0} k\sin\theta\sin^2\frac{\theta}{2}\sin\alpha\cos\alpha J_2(k\rho\sin\theta)\,d\theta \qquad \text{(a)}$$

Similarly, to find $\partial H_y/\partial z$ one uses the general expression for H_y,

$$H_y = jkfH_x^i e^{-j2kf}$$

$$\times\int_0^{\theta_0}\sin\theta\tan^2\frac{\theta}{2}J_2(k\rho\sin\theta)\sin 2\alpha e^{jkz\cos\theta}\,d\theta$$

which reduces in the focal plane to eqn (13.21).

Then

$$\left(\frac{\partial H_y}{\partial z}\right)_{z=0} = jkfH_x^i e^{-j2kf} \times \int_0^{\theta_0} \sin\theta \tan^2\frac{\theta}{2} J_2(k\rho\sin\theta)\sin 2\alpha \cdot jk\cos\theta\, d\theta \qquad \text{(b)}$$

Combining eqns (a) and (b) gives

$$E_x = \frac{k^2 f H_x^i e^{-j2kf}}{j\omega\varepsilon}\int_0^{\theta_0} \sin\theta\sin\alpha\cos\alpha J_2(k\rho\sin\theta) \times\left[-4\sin^2\frac{\theta}{2} + 2\cos\theta\tan^2\frac{\theta}{2}\right] d\theta$$

$$= jkfZ_0 H_x^i e^{-j2kf}\int_0^{\theta_0}\sin\theta\tan^2\frac{\theta}{2} J_2(k\rho\sin\theta)\sin 2\alpha\, d\theta \qquad (13.24)$$

$$= Z_0 H_y$$

Likewise E_y is found from Maxwell's equations,

$$j\omega\varepsilon E_y = \left(\frac{\partial H_x}{\partial z} - \frac{\partial H_z}{\partial x}\right)$$

Taking the general form of H_x as

$$H_x = jkfH_x^i e^{-jk2f}\int_0^{\theta_0}\left[\sin\theta J_0(k\rho\sin\theta) + \sin\theta\tan^2\frac{\theta}{2} J_2(k\rho\sin\theta)\cos 2\alpha\right] e^{jkz\cos\theta}\, d\theta$$

gives

$$\left(\frac{\partial H_x}{\partial z}\right)_{z=0} = -k^2 f H_x^i e^{-jk2f}\int_0^{\theta_0}\left[\sin\theta\cos\theta J_0(k\rho\sin\theta) + \sin\theta\cos\theta\tan^2\frac{\theta}{2} J_2(k\rho\sin\theta)\cos 2\alpha\right] d\theta \qquad \text{(c)}$$

Since

$$H_z = 4kfH_x^i e^{-jk2f}\int_0^{\theta_0}\sin^2\frac{\theta}{2} J_1(k\rho\sin\theta)\cos\alpha\, d\theta$$

then

$$\frac{\partial H_z}{\partial x} = 4kfH_x^i e^{-jk2f}\int_0^{\theta_0}\sin^2\frac{\theta}{2}\frac{\partial}{\partial x} \times [J_1(k\sin\theta\sqrt{(x^2+y^2)})\cdot x(x^2+y^2)^{-\frac{1}{2}}]\, d\theta$$

$$= 4kfH_x^i e^{-jk2f}\int_0^{\theta_0}\sin^2\frac{\theta}{2}$$

$$\times\left[J_1'(k\rho\sin\theta)\cdot k\sin\theta\cos^2\alpha + J_1(k\rho\sin\theta)\cdot\frac{1}{\rho}\right.$$

$$\left. - J_1(k\rho\sin\theta)\cdot\frac{\cos^2\alpha}{\rho}\right]d\theta \tag{d}$$

Combining eqns (c) and (d) gives

$$E_y = \frac{k^2 f H_x^i e^{-jk2f}}{j\omega\varepsilon}\int_0^{\theta_0}\left\{-\sin\theta\cos\theta J_0(k\rho\sin\theta)\right.$$

$$-\sin\theta\cos\theta\tan^2\frac{\theta}{2}J_2(k\rho\sin\theta)\cos 2\alpha$$

$$-4\sin^2\frac{\theta}{2}\sin\theta\cos^2\alpha\left[J_0(k\rho\sin\theta) - \frac{J_1(k\rho\sin\theta)}{k\rho\sin\theta}\right]$$

$$\left.-4\sin^2\frac{\theta}{2}\sin\theta\left[\frac{J_1(k\rho\sin\theta)}{k\rho\sin\theta} - \frac{J_1(k\rho\sin\theta)}{k\rho\sin\theta}\cos^2\alpha\right]\right\}d\theta$$

$$= \frac{k^2 f H_x^i e^{-jk2f}}{j\omega\varepsilon}\int_0^{\theta_0}\left\{-J_0(k\rho\sin\theta)\right.$$

$$\times\sin\theta\left[\cos\theta + 4\sin^2\frac{\theta}{2}\cos^2\alpha\right] + 4\frac{J_1(k\rho\sin\theta)}{k\rho\sin\theta}$$

$$\times\left[\sin^2\frac{\theta}{2}\sin\theta\cos^2\alpha - \sin^2\frac{\theta}{2}\sin\theta + \sin^2\frac{\theta}{2}\sin\theta\cos^2\alpha\right]$$

$$\left.-J_2(k\rho\sin\theta)\cdot\sin\theta\cos\theta\tan^2\frac{\theta}{2}\cos 2\alpha\right\}d\theta$$

$$= \frac{k^2 f H_x^i e^{-jk2f}}{j\omega\varepsilon}\int_0^{\theta_0}\left\{-J_0(k\rho\sin\theta)\cdot\left(\sin\theta\cos\theta + 2\sin^2\frac{\theta}{2}\sin\theta\right)\right.$$

$$+\left[\frac{2J_1(k\rho\sin\theta)}{k\rho\sin\theta} - J_0(k\rho\sin\theta)\right]2\sin^2\frac{\theta}{2}\sin\theta\cos 2\alpha$$

$$\left.-J_2(k\rho\sin\theta)\sin\theta\cos\theta\tan^2\frac{\theta}{2}\cos 2\alpha\right\}d\theta$$

$$= jkfZ_0 H_x^i e^{-jk2f}\int_0^{\theta_0}\left[J_0(k\rho\sin\theta)\cdot\sin\theta - J_2(k\rho\sin\theta)\right.$$

$$\left.\times\sin\theta\tan^2\frac{\theta}{2}\cos 2\alpha\right]d\theta \tag{13.25}$$

Thus the ratio of E_y/H_x is equal to Z_0 when α is equal to $\pi/4$, $3\pi/4$, etc.

The remaining component of electric field E_z is obtained from

$$j\omega\varepsilon E_z = \frac{\partial H_y}{\partial x} - \frac{\partial H_x}{\partial y}$$

Now

$$\frac{\partial H_y}{\partial x} = jkfH_x^i e^{-j2kf} \int_0^{\theta_0} \left\{ \sin\theta \tan^2\frac{\theta}{2} \right.$$

$$\times \left[J_2'(k\rho\sin\theta)\cdot k\sin\theta\cos\alpha\sin 2\alpha \right.$$

$$\left.\left. + 2J_2(k\rho\sin\theta)\frac{\sin\alpha}{\rho} - 4J_2(k\rho\sin\theta)\frac{\sin\alpha\cos^2\alpha}{\rho} \right]\right\} d\theta$$

and

$$\frac{\partial H_x}{\partial y} = jkfH_x^i e^{-j2kf} \int_0^{\theta_0} \left\{ -\sin\theta J_1(k\rho\sin\theta) \right.$$

$$\times k\sin\theta\sin\alpha + \sin\theta\tan^2\frac{\theta}{2} J_2'(k\rho\sin\theta)$$

$$\times k\sin\theta\sin\alpha\cos 2\alpha + \sin\theta\tan^2\frac{\theta}{2} J_2(k\rho\sin\theta)$$

$$\left. \times \left[-\frac{2\cos^2\alpha\sin\alpha}{\rho} - \frac{2\sin\alpha}{\rho} + \frac{2\sin^3\alpha}{\rho} \right] \right\} d\theta$$

Combining the two components gives

$$E_z = kfZ_0 H_x^i e^{-j2kf} \int_0^{\theta_0}$$

$$\times \left\{ J_2'(k\rho\sin\theta)\sin^2\theta\tan^2\frac{\theta}{2}(\cos\alpha\sin 2\alpha - \sin\alpha\cos 2\alpha) \right.$$

$$+ J_1(k\rho\sin\theta)\sin^2\theta\sin\alpha + \frac{J_2(k\rho\sin\theta)}{k\rho\sin\theta}$$

$$\times \left[2\sin^2\theta\tan^2\frac{\theta}{2}\sin\alpha - 4\sin^2\theta\tan^2\frac{\theta}{2}\sin\alpha\cos^2\alpha \right.$$

$$+ 2\sin^2\theta\tan^2\frac{\theta}{2}\sin\alpha\cos^2\alpha$$

$$\left.\left. + 2\sin^2\theta\tan^2\frac{\theta}{2}\sin\alpha - 2\sin^2\theta\tan^2\frac{\theta}{2}\sin^3\alpha \right]\right\} d\theta$$

i.e.

$$E_z = kfZ_0 H_x^i e^{-j2kf} \int_0^{\theta_0} \left\{ J_2'(k\rho\sin\theta) \right.$$

$$\times \sin^2\theta\tan^2\frac{\theta}{2}\sin\alpha + J_1(k\rho\sin\theta)\sin^2\theta\sin\alpha$$

$$+ 2\frac{J_2(k\rho\sin\theta)}{k\rho\sin\theta}\sin^2\theta\tan^2\frac{\theta}{2}\sin\alpha$$

$$\times [1 - 2\cos^2\alpha + \cos^2\alpha + 1 - \sin^2\alpha] \Big\} d\theta$$

$$= kfZ_0 H_x^i e^{-j2kf} \int_0^{\theta_0} \sin\theta \sin\alpha \Big\{ \sin\theta \tan^2\frac{\theta}{2} \frac{2J_0(k\rho\sin\theta)}{k\rho\sin\theta}$$

$$+ \sin\theta \tan^2\frac{\theta}{2}\left(1 - \frac{4}{k^2\rho^2\sin^2\theta}\right) J_1(k\rho\sin\theta)$$

$$+ \sin\theta J_1(k\rho\sin\theta) + 2\frac{\sin\theta\tan^2\dfrac{\theta}{2}}{k\rho\sin\theta}$$

$$\times \left[\frac{2J_1(k\rho\sin\theta)}{k\rho\sin\theta} - J_0(k\rho\sin\theta)\right]\Big\} d\theta$$

$$= kfZ_0 H_x^i e^{-j2kf} \int_0^{\theta_0} \sin\theta \sin\alpha$$

$$\times \Big\{ \frac{2J_0(k\rho\sin\theta)}{k\rho\sin\theta}\left[\sin\theta\tan^2\frac{\theta}{2} - \sin\theta\tan^2\frac{\theta}{2}\right]$$

$$+ J_1(k\rho\sin\theta)\left[\sin\theta\tan^2\frac{\theta}{2}\left(1 - \frac{4}{k^2\rho^2\sin^2\theta}\right) + \sin\theta\right.$$

$$\left. + \frac{4}{k^2\rho^2\sin^2\theta}\sin\theta\tan^2\frac{\theta}{2}\right]\Big\} d\theta$$

$$= 4kfZ_0 H_x^i e^{-j2kf} \int_0^{\theta_0} \sin^2\frac{\theta}{2} J_1(k\rho\sin\theta)\sin\alpha \, d\theta$$

$$= Z_0 \tan\alpha H_z \tag{13.26}$$

The expressions for these electric field components E_x, E_y and E_z for the same paraboloid, as before, have been computed and are shown in Fig. 13.11, normalised to the value at the focus. As has previously been stated the E_x-component is directly proportional to the magnetic component H_y and the E_z component is the same as the H_z, but rotated through 90°. Likewise the contours of the main E_y-component are virtually the same as for the H_x-component, but again rotated in space by 90°. It has been found numerically that along the $\alpha = 45°$ line there is a Z_0 relationship, within an accuracy of 1%, between E_y and H_x, out to a radius of approximately 0.45λ.

To illustrate the effect of different degrees of curvature on the focal plane pattern, contours of the electric field components for f/D ratios of 0.1 and 1.0 are shown in Figs. 13.12 and 13.13 for the same diameter of paraboloid, and at the same frequency. By comparing the results for the three different f/D ratios it is clear that the central spot becomes increasingly circular as the f/D ratio increases, and the orthogonal E_x- and E_z-components decrease.

These results are shown quantitatively in Fig. 13.14 in which the eccentricity e of the central spot, defined as the ratio of the distances S_y, S_x from the focus to the first minima in the y- and x-directions, is shown as a function of f/D. The Airy result, valid for large f/D ratio, is also shown for comparison. Also shown in Fig. 13.14 are the magnitudes and positions of the cross-polar E_x- and E_y-fields, and it can be seen that such fields can actually rise to exceed the y-component for very small f/D ratios.

13.11 Power flow across focal plane

The axial Poynting vector is given by

$$P_z = \tfrac{1}{2}\,\mathrm{Re}(E_x H_y^* - E_y H_x^*)$$

Fig. 13.11. Contours of scattered electric field in geometrical focal plane of paraboloid: $f/D = 0.25$: (*a*) E_x^S; (*b*) E_y^S; (*c*) E_z^S.

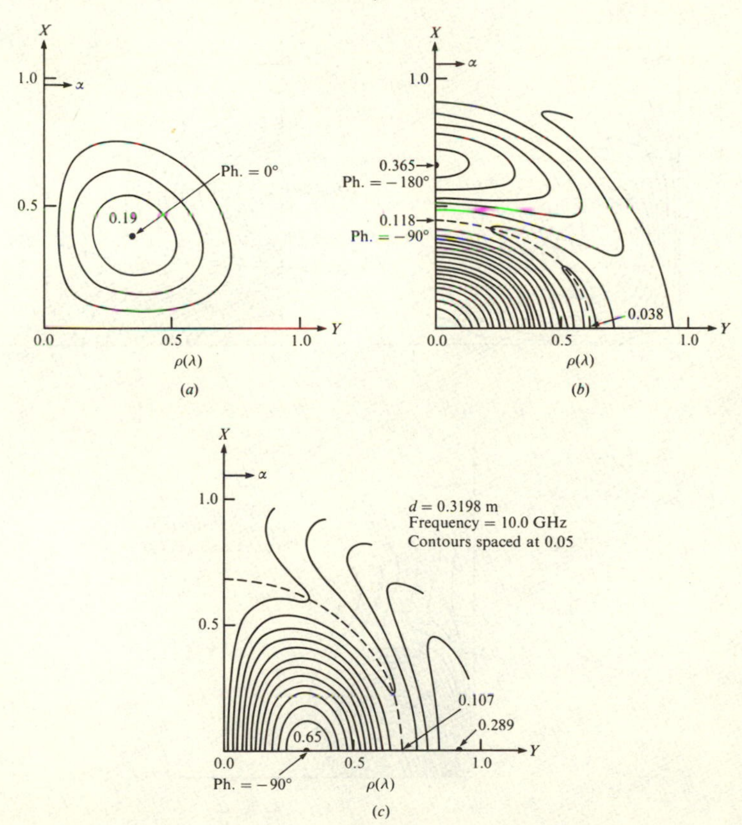

where

$$E_x = jkfZ_0H_x^i e^{-j2kf} \int_0^{\theta_0} \sin\theta \tan^2\frac{\theta}{2} J_2(k\rho\sin\theta)\sin 2\alpha \, d\theta$$

$$H_y = \frac{E_x}{Z_0}$$

$$E_y = jkfZ_0H_x^i e^{-j2kf} \int_0^{\theta_0} \times \sin\theta\left[J_0(k\rho\sin\theta) - J_2(k\rho\sin\theta)\tan^2\frac{\theta}{2}\cos 2\alpha\right] d\theta$$

$$H_x = jkfZ_0H_x^i e^{-jk2f} \int_0^{\theta_0} \times \sin\theta\left[J_0(k\rho\sin\theta) + J_2(k\rho\sin\theta)\tan^2\frac{\theta}{2}\cos 2\alpha\right] d\theta$$

Fig. 13.12. Contours of scattered electric field in geometrical focal plane of paraboloid: $f/D = 0.1$: (*a*) E_x^S; (*b*) E_y^S; (*c*) E_z^S.

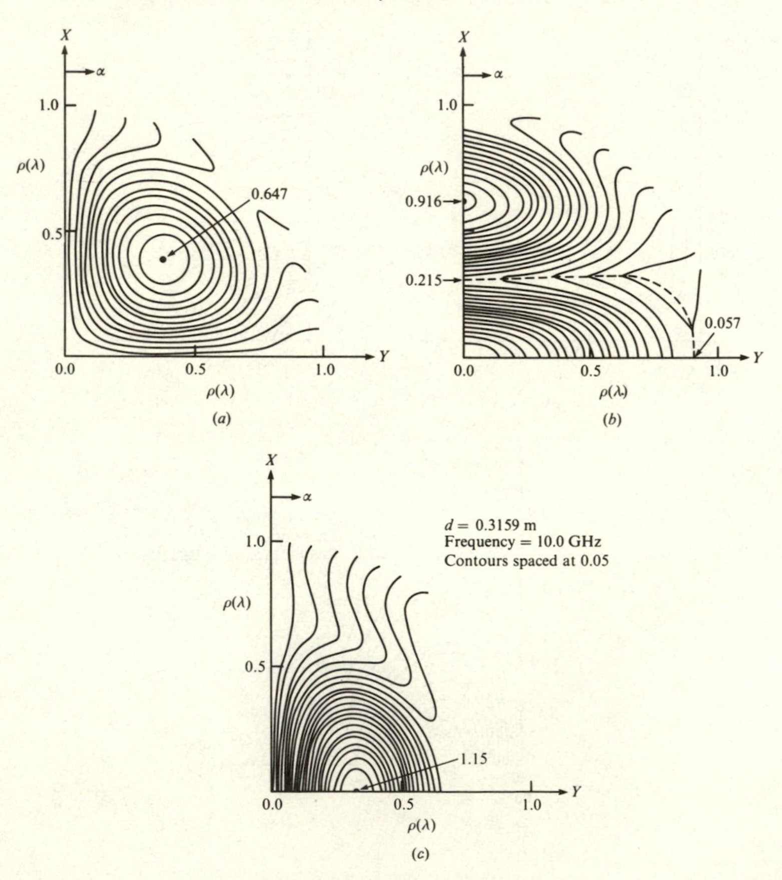

Defining

$$I_1 = \int_0^{\theta_0} \sin\theta J_0(k\rho\sin\theta)\,d\theta$$

$$I_2 = \int_0^{\theta_0} \sin\theta\tan^2\frac{\theta}{2} J_2(k\rho\sin\theta)\,d\theta$$

gives

$$\begin{aligned}
P_z &= \tfrac{1}{2}k^2 f^2 Z_0 H_x^{i2} \\
&\quad\times [I_2^2\sin^2 2\alpha - I_1^2 - I_1 I_2\cos 2\alpha + I_1 I_2\cos 2\alpha + I_2^2\cos^2 2\alpha] \\
&= \tfrac{1}{2}k^2 f^2 Z_0 H_x^{i2}(I_2^2 - I_1^2) \\
&= \frac{k^2 f^2 Z_0 H_x^{i2}}{2}\left\{\left[\int_0^{\theta_0} \sin\theta\tan^2\frac{\theta}{2}\cdot J_2(k\rho\sin\theta)\right]^2 \right. \\
&\quad \left. - \left[\int_0^{\theta_0} \sin\theta J_0(k\rho\sin\theta)\right]^2\right\} \qquad (13.27)
\end{aligned}$$

and it will be noted that this power density is axially symmetric since it is independent of α.

Fig. 13.13. Contours of scattered electric field in geometrical focal plane of paraboloid: $f/D = 1.0$: (*a*) E_x^S; (*b*) E_y^S; (*c*) E_z^S.

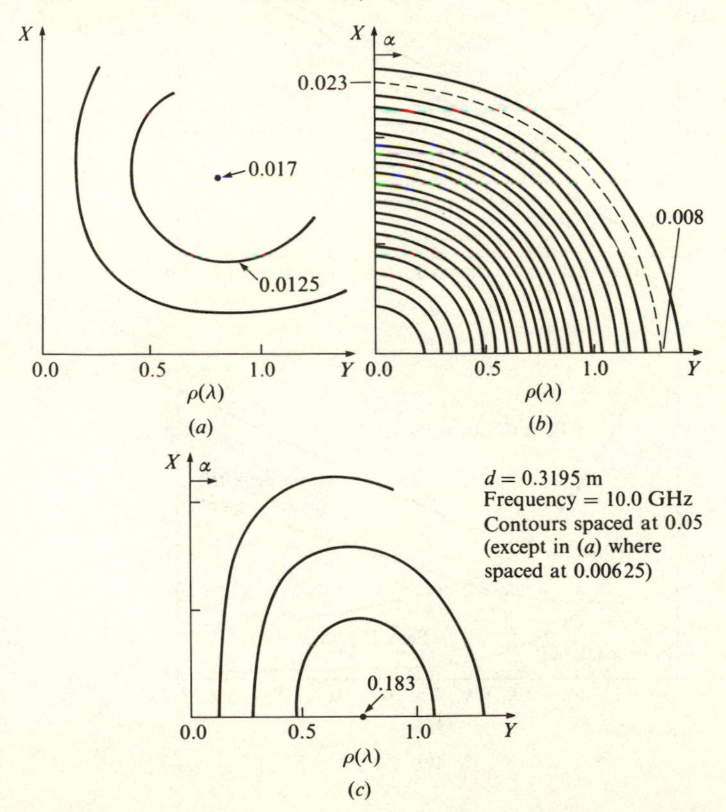

Assuming that all of this power can be collected by a circular aperture of radius a gives the available power as

$$W_1 = \int_0^a \tfrac{1}{2}k^2 f^2 Z_0 H_x^{i2}(I_2^2 - I_1^2)2\pi\rho \, d\rho$$

since P_z is symmetric in azimuth. But the incident power on the aperture of the paraboloid is

$$W_2 = \frac{\pi D^2}{4}\frac{H_x^{i2} Z_0}{2}$$

$$= 2\pi f^2 \tan^2\frac{\theta_0}{2} H_x^{i2} Z_0 \qquad (13.28)$$

Fig. 13.14. Focal plane electric field characteristics of paraboloid: (*a*) central spot eccentricity of E_y^S; (*b*) magnitude and position of peak cross-polar fields E_x^S and E_z^S.

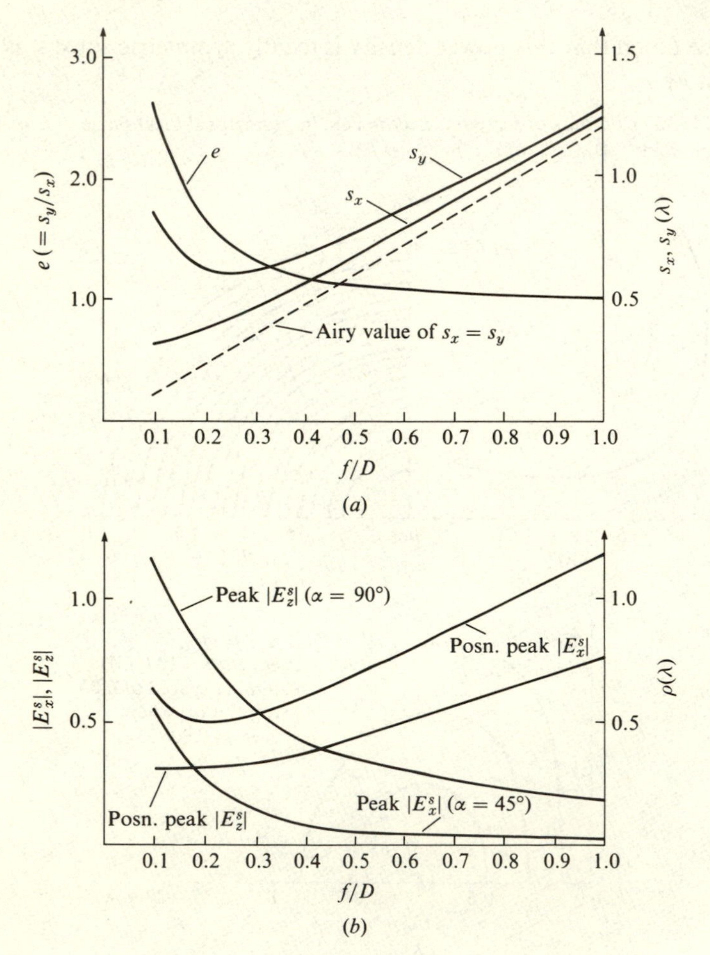

since $\tan\theta_0/2 = D/4f$, so that for this idealised feed the efficiency of the combined reflector–feed combination is

$$\eta = \frac{W_1}{W_2} = \frac{k^2}{2\tan^2\frac{\theta_0}{2}} \int_0^a [I_2^2 - I_1^2]\rho\, d\rho$$

For the case of the very long focal length paraboloid treated previously it was shown that

$$I_1 \approx \theta_0^2 \frac{J_1(k\rho\theta_0)}{k\rho\theta_0} \quad \text{and} \quad I_2 \approx \frac{\theta_0^4}{4}\frac{J_3(k\rho\theta_0)}{k\rho\theta_0} \approx 0$$

so that the efficiency becomes

$$\eta \approx -2 \int_0^{ka\theta_0} \frac{J_1^2(v)}{v}\, dv$$

Since

$$\frac{J_1^2(v)}{v} = -\frac{1}{2}\frac{d}{dv}[J_0^2(v) + J_1^2(v)]$$

this gives

$$\eta = 1 - J_0^2(ka\theta_0) - J_1^2(ka\theta_0) \tag{13.29}$$

Further reading

P.J.B. Clarricoats and G.T. Poulton: 'High efficiency microwave reflector antennas' *Proc. IEEE*, **65**, 1977, 1470–1504.

R.E. Collin and F.J. Zucker: *Antenna Theory*: McGraw-Hill, 1969.

R.C. Hansen: *Microwave Scanning Antennas, Vol. 1*: Academic Press, New York, 1964.

H. Jasik (Ed): *Antenna Engineering Handbook*: McGraw-Hill, New York, 1961.

G. Morris: Coupling between aperture antennas: PhD thesis, University of Birmingham, 1978.

J.F. Ramsay: *Tubular Beams from Radiating Apertures: Advances in Microwaves*: Academic Press, New York, 1968, Vol. 3, pp. 127–221.

W.V.T. Rusch and P.D. Potter: *Analysis of Reflector Antennas*: Academic Press, New York, 1970.

S. Silver: *Microwave Antenna Theory and Design*: McGraw-Hill, 1949.

FIG. 13.9.: M. Landry and Y. Chasse: 'Measurement of electromagnetic field, intensity in focal region of wide-angle paraboloid reflector: *Trans. IEEE* (*APG*), **AP-19**, 1971. (Copyright C, 1971, *IEEE*.)

14

+ − +

Receiving paraboloidal reflector with feed

In the previous chapter the receiving properties of an isolated paraboloidal reflector excited by a plane wave travelling along its axis were analysed, and numerical values were presented for the fields along that axis in the transverse focal plane. The purpose of a feed placed at the focus in a receiving situation is to abstract power from the concentrated fields set up in that region by the reflector, and the main part of this chapter will be concerned with quantifying this power for different types of feeds.

14.1 Infinitesimal dipole feed

First consider the simplest possible case of an infinitesimal dipole placed at the geometrical focus. It has previously been shown that the field magnitude there is related to the incident field by a factor $2kf\sin^2(\theta_0/2)$ when the high frequency approximation is used. The output power available from the infinitesimal dipole is then

$$W_1 = \frac{V_{oc}^2}{8R_r}$$

where V_{oc} is the open-circuited voltage at its terminals, and R_r is its radiation resistance. This radiation resistance can be considered to be unaffected by the presence of the reflector since in general it will be at least several wavelengths from it. Thus R_r has its isolated value given by

$$R_r = 5k^2dl^2$$

and the open-circuited voltage is

$$V_{oc} = E\frac{dl}{2}$$

where E is equal to $2kf\sin^2(\theta_0/2)$ times the incident field E^i. The available power from the dipole is then

$$W_1 = \frac{E^2 dl^2}{160k^2 dl^2} = \frac{\lambda^2 k^2 f^2}{160\pi^2} \sin^4 \frac{\theta_0}{2} E_i^2$$

By comparison an isotropic antenna would deliver a power W_2 given by

$$W_2 = \frac{E_1^2}{2Z_0} A_{eff}$$

where A_{eff} for an isotropic antenna, which has a power gain of unity by definition, is given by

$$A_{eff} = \frac{\lambda^2}{4\pi}$$

The power W_2 is then

$$W_2 = \frac{E_i^2}{2Z_0} \frac{\lambda^2}{4\pi}$$

and the power gain of this paraboloid with its infinitesimal dipole feed is

$$G = \frac{W_1}{W_2}$$

$$= \frac{\lambda^2 k^2 f^2}{160\pi^2} \frac{\sin^4 \frac{\theta_0}{2} 8\pi Z_0}{\lambda^2}$$

$$= 6k^2 f^2 \sin^4 \frac{\theta_0}{2}$$

The aperture efficiency is defined as the ratio of the output power W_1 to the power incident on the aperture of the paraboloid. This gives

$$\eta = \frac{\lambda^2 k^2 f^2 \sin^4 \frac{\theta_0}{2} E_i^2 2Z_0 \cdot 4}{160\pi^2 E_i^2 \pi D^2}$$

$$= 24 \left(\frac{f}{D}\right)^2 \sin^4 \frac{\theta_0}{2}$$

For a focal plane paraboloid with $f/D = 0.25$ and $\theta_0 = 90°$ this gives

$$\eta = 0.375$$

This is a very low value of aperture efficiency and it will be shown later how other types of feed enable much higher powers to be abstracted from the incoming wave. But before that is done an analysis will be carried out to show how the receiving radiation pattern of a paraboloidal reflector with an infinitesimal dipole feed may be found.

14.2 Receiving radiation pattern of paraboloid with infinitesimal dipole feed

Let the paraboloid be irradiated by a plane wave travelling horizontally in the yz-plane with its magnetic vector in the x-direction, and let the angle made by the propagation vector k with the positive z-axis be $(\pi - \beta)$. Thus when β is zero the wave is normally incident on the paraboloid and an infinitesimal dipole feed placed at the focus and aligned parallel to the y-axis would pick up maximum signal.

For other values of β the phase of the currents flowing in the reflector will be a function not only of the angular coordinate θ of the current element but also of its angular coordinate ϕ. Thus taking a phase reference at the vertex V of the paraboloid means, referring to Fig. 14.1, that the phase of the current at an element at $P(\theta, \phi)$ will be advanced on that at the vertex by $k(VP\cos\psi)$ where

$$\cos\psi = \cos\alpha\cos(\pi - \beta) + \sin\alpha\sin(\pi - \beta)\cos\left(\phi - \frac{\pi}{2}\right) \tag{14.1}$$

In eqn (14.1) α is the angle that VP makes the negative z-axis, such that

$$VP\sin\alpha = r\sin\theta$$

But VP, from the cosine rule for triangles, is given by

$$VP = [r^2 + f^2 - 2rf\cos\theta]^{\frac{1}{2}} = S$$

so that

$$\begin{aligned} S\cos\psi &= -S\cos\alpha\cos\beta + S\sin\alpha\sin\beta\sin\phi \\ &= -S\cos\beta\left[1 - \frac{r^2\sin^2\theta}{S^2}\right]^{\frac{1}{2}} + r\sin\theta\sin\beta\sin\phi \\ &= -\cos\beta[r^2\cos^2\theta + f^2 - 2rf\cos\theta]^{\frac{1}{2}} + r\sin\theta\sin\beta\sin\phi \\ &= -\cos\beta(f - r\cos\theta) + r\sin\theta\sin\beta\sin\phi \\ &= -f\tan^2\frac{\theta}{2}\cos\beta + 2f\tan\frac{\theta}{2}\sin\beta\sin\phi \end{aligned}$$

Fig. 14.1. Paraboloid illuminated by plane wave arriving off-axis.

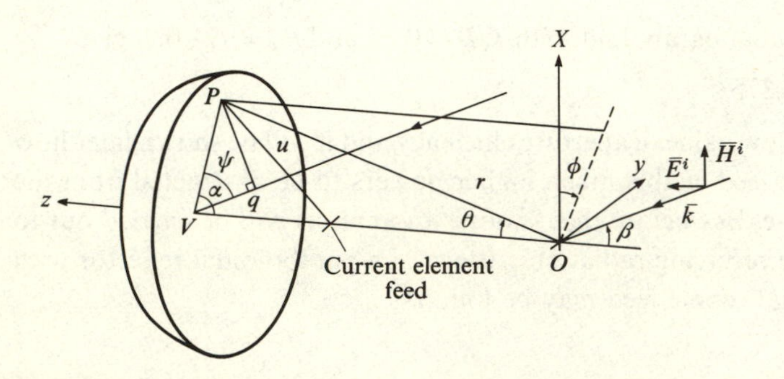

The reradiated field at a point of observation distant u from an element of current on the paraboloid will then be delayed with respect to the vertex by $k(-S\cos\psi+u)$, or with respect to the focus by $k(f-S\cos\psi+u)$. Apart from this the integrand for the scattered field components remains as for normal incidence, so that at the focus, assuming that the whole paraboloid is irradiated,

$$H_x=\frac{H_x^i}{2}e^{-jkf}\int_0^{\theta_0}\int_0^{2\pi} e^{-jk(r+f\tan^2(\theta/2)\cos\beta-2f\tan(\theta/2)\sin\beta\sin\phi)}$$
$$\times\left(-\frac{jk}{r^2}-\frac{1}{r^3}\right)\left\{-r\sin\theta\sin\phi\sin\frac{\theta}{2}\sin\phi-r\cos\theta\cos\frac{\theta}{2}\right\}$$
$$\times r^2\sin\theta\sec\frac{\theta}{2}\,d\theta\,d\phi$$

Then, using the standard integral,

$$\int_0^{2\pi}\cos n\phi\, e^{jv\sin\phi}\,d\phi=2\pi j^n\cos\frac{n\pi}{2}J_n(v)$$

gives

$$H_x=H_x^i e^{-j2kf}\int_0^{\theta_0}\left(-\frac{jk}{r^2}-\frac{1}{r^3}\right)e^{-jk(r+f\tan^2(\theta/2)\cos\beta)}$$
$$\times r^2\sin\theta\sec\frac{\theta}{2}\int_0^{2\pi} e^{jk2f\tan(\theta/2)\sin\beta\sin\phi}$$
$$\times\left(-\frac{r}{2}\sin\frac{\theta}{2}\sin\theta-r\cos\frac{\theta}{2}\cos\theta+\frac{r}{2}\sin\frac{\theta}{2}\sin\theta\cos 2\phi\right)d\theta\,d\phi$$
$$=H_x^i e^{-j2kf}\int_0^{\theta_0}\left(-\frac{jk}{r^2}-\frac{1}{r^3}\right)e^{-jk(r+f\tan^2(\theta/2)\cos\beta)}\cdot r^2\sin\theta\sec\frac{\theta}{2}$$
$$\times\left\{\left(-r\sin^2\frac{\theta}{2}\cos\frac{\theta}{2}-r\cos\frac{\theta}{2}\cos\theta\right)J_0\left(2kf\tan\frac{\theta}{2}\sin\beta\right)\right.$$
$$\left.+r\sin^2\frac{\theta}{2}\cos\frac{\theta}{2}J_2\left(2kf\tan\frac{\theta}{2}\sin\beta\right)\right\}d\theta$$

Fig. 14.2. Illustration of maximum value of pattern coordinate β before blockage occurs.

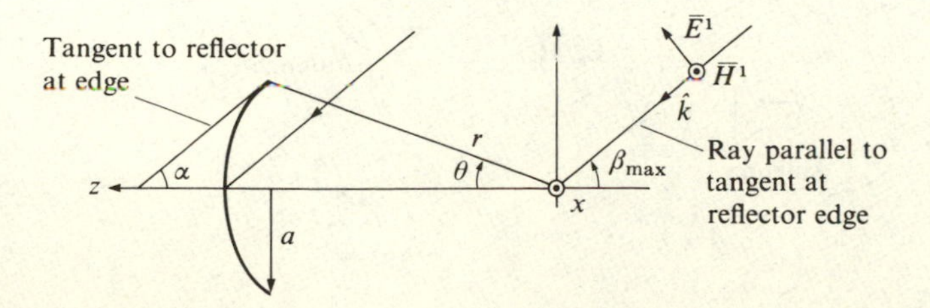

This is readily simplified to give

$$H_x = H_x^i e^{-j2kf} \int_0^{\theta_0} \left[J_0\left(2kf\tan\frac{\theta}{2}\sin\beta\right) \right.$$
$$\times \left(\frac{1}{r}+jk\right)f - J_2\left(2kf\tan\frac{\theta}{2}\sin\beta\right)(1+jkr)\sin^2\frac{\theta}{2}\left.\right]$$
$$\times \sin\theta e^{-jk(r+f\tan^2(\theta/2)\cos\beta)}\,d\theta \tag{14.2}$$

and the electric field may similarly be found in single integral form.

It should, however, be noted that the result applies only up to that angle for which irradiation of the whole inner surface by the incoming wave takes place. Beyond this angle β_{max}, shown in Fig. 14.2, an increasing part of the reflector is blocked off, in geometrical optics terms, and the integration on ϕ can no longer be performed analytically. The value of β_{max} is given by

$$\beta_{max} = \tan^{-1}\frac{4f}{D}$$

so that for an f/D of 0.25, β_{max} is equal to 45° in azimuth.

As β is varied between $\pm\beta_{max}$ the value of H_x given by eqn (14.2) represents the H-plane pattern within that range of the reflector plus y-directed infinitesimal dipole feed. Likewise the electric field E_y can be found in the same way and the variation of this with β represents the E-plane pattern of the antenna between the same limits.

14.3 Power coupling between apertures

Consider the antenna shown in Fig. 14.3. Let P_{1a} be the power radiated by antenna 1, and let P_{2a} be the corresponding receiving power into a matched load which is picked up by antenna 2 from antenna 1. The fields E_a, H_a at any observation point are related to a source current i_a associated with antenna 1 by

$$\nabla \times \mathbf{H}_a = \mathbf{i}_a + j\omega\varepsilon\mathbf{E}_a \tag{14.3}$$
$$\nabla \times \mathbf{E}_a = -j\omega\mu\mathbf{H}_a \tag{14.4}$$

Fig. 14.3. Two antennas positioned in a medium μ, ε.

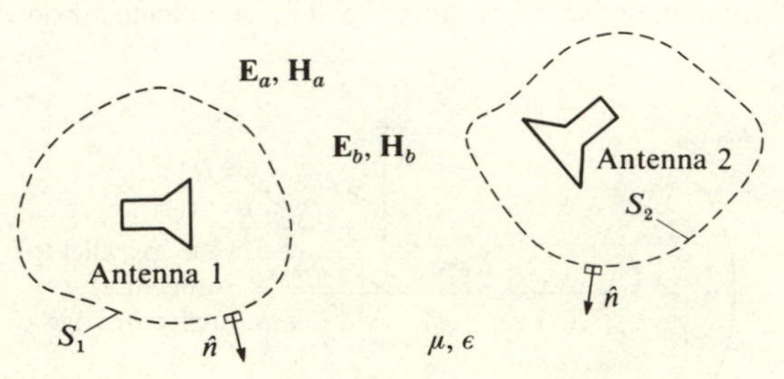

Now let the transmitting and receiving antennas be reversed so that P_{1b} is the input power to the transmitting antenna and P_{2b} is the corresponding received power into a matched load picked up by antenna 2 from antenna 1. Then, similarly, due to a source current i_b, the fields E_b, H_b are related by

$$\nabla \times \mathbf{H}_b = \mathbf{i}_b + j\omega\varepsilon\mathbf{E}_b \tag{14.5}$$

$$\nabla \times \mathbf{E}_b = -j\omega\mu\mathbf{H}_b \tag{14.6}$$

Multiplying eqn (14.3) scalarly by $\mathbf{E}_b$ and eqn (14.6) scalarly by $\mathbf{H}_a$ and subtracting gives

$$\mathbf{E}_b \cdot \nabla \times \mathbf{H}_a - \mathbf{H}_a \cdot \nabla \times \mathbf{E}_b = \mathbf{E}_b \cdot \mathbf{i}_a + j\omega\varepsilon\mathbf{E}_a \cdot \mathbf{E}_b + j\omega\mu\mathbf{H}_a \cdot \mathbf{H}_b$$

But since

$$\nabla \cdot (\mathbf{a} \times \mathbf{b}) = \mathbf{b} \cdot \nabla \times \mathbf{a} - \mathbf{a} \cdot \nabla \times \mathbf{b}$$

this gives

$$-\nabla \cdot (\mathbf{E}_b \times \mathbf{H}_a) = \mathbf{E}_b \cdot \mathbf{i}_a + j\omega\varepsilon\mathbf{E}_a \cdot \mathbf{E}_b + j\omega\mu\mathbf{H}_a \cdot \mathbf{H}_b \tag{14.7}$$

Similarly multiplying eqn (14.4) scalarly by $\mathbf{H}_b$ and eqn (14.5) scalarly by $\mathbf{E}_a$ and subtracting gives

$$\mathbf{H}_b \cdot \nabla \times \mathbf{E}_a - \mathbf{E}_a \cdot \nabla \times \mathbf{H}_b = -\mathbf{E}_a \cdot \mathbf{i}_b - j\omega\varepsilon\mathbf{E}_a \cdot \mathbf{E}_b - j\omega\mu\mathbf{H}_a \cdot \mathbf{H}_b$$

or

$$\nabla \cdot (\mathbf{E}_a \times \mathbf{H}_b) = -\mathbf{E}_a \cdot \mathbf{i}_b - j\omega\varepsilon\mathbf{E}_a \cdot \mathbf{E}_b - j\omega\mu\mathbf{H}_a \cdot \mathbf{H}_b \tag{14.8}$$

Combining eqns (14.7) and (14.8) gives

$$\nabla \cdot (\mathbf{E}_a \times \mathbf{H}_b) - \nabla \cdot (\mathbf{E}_b \times \mathbf{H}_a) = \mathbf{E}_b \cdot \mathbf{i}_a - \mathbf{E}_a \cdot \mathbf{i}_b$$

Returning to the situation when antenna 1 radiates the power P_{1a} and integrating over a surface S_1 which encloses this antenna gives

$$\oint\!\!\oint (\mathbf{E}_a \times \mathbf{H}_b - \mathbf{E}_b \times \mathbf{H}_a) \cdot \mathbf{ds} = \iiint (\mathbf{E}_b \cdot \mathbf{i}_a)\, dv$$

Likewise taking a surface S_2 surrounding the antenna transmitting the power P_{1b} gives

$$\oint\!\!\oint (\mathbf{E}_a \times \mathbf{H}_b - \mathbf{E}_b \times \mathbf{H}_a) \cdot \mathbf{ds} = -\iiint \mathbf{E}_a \cdot \mathbf{i}_b\, dv$$

Taking still another closed surface S_3 which does not enclose either of the antennas gives

$$\oint\!\!\oint (\mathbf{E}_a \times \mathbf{H}_b - \mathbf{E}_b \times \mathbf{H}_a) \cdot \mathbf{ds} = 0$$

Finally taking any closed surface S_4 which encloses both sources gives

$$\iint (\mathbf{E}_a \times \mathbf{H}_b - \mathbf{E}_b \times \mathbf{H}_a) \cdot \mathbf{ds} = \iiint \mathbf{E}_b \cdot \mathbf{i}_a\, dv - \iiint \mathbf{E}_a \cdot \mathbf{i}_b\, dv \tag{a}$$

irrespective of the shape or size of S_4, provided the sources are enclosed.

It may be shown that the left-hand side of eqn (a) is identically zero so that

$$\iiint \mathbf{E}_b \cdot \mathbf{i}_a \, dv = \iiint \mathbf{E}_a \cdot \mathbf{i}_b \, dv$$

or, in network terms under power matched conditions,

$$2V_{1b}I_{1a} = 2V_{2a}I_{2b}$$

Returning now to the case when the integration is taken over S_1, for which the left-hand side is no longer zero gives

$$\oiint (\mathbf{E}_a \times \mathbf{H}_b - \mathbf{E}_b \times \mathbf{H}_a) \cdot \mathbf{ds} = 2V_{1b}I_{1a} = 2V_{2a}I_{2b}$$

Likewise the integration could have been taken over S_2 to give the same result. Now, under power matched conditions,

$$V_{1b} = I_{1b}R_{r1}$$

and similarly

$$V_{2a} = I_{2a}R_{r2}$$

Therefore

$$\oiint (\mathbf{E}_a \times \mathbf{H}_b - \mathbf{E}_b \times \mathbf{H}_a) \cdot \mathbf{ds} = 2I_{1a}I_{1b}R_{r1} = 2I_{2a}I_{2b}R_{r2} \tag{14.9}$$

Taking complex conjugates of both sides gives

$$\oiint (\mathbf{E}_a \times \mathbf{H}_b - \mathbf{E}_b \times \mathbf{H}_a) \cdot \mathbf{ds}^* = 2I_{1a}^*I_{1b}^*R_{r1} = 2I_{2a}^*I_{2b}^*R_{r2} \tag{14.10}$$

Multiplying both sides of eqns (14.9) and (14.10) gives

$$\left| \oiint (\mathbf{E}_a \times \mathbf{H}_b - \mathbf{E}_b \times \mathbf{H}_a) \cdot ds \right|^2 = 4|I_{1a}|^2|I_{1b}|^2 R_{r1}^2$$
$$= 4|I_{2a}|^2|I_{2b}|^2 R_{r2}^2$$

The ratio of the received to transmitted power in either direction is

$$\frac{P_{2a}}{P_{1a}} = \frac{P_{2b}}{P_{1b}} = \frac{\frac{1}{2}|I_{2a}|^2 R_{r2}}{\frac{1}{2}|I_{1a}|^2 R_{r1}} = \frac{\frac{1}{2}|I_{1b}|^2 R_{r1}}{\frac{1}{2}|I_{2b}|^2 R_{r2}}$$
$$= \frac{|\oiint (\mathbf{E}_a \times \mathbf{H}_b - \mathbf{E}_b \times \mathbf{H}_a) \cdot ds|^2}{4|I_{1a}|^2 R_{r1} |I_{2b}|^2 R_{r2}}$$
$$= \frac{|\oiint (\mathbf{E}_a \times \mathbf{H}_b - \mathbf{E}_b \times \mathbf{H}_a) \cdot ds|^2}{16 P_{1a} P_{1b}} \tag{14.11}$$

14.4 Coupling between paraboloid and TE_{11} circular cylindrical waveguide feed

Applying the power coupling theorem to a paraboloid which reradiates the fields incident on it and thereby acts as a transmitting antenna 1, and a circular waveguide feed delivering power to a matched load which acts as antenna 2, gives for the coupling factor or aperture efficiency,

$$\eta = \frac{|\int(\mathbf{H}^s \times \mathbf{E}_g + \mathbf{E}^s \times \mathbf{H}_g)\cdot\hat{z}\,ds|^2}{16P_sP_g} \tag{14.12}$$

where $\mathbf{H}^s$, $\mathbf{E}^s$ are the fields produced by the paraboloid over the mouth of the waveguide feed placed at the focus, and $\mathbf{E}_g$, $\mathbf{H}_g$ are the fields over the same surface when the waveguide is transmitting. The power scattered by the paraboloid, P_s, for the case of a plane wave normally incident on it is given by

$$P_s = \frac{E_i^2 \pi a^2}{2Z_0}$$

where a is the radius of the paraboloid. The power P_g radiated by the waveguide when it is transmitting is given, for TE_{11} mode propagation, by

$$P_g = \frac{\pi Z_0 \beta k A^2}{4k_c^4}(\delta^2 - 1)J_1^2(k_c b)$$

where $\delta = 1.841$, $k_c = \delta/b$ where b is the radius of the waveguide and A is a constant.

The aperture efficiency then becomes

$$\eta = \frac{|\int_0^b\int_0^{2\pi}[H_{1\rho}E_{2\alpha} - H_{1\alpha}E_{2\rho} + E_{1\rho}H_{2\alpha} - E_{1\alpha}H_{2\rho}]|^2}{16P_sP_g} \tag{14.13}$$

where the subscripts 1, 2 refer to fields associated with the paraboloid and waveguide respectively. But the fields associated with the paraboloid are themselves given by single integrals and have been given in rectangular form by eqns (13.18)–(13.26). The four of these required in the above expression for η are, in cylindrical form,

$$\begin{aligned} H_{1\rho} &= H_{1x}\cos\alpha + H_{1y}\sin\alpha \\ &= jkfH_x^i e^{-j2kf}\cos\alpha \\ &\quad\times \int_0^{\theta_0}\left[\sin\theta J_0(k\rho\sin\theta) + \sin\theta\tan^2\frac{\theta}{2}J_2(k\rho\sin\theta)\right]d\theta \end{aligned}$$

$$
\begin{aligned}
H_{1\alpha} &= -H_{1x}\sin\alpha + H_{1y}\cos\alpha \\
&= jkfH_x^i e^{-j2kf}\sin\alpha \\
&\quad\times \int_0^{\theta_0}\left[-\sin\theta J_0(k\rho\sin\theta) + \sin\theta\tan^2\frac{\theta}{2}J_2(k\rho\sin\theta)\right]d\theta
\end{aligned}
$$

$$
\begin{aligned}
E_{1\rho} &= jkfZ_0H_x^i e^{-j2kf}\sin\alpha \\
&\quad\times \int_0^{\theta_0}\sin\theta\left[J_0(k\rho\sin\theta) + \tan^2\frac{\theta}{2}J_2(k\rho\sin\theta)\right]d\theta
\end{aligned}
$$

and

$$
\begin{aligned}
E_{1\alpha} &= jkfZ_0H_x^i e^{-j2kf}\cos\alpha \\
&\quad\times \int_0^{\theta_0}\sin\theta\left[J_0(k\rho\sin\theta) - \tan^2\frac{\theta}{2}J_2(k\rho\sin\theta)\right]d\theta
\end{aligned}
$$

Likewise the corresponding waveguide fields are

$$E_{2\rho} = \frac{jkZ_0A}{k_c}\frac{J_1(k_c\rho)}{k_c\rho}\sin\alpha$$

$$E_{2\alpha} = \frac{jkZ_0A}{k_c}J_1'(k_c\rho)\cos\alpha$$

$$H_{2\rho} = -\frac{j\beta A}{k_c}J_1'(k_c\rho)\cos\alpha$$

$$H_{2\alpha} = \frac{j\beta A}{k_c}\frac{J_1(k_c\rho)}{k_c\rho}\sin\alpha$$

Substituting in the equation for η gives

$$
\begin{aligned}
\eta = \frac{1}{16P_sP_g}\Bigg|\int_0^b\int_0^{2\pi}\int_0^{\theta_0}\Bigg[&-\frac{k^2fZ_0AH_x^i}{k_c}e^{-j2kf} \\
&\times\left\{\cos^2\alpha\sin\theta J_1'(k_c\rho)\left[J_0(k\rho\sin\theta)\right.\right. \\
&\left.+\tan^2\frac{\theta}{2}J_2(k\rho\sin\theta)\right] - \sin^2\alpha\sin\theta\frac{J_1(k_c\rho)}{k_c\rho} \\
&\left.\times\left[-J_0(k\rho\sin\theta) + \tan^2\frac{\theta}{2}J_2(k\rho\sin\theta)\right]\right\} \\
&+\frac{k\beta fZ_0AH_x^i}{k_c}e^{-j2kf}\left\{\sin^2\alpha\sin\theta\frac{J_1(k_c\rho)}{k_c\rho}\right. \\
&\times\left[J_0(k\rho\sin\theta) + \tan^2\frac{\theta}{2}J_2(k\rho\sin\theta)\right] - \cos^2\alpha\sin\theta J_1'(k_c\rho) \\
&\left.\times\left[J_0(k\rho\sin\theta) - \tan^2\frac{\theta}{2}J_2(k\rho\sin\theta)\right]\right\}\Bigg]\rho\,d\rho\,d\theta\,d\alpha\Bigg|^2
\end{aligned}
\qquad (14.14)
$$

The integration on α can be performed analytically since

$$\int_0^{2\pi} \cos^2 \alpha \, d\alpha = \int_0^{2\pi} \sin^2 \alpha \, d\alpha = \pi$$

and the aperture efficiency then becomes

$$\begin{aligned}\eta = \frac{k_c^2 k^2 f^2}{\pi a^2 k\beta(\delta^2 - 1)J_1^2(\delta)} \Bigg| \int_0^b \int_0^{\theta_0} \Bigg[& -k\cos^2\alpha \sin\theta J_1'(k_c\rho) \\ & + \beta \sin^2\alpha \sin\theta \frac{J_1(k_c\rho)}{k_c\rho} \Bigg] \\ & \times \left[J_0(k\rho\sin\theta) + \tan^2\frac{\theta}{2} J_2(k\rho\sin\theta) \right] \\ & + \left[k\sin^2\alpha \sin\theta \frac{J_1(k_c\rho)}{k_c\rho} - \beta\cos^2\alpha \sin\theta J_1'(k_c\rho) \right] \\ & \times \left[-J_0(k\rho\sin\theta) + \tan^2\frac{\theta}{2} J_2(k\rho\sin\theta) \right] \rho \, d\rho \, d\theta \Bigg|^2 \end{aligned} \quad (14.15)$$

which can be evaluated numerically on a computer.

14.5 Computed aperture efficiency for cylindrical waveguide carrying TE_{11} mode placed at focus

Numerical results of aperture efficiency for a paraboloidal reflector of 31.95 cm diameter, together with cylindrical waveguide feeds of radii 0.383λ, 0.641λ and 1.056λ at a frequency of 10 GHz, have been computed from eqn (14.15) for a range of f/D values and are shown in Fig. 14.4. It

Fig. 14.4. Aperture efficiency of paraboloid with focussed TE_{11} mode circular waveguide feed.

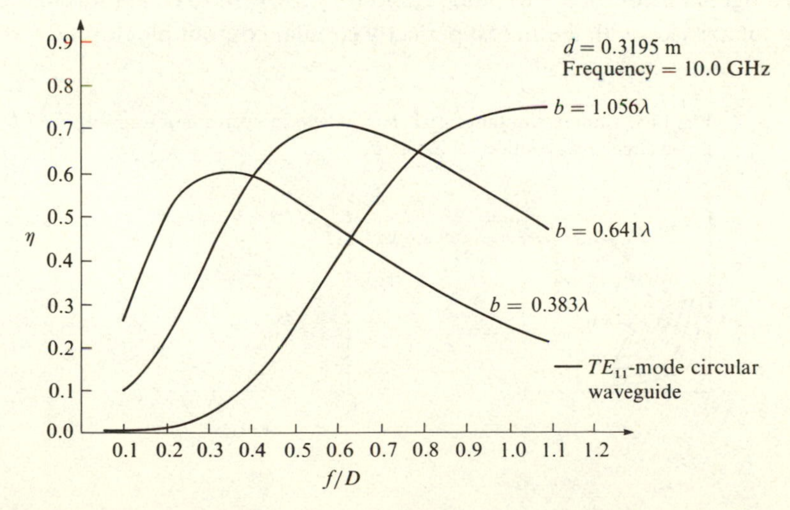

should be noted that the ordinate of aperture efficiency is applicable to any paraboloid with a diameter greater than 5λ, since the focal plane fields are unchanged in distribution for large diameters.

From the diagram it can be seen that for a given f/D ratio of the paraboloid there is a particular radius of waveguide which gives maximum aperture efficiency, and that this maximum efficiency is itself a function of the f/D ratio. Over the normal range of f/D used the maximum efficiency is of the order of 60–70% for this type of feed.

To understand why this should be the case consider first a paraboloid with a f/D of 1.0. From Fig. 13.13 it will be seen that the radius of the central spot formed when it is irradiated by a uniform plane wave is approximately 1.25λ. Hence a feed waveguide with a radius less than this is not going to abstract all the energy within this central spot, and the feed waveguide of 0.383λ radius in particular is only going to cover 9.4% of its central area. The efficiency of such a feed will therefore be small, though larger than the percentage area covered because the spot has maximum intensity at its centre.

But to achieve maximum efficiency there are two additional requirements beyond simply covering the central spot of the focal plane distribution. Firstly, the focal plane contours should match the corresponding contours of the TE_{11} mode in the feed guide and, secondly, the phase distribution across that part of the spot covered by the guide should match that of the TE_{11} mode. As far as the phase requirement for a 0.383λ guide is concerned, this is almost perfectly satisfied since the phase of the central spot changes by less than 1° from the centre to a radius of 0.383λ, and that of the guide is constant across a radius. But the amplitude requirements differ significantly. Fig. 14.5 shows the contour plots for the main and transverse orthogonal fields for a TE_{11} mode, and they are seen to be highly elliptical by comparison with the almost perfectly circular contour plot for the f/D of

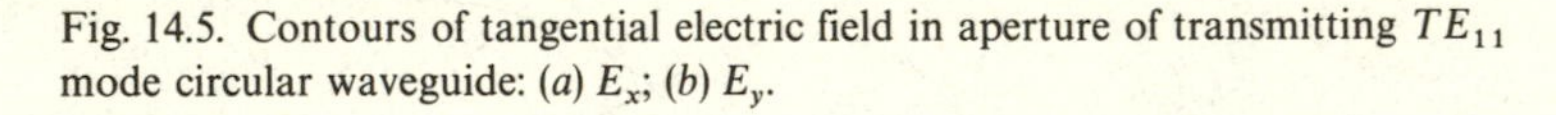

Fig. 14.5. Contours of tangential electric field in aperture of transmitting TE_{11} mode circular waveguide: (*a*) E_x; (*b*) E_y.

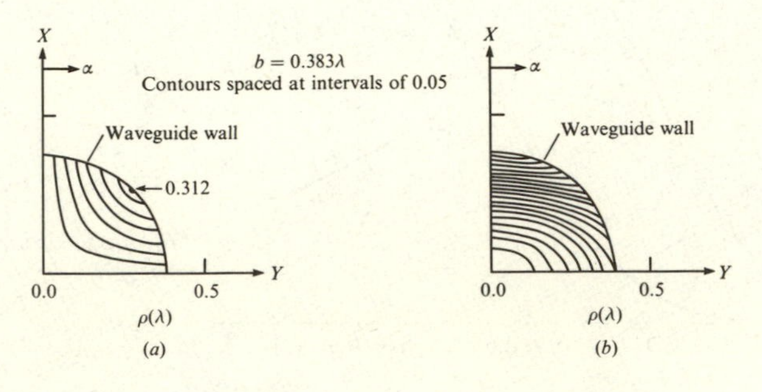

1.0 shown in Fig. 13.13. Hence the 0.383λ radius guide satisfies only one of the three requirements for producing a high aperture efficiency.

On the other hand for a f/D ratio of 0.25, the radii to first minima along and perpendicular to the direction of the electric field are approximately 0.61λ and 0.42λ. Thus the 0.383λ waveguide almost covers the central spot in the smaller of these two directions, and a comparison of its contour plot shows it to be similar to that of the TE_{11} mode. The phase of the central spot at the edge of the waveguide is 4.1° in the direction of the electric field, but it is 48° in the transverse direction where the waveguide comes close to the edge of the spot. The efficiency is therefore much higher overall when this size of guide is used with a paraboloid of $0.25f/D$ ratio.

14.6 Coupling between paraboloid and HE_{11} corrugated waveguide feed

The tangential field over the aperture of the corrugated waveguide due to scattering from the paraboloid remains as in Chapter 13. The corresponding waveguide fields for a predominantly y-directed electric field in the guide are derived from

$$\left.\begin{aligned} E_z &= A_1 J_1(h\rho)\sin\alpha e^{-j\beta z} \\ H_z &= -A_1 Y_0 J_1(h\rho)\cos\alpha e^{-j\beta z} \end{aligned}\right\} \tag{14.16}$$

where it will be noted that a minus sign is required for the H_z-component. It may be shown that the tangential fields in the aperture are

$$E_\rho = -\frac{jA_1}{2h}[(k+\beta)J_0(h\rho) + (k-\beta)J_2(h\rho)]\sin\alpha$$

$$E_\alpha = -\frac{jA_1}{2h}[(k+\beta)J_0(h\rho) - (k-\beta)J_2(h\rho)]\cos\alpha$$

$$H_\rho = \frac{jY_0A_1}{2h}[(k+\beta)J_0(h\rho) + (k-\beta)J_2(h\rho)]\cos\alpha$$

$$H_\alpha = -\frac{jY_0A_1}{2h}[(k+\beta)J_0(h\rho) - (k-\beta)J_2(h\rho)]\sin\alpha$$

From these, the expression for the z-directed Poynting vector becomes

$$\begin{aligned} P_z &= \tfrac{1}{2}\mathrm{Re}[E_\rho H_\alpha^* - E_\alpha H_\rho^*] \\ &= \frac{A_1^2 Y_0}{8h^2}[(k+\beta)^2 J_0^2(h\rho) - (k-\beta)^2 J_2^2(h\rho)] \end{aligned} \tag{14.17}$$

and it will be noted that this is independent of the angle α, so that the power density is circularly symmetric. Integrating over the cross-section of the waveguide gives the total axial power as

$$W = \frac{A_1^2 Y_0}{8h^2} \int_0^a \int_0^{2\pi} [(k+\beta)^2 J_0^2(h\rho) - (k-\beta)^2 J_2^2(h\rho)]\rho \, d\rho \, d\alpha$$

$$= \frac{\pi A_1^2 Y_0}{4h^2} \int_0^a [(k+\beta)^2 J_0^2(h\rho) - (k-\beta)^2 J_2^2(h\rho)]\rho \, d\rho \tag{14.18}$$

The Bessel function integrals are evaluated using

$$\int_0^a J_n^2(h\rho)\rho \, d\rho = \frac{a^2}{2}\left[J_n'^2(ha) + \left(1 + \frac{n^2}{h^2a^2}\right) J_n^2(ha)\right]$$

to give, with the arguments of the Bessel functions being dropped,

$$W = \frac{\pi A_1^2 Y_0 a^2}{8h^2} \left\{ (k+\beta)^2 (J_0'^2 + J_0^2) - (k^2 - \beta^2) \right.$$
$$\left. \times \left[J_2'^2 + \left(1 - \frac{4}{h^2a^2}\right) J_2^2 \right] \right\}$$

But

$$J_0' = -J_1$$

$$J_2' = \frac{2J_0}{ha} + \left(1 - \frac{4}{h^2a^2}\right) J_1$$

and

$$J_2 = \frac{2J_1}{ha} - J_0$$

so that

$$(J_2')^2 = \frac{4J_0^2}{h^2a^2} + \frac{4J_0J_1}{ha}\left(1 - \frac{4}{h^2a^2}\right) + \left(1 - \frac{4}{h^2a^2}\right)^2 J_1^2$$

and

$$\left(1 - \frac{4}{h^2a^2}\right) J_2^2 = \frac{4}{h^2a^2}\left(1 - \frac{4}{h^2a^2}\right) J_1^2 + \left(1 - \frac{4}{h^2a^2}\right) J_0^2$$
$$- \frac{4}{ha}\left(1 - \frac{4}{h^2a^2}\right) J_0 J_1$$

Consequently the power transmitted along the guide is

$$W_g = \frac{\pi A_1^2 Y_0 a^2}{8h^2} \left\{ (k^2 + \beta^2)(J_0^2 + J_1^2) - (k^2 - \beta^2) \right.$$
$$\left. \times \left[J_0^2 + \left(1 - \frac{4}{h^2a^2}\right) J_1^2 \right] \right\}$$

$$= \frac{\pi A_1^2 Y_0 a^2}{4h^2} \left\{ \beta^2 J_0^2 + \left(\beta^2 + \frac{2k^2}{h^2a^2} - \frac{2\beta^2}{h^2a^2}\right) J_1^2 \right\} \tag{14.19}$$

where the argument of the Bessel functions is ha.

Substituting in the equation for η gives

$$\eta = \frac{1}{16W_sW_g}\left|\int_0^a\int_0^{2\pi}\int_0^{\theta_0}\frac{kfA_1H_x^i}{2h}e^{-j2kf}\right.$$
$$\times\left\{\cos^2\alpha\sin\theta[(k+\beta)J_0(h\rho)-(k-\beta)J_2(h\rho)]\right.$$
$$\times\left[J_0(k\rho\sin\theta)+\tan^2\frac{\theta}{2}J_2(k\rho\sin\theta)\right]$$
$$+\sin^2\alpha\sin\theta[(k+\beta)J_0(h\rho)+(k-\beta)J_2(h\rho)]$$
$$\times\left[J_0(k\rho\sin\theta)-\tan^2\frac{\theta}{2}J_2(k\rho\sin\theta)\right]$$
$$+\sin^2\alpha\sin\theta[(k+\beta)J_0(h\rho)-(k-\beta)J_2(h\rho)]$$
$$\times\left[J_0(k\rho\sin\theta)+\tan^2\frac{\theta}{2}J_2(k\rho\sin\theta)\right]$$
$$+\cos^2\alpha\sin\theta[(k+\beta)J_0(h\rho)+(k-\beta)J_2(h\rho)]$$
$$\left.\left.\times\left[J_0(k\rho\sin\theta)-\tan^2\frac{\theta}{2}J_2(k\rho\sin\theta)\right]\right\}\rho\,d\rho\,d\theta\,d\alpha\right|^2 \quad (14.20)$$

The integration on α can be performed analytically since

$$\int_0^{2\pi}\cos^2\alpha\,d\alpha = \int_0^{2\pi}\sin^2\alpha\,d\alpha = \pi$$

and the aperture efficiency then becomes

$$\eta = \frac{kfA_1H_x^i}{32hW_sW_g}\left|\int_0^a\int_0^{\theta_0}\sin\theta\left\{[(k+\beta)J_0(h\rho)-(k-\beta)J_2(h\rho)]\right.\right.$$
$$\times\left[J_0(k\rho\sin\theta)+\tan^2\frac{\theta}{2}J_2(k\rho\sin\theta)\right]$$
$$+[(k+\beta)J_0(h\rho)+(k-\beta)J_2(h\rho)]$$
$$\left.\times\left[J_0(k\rho\sin\theta)-\tan^2\frac{\theta}{2}J_2(k\rho\sin\theta)\right]\right\} \quad (14.21)$$

which can be evaluated numerically on a computer and gives a higher aperture efficiency than for the TE_{11} mode in a smooth walled guide. A maximum value of efficiency of around 80% may be achieved.

14.7 Signal to noise ratio of receiving paraboloidal reflector plus feed

In a receiving situation the output from the feed terminals will consist of both the desired signal and undesired noise. It is necessary not only that the signal there should be large enough to operate the receiver, but that the signal to noise ratio at these terminals should be as large as possible. Some of the noise picked up will be galactic radiation, but a large amount of it will be ground noise which will enter through the back and side lobes of the radiation pattern of the antenna. The sky noise will normally, at microwave frequencies, be much less than this ground noise. Typical values of average sky temperature might be 25° K at 1GHz, whereas the ground temperature will be approximately 290° K for all propagation frequencies.

Consider for simplicity a paraboloid reflector plus feed which is pointed towards the zenith. Let the average sky temperature in forward half-space, i.e. for $0 < \theta < \pi/2$, be denoted by T_s so that if this surrounded the whole antenna the available noise power would be $KT_s df$ in a bandwidth df. Likewise if the ground temperature T_g surrounded the whole antenna the available noise power would be $KT_g df$. But since the two acceptance angles of the different temperature sources are equal, the contribution accepted from each of the two sources is proportional to the average power gain of the complete antenna in the forward and backward half-spaces respectively.

Since this measured information is not normally available at the design stage, some other method of calculation is required. Consider, therefore, the power picked up by the feed as the sum of a component which arrives via the parabolic reflector and a component which arrives directly from the ground. In the corresponding transmitting situation let α be the proportion of the power radiated by the feed which falls on the reflector so that $(1 - \alpha)$ is the proportion which falls outside the paraboloid. This is the same as falls on the ground if one assumes as an approximation that there is no backward radiation from the feed. Then the total noise power picked up by the antenna is

$$P_N = K(\alpha T_s + (1 - \alpha) T_g) df \tag{14.22}$$

The signal power delivered by the antenna is equal to the incident power density multiplied by its effective receiving area, i.e.

$$P_s = \frac{E_i^2}{2Z_0} A_e \tag{14.23}$$

where A_e is related to the receiving power gain through the equation

$$G = \frac{4\pi}{\lambda^2} A_e$$

Hence the signal power becomes

$$P_s = \frac{E_i^2}{2Z_0} \frac{G\lambda^2}{4\pi} \tag{14.24}$$

The associated signal to noise power ratio is then

$$S{:}N = \frac{E_i^2 G \lambda^2}{8\pi Z_0 K(\alpha T_s + (1-\alpha)T_g)df} \tag{14.25}$$

It might appear from eqn (14.25) that the best signal to noise ratio is obtained when the proportion of the power radiated by the feed which falls on the reflector is unity. But if this were to be taken as the sole design aim it would result in the receiving power gain G being abnormally low because the edge taper of the illumination on the paraboloid would be excessively great.

This points to the fact that the design of a paraboloid for maximum power gain is different from the design for maximum signal to noise ratio. Maximum power gain is achieved with uniform illumination of the aperture when transmitting, whereas maximum signal to noise ratio is achieved when the ratio $G/(\alpha T_s + (1-\alpha)T_g)$ is a maximum. In practice this means that different feeds will be required if the designs are to be optimised.

However, one factor which has been neglected in the above analysis is the effective input temperature of the waveguide leading to the receiver. If this is denoted by T_e then this forms an additional term in the brackets in eqn (14.25). The effect of this could be dominant if the waveguide were lossy or if the receiver had a high noise figure, but normally this would be designed not to be the case.

Further reading

M.K. Hu: *Near Zone Power Transmission Formulas*: *IRE* National Convention Record, Part 8, 1958, pp. 128–35.

G. Morris: Coupling between aperture antennas: PhD thesis University of Birmingham, 1978.

R. Neri: Low frequency operation of grid reflector antennas: PhD thesis, University of Birmingham, 1979.

J. Robieux: 'General theorems on the transmission coefficient from a transmitting to a receiving system': *IRE Trans. Antennas and Propagation*, **AP-7** (Special Supplement), 1959, p. S118.

A.W. Rudge, K. Milne, A.D. Olver, P. Knight (Ed.): *The Handbook of Antenna Design*: Peter Peregrinus, 1982.

S. Silver: *Microwave Antenna Theory and Design*: McGraw-Hill, 1949.

15

+ – +

Analysis of transmitting paraboloids

The study of paraboloidal reflectors from a receiving point of view in the preceding chapter led to a physical understanding of their operation, and by the use of the power coupling theorem enabled a quantitative assessment of their aperture efficiency to be made. This calculation, however, involved double numerical integration, whereas the approach to be adopted in this chapter will enable this calculation to be carried out with a single numerical integration, though at the expense of the same physical understanding.

A study will first be carried out of a paraboloid which uses a current element feed, in order to allow a comparison with the receiving approach developed in the preceding chapter. It will be shown that the radiation field at any point in space can be found by either of two methods, the aperture field approach or the current distribution approach, which give identical results in the forward direction and very similar results out to several beamwidths of the main beam. By contrast it will be recalled that the receiving analysis, as given in the previous chapter, produced a radiation pattern only in that part of forward half-space corresponding to values of the angle θ where the curvature of the reflector allowed full circumferential flow of current to exist. Thus for a f/D of 0.25 the receiving pattern was available in the form of a single integral from the forward ($\theta = 0°$) direction out to $\theta = 45°$ only.

In both the aperture field and the current distribution approaches it will be assumed that the field from the feed which is incident on the reflector is the same as that which would exist at that location if the reflector were absent. Both infinitesimal dipole and current element feeds will give the same relative field distributions over the reflector.

15.1 Radiation pattern of paraboloid with current element feed – aperture field approach

Let the current element dipole be directed along the y-axis, as shown in Fig. 15.1. It has been shown in an earlier chapter that the fields radiated by such a dipole are given, in the far field, by

$$E_\theta = -j\frac{Z_0 I dy}{2\lambda}\frac{e^{-jkr}}{r}\cos\theta\sin\phi = B\cos\theta\sin\phi \tag{15.1}$$

and

$$E_\phi = -j\frac{Z_0 I dy}{2\lambda}\frac{e^{-jkr}}{r}\cos\phi = B\cos\phi \tag{15.2}$$

These are the fields incident on the paraboloid, and converting these into rectangular coordinates gives

$$\begin{aligned} E_x^i &= E_\theta\cos\theta\cos\phi - E_\phi\sin\phi \\ &= B\sin\phi\cos\phi(\cos^2\theta - 1) \\ &= -\frac{B}{2}\sin^2\theta\sin 2\phi \end{aligned} \tag{15.3}$$

$$\begin{aligned} E_y^i &= E_\theta\cos\theta\sin\phi + E_\phi\cos\phi \\ &= B(\cos^2\theta\sin^2\phi + \cos^2\phi) \end{aligned} \tag{15.4}$$

$$E_z^i = -E_\theta\sin\theta = -\frac{B}{2}\sin 2\theta\sin\phi \tag{15.5}$$

These incident fields are converted at the paraboloid surface into reflected fields whose magnitudes and directions may be found from the following considerations. Assuming that the paraboloid surface is locally plane the tangential component of reflected electric field will be equal and opposite to the tangential component of the incident electric field, i.e.

$$\mathbf{n}\times\mathbf{E}^r = -(\mathbf{n}\times\mathbf{E}^i) \tag{15.6}$$

where $\mathbf{n}$ is the unit normal at the reflector surface. Similarly the normal components will be equal and codirected for optical type reflection, which gives

$$\mathbf{n}\cdot\mathbf{E}^r = \mathbf{n}\cdot\mathbf{E}^i \tag{15.7}$$

Fig. 15.1. y-directed current element feed at focus of paraboloid.

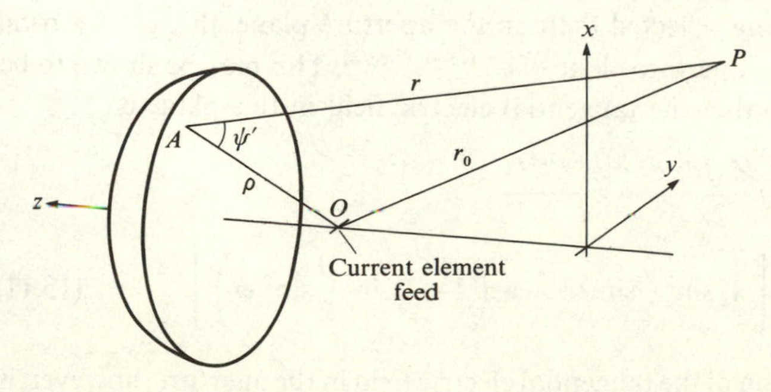

Now using the vector identity,

$$\mathbf{n} \times (\mathbf{n} \times \mathbf{E}) = (\mathbf{n} \cdot \mathbf{E})\mathbf{n} - (\mathbf{n} \cdot \mathbf{n})\mathbf{E}$$

gives

$$\mathbf{n} \times (\mathbf{n} \times \mathbf{E}^i) = (\mathbf{n} \cdot \mathbf{E}^i)\mathbf{n} - \mathbf{E}^i$$

and

$$\mathbf{n} \times (\mathbf{n} \times \mathbf{E}^r) = (\mathbf{n} \cdot \mathbf{E}^r)\mathbf{n} - \mathbf{E}^r$$

Using eqns (15.6) and (15.7) gives

$$\mathbf{E}^r = 2(\mathbf{n} \cdot \mathbf{E}^i)\mathbf{n} - \mathbf{E}^i \tag{15.8}$$

Since the unit normal to the reflector is given by

$$\mathbf{n} = -\hat{\mathbf{x}} \sin\frac{\theta}{2}\cos\phi - \hat{\mathbf{y}}\sin\frac{\theta}{2}\sin\phi - \hat{\mathbf{z}}\cos\frac{\theta}{2}$$

this means that the scalar product can be written as

$$(\mathbf{n} \cdot \mathbf{E}^i) = \frac{B}{2}\left[-2\sin\frac{\theta}{2}\sin\phi(\cos^2\phi + \cos^2\theta\sin^2\phi) + 2\sin\theta\sin\phi\left(\cos\frac{\theta}{2}\cos\theta + \sin\frac{\theta}{2}\sin\theta\cos^2\phi\right)\right]$$

and hence using

$$E_x^r = 2(\mathbf{n} \cdot \mathbf{E}^i)n_x - E_x^i$$

gives, after simplifying,

$$E_x^r = B\sin^2\frac{\theta}{2}\sin 2\phi \tag{15.9}$$

Similarly

$$E_y^r = -B\left(1 - 2\sin^2\frac{\theta}{2}\sin^2\phi\right) \tag{15.10}$$

These are the fields at the reflector surface, which, continuing with the optical ray treatment, travel as parallel rays to the reflector aperture, arriving in phase at this plane. Since the total electrical distance travelled from the focus corresponds to that of a ray travelling to the outer rim of the reflector and being reflected there in the aperture plane, this gives a total phase delay in the aperture plane of $e^{-jkf\sec^2(\psi_0/2)}$. This may be shown to be $e^{-jk[f+(a^2/4f)]}$, so that the tangential electric field in this plane is

$$\mathbf{E}_a = -j\frac{Z_0 I dy}{2\lambda}\frac{e^{-jk(f+a^2/4f)}}{\rho} \times \left[\mathbf{a}_x \sin^2\frac{\theta}{2}\sin 2\phi - \mathbf{a}_y\left(1 - 2\sin^2\frac{\theta}{2}\sin^2\phi\right)\right] \tag{15.11}$$

This description of the tangential electric field in the aperture, however, is based on the coordinate system centred on the focus, with its positive z-axis directed towards the vertex of the paraboloid. To describe the radiated field from the paraboloid it is preferable to transform this coordinate system so that the positive z-direction is reversed and, in order to maintain a right-handed system, y will also be reversed and the direction of x left unchanged. Also the positive azimuthal direction ϕ' is the negative of ϕ in the previous coordinate system and the angle θ in that system is related to the radius ρ' in the aperture plane by

$$\rho' = 2f\tan\frac{\theta}{2}$$

so that

$$\cos\theta = \frac{p^2 - \rho'^2}{p^2 + \rho'^2} \quad \text{where} \quad p = 2f$$

and

$$\rho = f\sec^2\frac{\theta}{2} = f\left[1 + \left(\frac{\rho'}{2f}\right)^2\right] = \frac{p^2 + \rho'^2}{2p}$$

Hence the tangential electric field in the aperture plane, referred to the new coordinate system, can be written as

$$\begin{aligned}\mathbf{E}_a = & -j\frac{60\pi I dy}{2\lambda\rho} \\ & \times e^{-jk(f + a^2/4f)}\{-\mathbf{a}_x(1 - (p^2 - \rho'^2)/(p^2 + \rho'^2))\sin 2\phi' \\ & - \mathbf{a}_y[2 - 2(1 - (p^2 - \rho'^2)/(p^2 + \rho'^2))((1 - \cos 2\phi')/2)]\} \\ = & A\{-\mathbf{a}_x f_1(\rho')\sin 2\phi' + \mathbf{a}_y[f_1(\rho')\cos 2\phi' + f_2(\rho')]\} \qquad (15.12)\end{aligned}$$

where

$$A = -j\frac{60\pi p I dy}{\lambda} e^{-jk(f + a^2/4f)}$$

$$f_1(\rho') = \frac{1}{p^2 + \rho'^2}\left(1 - \frac{p^2 - \rho'^2}{p^2 + \rho'^2}\right)$$

and

$$f_2(\rho') = \frac{1}{p^2 + \rho'^2}\left(1 + \frac{p^2 - \rho'^2}{p^2 + \rho'^2}\right)$$

It has previously been shown that for tangential electric field components E_x, E_y in a circular aperture the radiated electric field components are given by

$$\begin{aligned}E_\theta = \frac{j}{\lambda r_0} e^{-jkr_0}\Big\{ & \cos\phi \iint_{aperture} E_x e^{jk\rho'\sin\theta\cos(\phi - \phi')} \rho'\, d\rho'\, d\phi' \\ & + \sin\phi \iint_{aperture} E_y e^{jk\rho'\sin\theta\cos(\phi - \phi')} \rho'\, d\rho'\, d\phi' \Big\} \qquad (15.13)\end{aligned}$$

$$E_\phi = \frac{j}{\lambda r_0} e^{-jkr_0} \cos\theta \left\{ -\sin\phi \iint_{aperture} E_x \right.$$
$$\times e^{jk\rho' \sin\theta \cos(\phi-\phi')} \rho' \, d\rho' \, d\phi'$$
$$\left. + \cos\phi \iint_{aperture} E_y e^{jk\rho' \sin\theta \cos(\phi-\phi')} \rho' \, d\rho' \, d\phi' \right\} \tag{15.14}$$

Substituting from eqn (15.12) into eqn (15.13) gives

$$E_\theta = \frac{jA}{\lambda r_0} e^{-jkr_0} \left\{ \cos\phi \iint_{aperture} -f_1(\rho') \sin 2\phi' \right.$$
$$\times e^{jk\rho' \sin\theta \cos(\phi-\phi')} \rho' \, d\rho' \, d\phi'$$
$$+ \sin\phi \iint_{aperture} [f_1(\rho') \cos 2\phi' + f_2(\rho')]$$
$$\left. \times e^{jk\rho' \sin\theta \cos(\phi-\phi')} \rho' \, d\rho' \, d\phi' \right\}$$
$$= \frac{jA^{-jkr_0}}{\lambda r_0} \int_0^a \left\{ f_1(\rho^1) \int_0^{2\pi} \sin(\phi - 2\phi') \right.$$
$$\times e^{jk\rho' \sin\theta \cos(\phi-\phi')} \rho' \, d\rho' \, d\phi'$$
$$\left. + \sin\phi f_2(\rho') \int_0^{2\pi} e^{jk\rho' \sin\theta \cos(\phi-\phi')} \rho' \, d\rho' \, d\phi' \right\}$$
$$= j\frac{2\pi A}{\lambda r_0} e^{-jkr_0} \left\{ \int_0^a (-\sin\phi \cos 2\phi + \cos\phi \sin 2\phi) \right.$$
$$\times J_2(k\rho' \sin\theta) f_1(\rho') \rho' \, d\rho'$$
$$\left. + \int_0^a \sin\phi f_2(\rho') J_0(k\rho' \sin\theta) \rho' \, d\rho' \right\}$$
$$= j\frac{kA}{r_0} e^{-jkr_0} \sin\phi \left\{ \int_0^a J_0(k\rho' \sin\theta) f_2(\rho') \right.$$
$$\left. + J_2(k\rho' \sin\theta) f_1(\rho') \right\} \rho' \, d\rho' \tag{15.15}$$

In the same way it can be shown that

$$E_\phi = j\frac{kA}{r_0} e^{-jkr_0} \cos\theta \cos\phi \left\{ \int_0^a J_0(k\rho' \sin\theta) f_2(\rho') \right.$$
$$\left. - J_2(k\rho' \sin\theta) f_1(\rho') \right\} \rho' \, d\rho' \tag{15.16}$$

The form of eqns (15.15) and (15.16) will now be modified to allow a more direct comparison with the results to be derived later from the current distribution approach. First let E_θ, E_ϕ be transformed to rectangular coordinate form using

$$E_x = E_\theta \cos\theta \cos\phi - E_\phi \sin\phi$$
$$E_y = E_\theta \cos\theta \sin\phi + E_\phi \cos\phi$$

This gives

$$E_x = j\frac{kA}{r_0} e^{-jkr_0} \cos\theta \sin 2\phi \int_0^a J_2(k\rho' \sin\theta) f_1(\rho')\rho'\, d\rho'$$

$$E_y = j\frac{kA}{r_0} e^{-jkr_0} \cos\theta \left\{ \int_0^a J_0(k\rho' \sin\theta) f_2(\rho')\rho'\, d\rho' - \cos 2\phi \int_0^a J_2(k\rho' \sin\theta) f_1(\rho')\rho'\, d\rho' \right\}$$

Now let the variable of integration ρ' be changed to θ' where

$$\theta' = (\pi - \theta) \quad \text{and} \quad \rho' = 2f \cot\frac{\theta'}{2}$$

Then

$$f_1(\rho') = \frac{1 + \cos\theta'}{4f^2\left(1 + \cot^2\dfrac{\theta'}{2}\right)} = \frac{\cos^2\dfrac{\theta'}{2}}{2f^2 \operatorname{cosec}^2\dfrac{\theta'}{2}} = \frac{1}{8f^2}\sin^2\theta'$$

and

$$f_2(\rho') = \frac{1 - \cos\theta'}{4f^2\left(1 + \cot^2\dfrac{\theta'}{2}\right)} = \frac{1}{2f^2}\sin^4\frac{\theta'}{2}$$

Substituting in the expression for the rectangular components of electric field gives

$$E_x = j\frac{kA}{8f^2 r_0} e^{-jkr_0} \cos\theta \sin 2\phi \int_\pi^{\theta_0'} J_2\left(2kf \sin\theta \cot\frac{\theta'}{2}\right) \times \sin^2\theta' \cdot 2f\cot\frac{\theta'}{2}(-)f \operatorname{cosec}^2\frac{\theta'}{2}\, d\theta'$$

$$= -\frac{Z_0 I\, dy}{\lambda}\frac{kf}{r_0} e^{-jkr_0} e^{-jk(f + a^2/4f)} \cos\theta \sin 2\phi \times \int_\pi^{\theta_0'} J_2\left(2kf \sin\theta \cot\frac{\theta'}{2}\right) \cos^2\frac{\theta'}{2} \cot\frac{\theta'}{2}\, d\theta' \tag{15.17}$$

Similarly

$$E_y = \frac{Z_0 I dy}{\lambda} \frac{kf}{r_0} e^{-jkr_0} e^{-jk(f+a^2/4f)}$$

$$\times \cos\theta \int_\pi^{\theta'_0} \left[-J_0\left(2kf \sin\theta \cot\frac{\theta'}{2}\right) \sin^2\frac{\theta'}{2} \cot\frac{\theta'}{2}\right.$$

$$\left. + \cos 2\phi \int_\pi^{\theta'_0} J_2\left(2kf \sin\theta \cot\frac{\theta'}{2}\right) \cos^2\frac{\theta'}{2} \cot\frac{\theta'}{2}\right] d\theta' \qquad (15.18)$$

15.2 Radiation pattern of paraboloid with current element feed – current distribution approach

Consider the current element $I dy$ aligned along the y-axis of the original coordinate system with the positive z-axis directed towards the vertex of the paraboloid. This current element is situated at the geometrical focus of the paraboloid and produces a vector potential

$$dA_y = \frac{\mu I dy}{4\pi} \frac{e^{-jkr}}{r} \qquad (15.19)$$

Since this is the only component of vector potential it follows that

$$dH_x = -\frac{1}{\mu} \frac{\partial(dA_y)}{\partial z}$$

$$dH_y = 0$$

$$dH_z = \frac{1}{\mu} \frac{\partial(dA_y)}{\partial x}$$

Now

$$r = [x^2 + y^2 + z^2]^{\frac{1}{2}}$$

so that

$$\frac{\partial}{\partial z}\left(\frac{e^{-jkr}}{r}\right) = -\frac{z}{r^2}\left(jk + \frac{1}{r}\right) e^{-jkr}$$

$$\frac{\partial}{\partial x}\left(\frac{e^{-jkr}}{r}\right) = -\frac{x}{r^2}\left(jk + \frac{1}{r}\right) e^{-jkr}$$

and hence

$$dH_x = \frac{I dy}{4\pi} \frac{e^{-jkr}}{r}\left(\frac{1}{r} + jk\right) \cos\theta \qquad (15.20)$$

$$dH_y = 0 \qquad (15.21)$$

$$dH_z = -\frac{I dy}{4\pi} \frac{e^{-jkr}}{r}\left(\frac{1}{r} + jk\right) \sin\theta \cos\phi \qquad (15.22)$$

Alternatively these results could have been derived from their spherical components as given in an earlier chapter.

The current distribution of the reflector is then obtained from

$$\mathbf{J} = 2(\mathbf{n} \times \mathbf{H})$$

with $\bar{\mathbf{n}}$ given by

$$\mathbf{n} = -\hat{\mathbf{x}} \sin\frac{\theta}{2}\cos\phi - \hat{\mathbf{y}}\sin\frac{\theta}{2}\sin\phi - \hat{\mathbf{z}}\cos\frac{\theta}{2}$$

Thus

$$J_x = 2(n_y H_z - n_z H_y)$$

$$= \frac{Idy}{2\pi}\frac{e^{-jk\rho}}{\rho}\left(\frac{1}{\rho} + jk\right)\sin\frac{\theta}{2}\sin\theta\sin\phi\cos\phi \qquad (15.23)$$

Similarly

$$J_y = 2(n_z H_x - n_x H_z)$$

$$= -\frac{Idy}{2\pi}\frac{e^{-jk\rho}}{\rho}\left(\frac{1}{\rho} + jk\right)\cos\frac{\theta}{2}\left(\cos\theta + 2\sin^2\frac{\theta}{2}\cos^2\phi\right) \qquad (15.24)$$

and

$$J_z = 2(n_x H_y - n_y H_x)$$

$$= \frac{Idy}{2\pi}\frac{e^{-jk\rho}}{\rho}\left(\frac{1}{\rho} + jk\right)\sin\frac{\theta}{2}\cos\theta\sin\phi \qquad (15.25)$$

These expressions for the current density will now be transformed into the new coordinate system with y, z reversed while x is left the same. This implies also that θ is replaced by $\theta' = (\pi - \theta)$ and ϕ by $\phi' = -\phi$. Then the new equations become

$$J_x = -\frac{Idy}{2\pi}\frac{e^{-jk\rho}}{\rho}\left(\frac{1}{\rho} + jk\right)\cos\frac{\theta'}{2}\sin\theta'\sin\phi'\cos\phi' \qquad (15.26)$$

$$J_y = \frac{Idy}{2\pi}\frac{e^{-jk\rho}}{\rho}\left(\frac{1}{\rho} + jk\right)\sin\frac{\theta'}{2}\left(-\cos\theta' + 2\cos^2\frac{\theta'}{2}\cos^2\phi'\right) \qquad (15.27)$$

$$J_z = -\frac{Idy}{2\pi}\frac{e^{-jk\rho}}{\rho}\left(\frac{1}{\rho} + jk\right)\cos\frac{\theta'}{2}\cos\theta'\sin\phi' \qquad (15.28)$$

The corresponding vector potentials are

$$dA_x = -\frac{\mu I dy}{8\pi^2}\frac{e^{-jkf\,\mathrm{cosec}^2(\theta'/2)}}{f\,\mathrm{cosec}^2\dfrac{\theta'}{2}}\left(\frac{1}{f\,\mathrm{cosec}^2\dfrac{\theta'}{2}} + jk\right)\frac{e^{-jkr}}{r}\cos\frac{\theta'}{2}$$

$$\times \sin\theta'\sin\phi'\cos\phi'\,dS$$

$$dA_y = \frac{\mu I dy}{8\pi^2} \frac{e^{-jkf\,\mathrm{cosec}^2(\theta'/2)}}{f\,\mathrm{cosec}^2\dfrac{\theta'}{2}} \left(\frac{1}{f\,\mathrm{cosec}^2\dfrac{\theta'}{2}} + jk \right) \frac{e^{-jkr}}{r} \sin\frac{\theta'}{2}$$

$$\times \left(-\cos\theta' + 2\cos^2\frac{\theta'}{2}\cos^2\phi' \right) dS$$

$$dA_z = -\frac{\mu I dy}{8\pi^2} \frac{e^{-jkf\,\mathrm{cosec}^2(\theta'/2)}}{f\,\mathrm{cosec}^2\dfrac{\theta'}{2}} \left(\frac{1}{f\,\mathrm{cosec}^2\dfrac{\theta'}{2}} + jk \right) \frac{e^{-jkr}}{r}$$

$$\times \cos\frac{\theta'}{2}\cos\theta'\sin\phi' dS$$

so that the contribution to the magnetic field at an observation point (x, y, z) from an element of current over the area dS is given by the components

$$dH_x = \frac{1}{\mu}\left[\frac{\partial}{\partial y}(dA_z) - \frac{\partial}{\partial z}(dA_y) \right]$$

$$dH_y = \frac{1}{\mu}\left[\frac{\partial}{\partial z}(dA_x) - \frac{\partial}{\partial x}(dA_z) \right]$$

$$dH_z = \frac{1}{\mu}\left[\frac{\partial}{\partial x}(dA_y) - \frac{\partial}{\partial y}(dA_x) \right]$$

The derivatives are taken at the point of observation and involve only the term e^{-jkr}/r, for which

$$\frac{\partial}{\partial x}\left(\frac{e^{-jkr}}{r} \right) = \frac{\partial}{\partial r}\left(\frac{e^{-jkr}}{r} \right)\frac{\partial r}{\partial x} = -\frac{e^{-jkr}}{r^2}\left(\frac{1}{r} + jk \right)$$

$$\times \left(x - f\,\mathrm{cosec}^2\frac{\theta'}{2}\cdot\sin\theta'\cos\phi' \right)$$

Similarly

$$\frac{\partial}{\partial y}\left(\frac{e^{-jkr}}{r} \right) = -\frac{e^{-jkr}}{r^2}\left(\frac{1}{r} + jk \right)\left(y - f\,\mathrm{cosec}^2\frac{\theta'}{2}\sin\theta'\sin\phi' \right)$$

$$\frac{\partial}{\partial z}\left(\frac{e^{-jkr}}{r} \right) = -\frac{e^{-jkr}}{r^2}\left(\frac{1}{r} + jk \right)\left(z - f\,\mathrm{cosec}^2\frac{\theta'}{2}\cos\theta' \right)$$

The distance r from the source point to the point of observation is given for the far field, referring to Fig. 15.1, by

$$r \approx r_0 + \rho\cos\psi'$$

$$r = r_0 + \rho\cos(\pi - A\hat{O}P)$$

$$r = r_0 - \rho[\cos\theta\cos\theta' + \sin\theta\sin\theta'\cos(\phi - \phi')]$$

$$= r_0 - f\,\mathrm{cosec}^2\frac{\theta'}{2}[\cos\theta\cos\theta' + \sin\theta\sin\theta'\cos(\phi - \phi')]$$

so that

$$dH_x = -\frac{Idy}{8\pi^2}\frac{e^{-jkf\,\mathrm{cosec}^2(\theta'/2)}}{f\,\mathrm{cosec}^2\dfrac{\theta'}{2}}\left(\frac{1}{f\,\mathrm{cosec}^2\dfrac{\theta'}{2}}+jk\right)(-)$$

$$\times\frac{e^{-jkr}}{r^2}\left(\frac{1}{r}+jk\right)\left[\left(y-f\,\mathrm{cosec}^2\frac{\theta'}{2}\sin\theta'\sin\phi'\right)\right.$$

$$\times\cos\frac{\theta'}{2}\cos\theta'\sin\phi'+\left(z-f\,\mathrm{cosec}^2\frac{\theta'}{2}\cos\theta'\right)\sin\frac{\theta'}{2}$$

$$\left.\times\left(-\cos\theta'+2\cos^2\frac{\theta'}{2}\cos^2\phi'\right)\right]dS \tag{15.29}$$

In the far field the following approximations can be made

$$jk \gg \frac{1}{r}$$

$$y \gg f\,\mathrm{cosec}^2\frac{\theta'}{2}\sin\theta'\sin\phi'$$

$$z \gg f\,\mathrm{cosec}^2\frac{\theta'}{2}\cos\theta'$$

Hence integrating over the total front surface of the paraboloid gives, for the total magnetic field in the x-direction,

$$H_x = +j\frac{Idy}{8\pi^2}k\frac{e^{-jkr_0}}{r_0^2}\int_{\theta'_0}^{\pi}\int_0^{2\pi}\frac{e^{-jkf\,\mathrm{cosec}^2(\theta'/2)}}{f\,\mathrm{cosec}^2\dfrac{\theta'}{2}}$$

$$\times\left(\frac{1}{f\,\mathrm{cosec}^2\dfrac{\theta'}{2}}+jk\right)e^{jkf\,\mathrm{cosec}^2(\theta'/2)[\cos\theta\cos\theta'+\sin\theta\sin\theta'\cos(\phi-\phi')]}$$

$$\times\left[y\cos\frac{\theta'}{2}\cos\theta'\sin\phi'+z\sin\frac{\theta'}{2}\right.$$

$$\left.\times\left(-\cos\theta'+2\cos^2\frac{\theta'}{2}\cos^2\phi'\right)\right]dS$$

$$=j\frac{Idy}{8\pi^2}k\frac{e^{-jkr_0}}{r_0}\int_{\theta'_0}^{\pi}\frac{e^{-jkf\,\mathrm{cosec}^2(\theta'/2)(1-\cos\theta\cos\theta')}}{f\,\mathrm{cosec}^2\dfrac{\theta'}{2}}\left(\frac{1}{f\,\mathrm{cosec}^2\dfrac{\theta'}{2}}+jk\right)$$

$$\times\int_0^{2\pi}e^{j2kf\sin\theta\cot(\theta'/2)\cos(\theta-\phi')}.\left[\sin\theta\sin\phi\cos\frac{\theta'}{2}\cos\theta'\sin\phi'\right.$$

$$\left.+\cos\theta\sin\frac{\theta'}{2}\left(-\cos\theta'+\cos^2\frac{\theta'}{2}\right)(1+\cos 2\phi')\right]dS$$

Now

$$dS = f^2 \operatorname{cosec}^5 \frac{\theta'}{2} \sin\theta' \, d\theta' \, d\phi'$$

and

$$\int_0^{2\pi} e^{ju\cos(\phi-\phi')} \begin{matrix}\cos n\phi' \\ \sin n\phi'\end{matrix} \, d\phi' = 2\pi j^n \begin{matrix}\cos n\phi \\ \sin n\phi\end{matrix} J_n(u)$$

so that

$$H_x = j\frac{Idy}{4\pi} kf \frac{e^{-jkr_0}}{r_0} \int_{\theta_0'}^{\pi} e^{-jkf\operatorname{cosec}^2(\theta'/2)(1-\cos\theta\cos\theta')}$$
$$\times\left(\frac{1}{f\operatorname{cosec}^2\frac{\theta'}{2}} + jk\right)\left\{J_1\left(2kf\sin\theta\cot\frac{\theta'}{2}\right)\right.$$
$$\times j\sin^2\frac{\theta'}{2}\sin\theta\cos\frac{\theta'}{2}\cos\theta'\sin^2\phi + J_0\left(2kf\sin\theta\cot\frac{\theta'}{2}\right)$$
$$\times\left[\cos\theta\sin^3\frac{\theta'}{2}\left(\cos^2\frac{\theta'}{2} - \cos\theta'\right)\right]$$
$$\left. - J_2\left(2kf\sin\theta\cot\frac{\theta'}{2}\right)\cos\theta\sin^3\frac{\theta'}{2}\cos^2\frac{\theta'}{2}\cos 2\phi\right\}$$
$$\times\operatorname{cosec}^5\frac{\theta'}{2}\sin\theta' \, d\theta' \qquad (15.30)$$

If the reflector is in the far field of the current element so that $jk \gg (1/f\operatorname{cosec}^2)(\theta'/2)$, and if attention is restricted to the forward region where θ is close to zero this gives

$$H_x = -\frac{Idy}{2\lambda}\frac{kf}{r_0} e^{-jkr_0} e^{-j2kf}$$
$$\times\left\{\int_{\theta_0'}^{\pi} J_0\left(2kf\sin\theta\cot\frac{\theta'}{2}\right)\sin\theta'\right.$$
$$\times\cos\theta\, d\theta' - \cos 2\phi \int_{\theta_0'}^{\pi} J_2\left(2kf\sin\theta\cot\frac{\theta'}{2}\right)$$
$$\left.\times\cot^2\frac{\theta'}{2}\sin\theta'\cos\theta\, d\theta'\right\}$$
$$= \frac{Idy}{\lambda}\frac{kf}{r_0} e^{-jkr_0} e^{-j2kf}\cos\theta$$
$$\times\left\{\int_{\pi}^{\theta_0'} J_0\left(2kf\sin\theta\cot\frac{\theta'}{2}\right)\sin\frac{\theta'}{2}\right.$$

$$\times \cos\frac{\theta'}{2}\,d\theta' - \cos 2\phi \int_{\pi}^{\theta'_0} J_2\left(2kf\sin\theta\cot\frac{\theta'}{2}\right)$$
$$\times \cos^2\frac{\theta'}{2}\cot\frac{\theta'}{2}\,d\theta'\Bigg\} \tag{15.31}$$

Comparison with the equation for E_y obtained from the aperture field approach shows that apart from a phase factor term $e^{-jk[f-(a^2/4f)]}$ this is exactly $(-Y_0)$ times the orthogonal E_y at the same observation point. The minus sign arises from the fact that in the far field $E_y/H_x = -Z_0$ for an outward travelling wave.

If a comparison is made for values of θ such that the product involving the J_1 term times $\sin\theta$ is small compared with the products involving J_0 and J_2 times $\cos\theta$, then the difference between the two approaches centres on the term $e^{-jk\,\mathrm{cosec}^2(\theta'/2)(1-\cos\theta\cos\theta')}$ in eqn (15.30). It is convenient to write this as

$$e^{-jkf\,\mathrm{cosec}^2(\theta'/2)[1-\cos\theta'+\cos\theta'(1-\cos\theta)]}$$

or

$$e^{-j2kf}e^{-j2kf\cos\theta'\,\mathrm{cosec}^2(\theta'/2)\sin^2(\theta/2)}$$

Thus the phase difference at an observation point in the direction θ, between sources located at $\theta' = \pi$ and at $\theta' = \theta_0$ is

$$2kf\sin^2\frac{\theta}{2}\left(1+\cos\theta'_0\,\mathrm{cosec}^2\frac{\theta'_0}{2}\right)$$

If this maximum phase difference is restricted to $\pi/8$, which corresponds to the phase difference associated with an aperture antenna of maximum linear dimension D at a range of $2D^2/\lambda$, then the angle at which this occurs is given by

$$\sin^2\frac{\theta}{2} = \frac{\lambda}{32f\left(1+\cos\theta'_0\,\mathrm{cosec}^2\dfrac{\theta'_0}{2}\right)} \tag{15.32}$$

As an example for a paraboloid with f/D of 0.25, corresponding to an angle θ_0 of 90°, $\theta = 9°$ for a diameter D/λ of 20, and 5.7° for D/λ of 50. These angles compare with $\theta = 3.5°$ and $\theta = 1.4°$, respectively, to the first nulls of the mean beam, so that the current distribution approach for which this phase difference occurs, and the aperture field approach, which is cophasal, can be expected to agree out to several beamwidths.

In the same way the orthogonal component of magnetic field **H** in the radiation zone may be shown to be given by

$$H_y = -\frac{Idy}{\lambda}\frac{kf}{r_0}e^{-jkr_0}\int_{\pi}^{\theta_0'} e^{-jkf\,\mathrm{cosec}^2(\theta'/2)(1-\cos\theta\cos\theta')}$$

$$\times\left\{\cos\theta\sin 2\phi J_2\left(2kf\sin\theta\cot\frac{\theta'}{2}\right)\right.$$

$$\times\cos^2\frac{\theta'}{2}\cot\frac{\theta'}{2}+\frac{j}{2}\sin\theta\sin 2\phi$$

$$\left.\times J_1\left(2kf\sin\theta\cot\frac{\theta'}{2}\right)\cos\theta'\cot^2\frac{\theta'}{2}\right\}d\theta'$$

which equals the result for the aperture field method in the forward direction along the axis, and agrees with it out to several beamwidths away from the axis, as before.

It will be noted that along the z-axis the only component of $\mathbf{H}$ is H_x given by

$$H_x = \frac{Idy}{2\lambda}\frac{kf}{r_0}e^{-jkr_0}e^{-j2kf}\int_{\pi}^{\theta_0'}\sin\theta'\,d\theta'$$

$$= -\frac{Idy}{\lambda}\frac{kf}{r_0}e^{-jk(r_0+2f)}\cos^2\frac{\theta_0'}{2}$$

$$= -\frac{Idy}{\lambda}\frac{k}{4r_0}D\cot\frac{\theta_0}{2}\sin^2\frac{\theta_0}{2}e^{-jk(r_0+2f)}$$

$$|H_x| = \frac{Idy}{8\lambda r_0}kD\sin\theta_0 \tag{15.33}$$

Hence the magnification of the field of the feed current element is $(\pi/2)(D/\lambda)\sin\theta_0$ which has a maximum value of $(\pi/2)(D/\lambda)$ for the focal plane paraboloid, as found from the receiving approach in Chapter 13.

15.3 Power gain on axis of paraboloid excited by *y*-directed current element

The power gain is defined as

$$G = \frac{\text{maximum power density}}{\text{average power density}} = \frac{\Phi_{max}}{\Phi_{av}}$$

where

$$\Phi_{max} = |H_{x\,max}|^2\frac{Z_0}{2}$$

$$= \frac{I^2dy^2k^2D^2\sin^2\theta_0}{64\lambda^2r^2}\frac{Z_0}{2}$$

and

$$\Phi_{av} = \frac{1}{2}\frac{I^2 R_r}{4\pi r^2}$$

$$= \frac{I^2 80\pi^2 dy^2}{2\lambda^2 4\pi r^2}$$

Hence

$$G = \frac{Z_0 k^2 D^2 \sin^2\theta_0}{1280\pi}$$

$$= \frac{3}{8}\left(\frac{\pi D}{\lambda}\right)^2 \sin^2\theta_0$$

The aperture efficiency η is then found from

$$G = \frac{4\pi}{\lambda^2}\frac{\pi D^2}{4}\eta$$

giving

$$\eta = 0.375 \sin^2\theta_0 \qquad (15.34)$$

with a maximum value of 0.375 for the focal plane case.

15.4 Radiation pattern of paraboloid excited by current element feed

The radiation pattern of the paraboloid excited by the current element feed has been shown in eqns (15.15) and (15.16) to be given by

$$E_\theta = j\frac{kA}{r_0}e^{-jkr_0}\sin\phi\left\{\int_0^a J_0(k\rho'\sin\theta)f_2(\rho')\right.$$
$$\left. + J_2(k\rho'\sin\theta)\cdot f_1(\rho')\right\}\rho'\,d\rho' \qquad (15.35)$$

Fig. 15.2. *E*-plane radiation pattern for 10λ diameter paraboloid: $f/D = 0.25$.

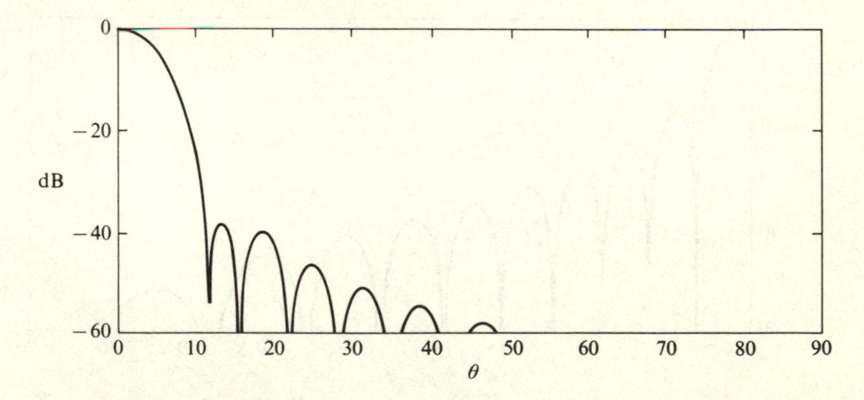

and

$$E_\phi = j\frac{kA}{r_0}e^{-jkr_0}\cos\theta\cos\phi\left\{\int_0^a J_0(k\rho'\sin\theta)\cdot f_2(\rho')\right.$$
$$\left. - J_2(k\rho'\sin\theta)\cdot f_1(\rho')\right\}\rho'\,d\rho' \quad (15.36)$$

These equations are readily integrated numerically to produce the principal E- and H-plane patterns in forward half-space, and examples to illustrate this are given in Figs. 15.2 and 15.3 for a paraboloid diameter of 10λ and f/D ratio 0.25. It will be seen that the E-plane pattern has a broader beamwidth and lower sidelobes than that of the H-plane, which is in keeping with the tapered distribution produced in the E-plane of the dipole over the paraboloid aperture.

15.5 Reflected fields from paraboloid with incident field $(E_\theta(\theta,\phi) + E_\phi(\theta,\phi))$

It has previously been shown that the reflected field E^r at the surface of the paraboloid is related to the incident field E^i by the equation

$$\mathbf{E}^r = 2(\mathbf{n}\cdot\mathbf{E}^i)\mathbf{n} - \mathbf{E}^i$$

where $\mathbf{n}$ is the inward unit normal at the surface of the paraboloid given by

$$\mathbf{n} = -\mathbf{x}\sin\frac{\theta}{2}\cos\phi - \mathbf{y}\sin\frac{\theta}{2}\sin\phi - \mathbf{z}\cos\frac{\theta}{2}$$

This gives

$$E_x^r = 2(\mathbf{n}\cdot\mathbf{E}^i)n_x - E_x^i$$

which reduces to

$$E_x^r = -E_\theta\cos\phi + E_\phi\sin\phi$$

Fig. 15.3. H-plane radiation pattern for 10λ diameter paraboloid: $f/D = 0.25$.

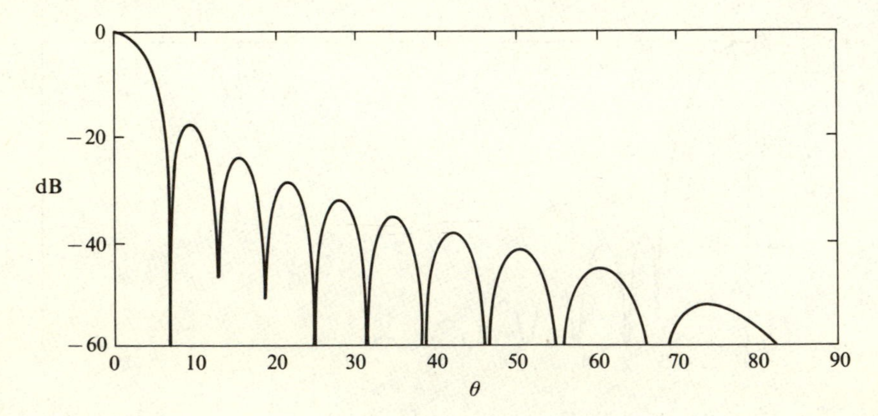

Similarly

$$E_y^r = -E_\theta \sin\phi - E_\phi \cos\phi$$

and

$$E_z^r = 0$$

In the aperture plane the phase is $e^{-jk(f+z_0)}$ where

$$z_0 = \frac{a^2}{4f}$$

and the electric field in that plane is given, in the primary coordinate system, by

$$\mathbf{E}_a = \frac{e^{-jk(f+z_0)}}{\rho}\{\mathbf{a}_{xF}[-E_\theta \cos\phi + E_\phi \sin\phi] + \mathbf{a}_{yF}[-E_\theta \sin\phi - E_\phi \cos\phi]\} \tag{15.37}$$

Changing to a new coordinate system with

$$\hat{x}_F = \hat{x}, \hat{y}_F = -\hat{y}, \hat{z}_F = -\hat{z}$$
$$\theta' = (\pi - \theta); \quad \phi' = -\phi$$

gives

$$\mathbf{E}_a = \frac{e^{-jk(f+z_0)}}{\rho}\{\mathbf{a}_x[-E_\theta \cos\phi' - E_\phi \sin\phi'] + \mathbf{a}_y[-E_\theta \sin\phi' + E_\phi \cos\phi']\} \tag{15.38}$$

15.6 Radiation from paraboloid using TE_{11} mode cylindrical waveguide feed

The radiation field of a TE_{11} mode circular waveguide feed is given, in the primary coordinate system, for a predominantly y-directed electric field, by

$$\left.\begin{aligned} E_\theta &= A\frac{e^{-jkr}}{r}\frac{J_1(kb\sin\theta)}{kb\sin\theta}\sin\phi \\ E_\phi &= A\frac{e^{-jkr}}{r}\frac{\delta^2 J_1'(kb\sin\theta)}{\delta^2 - k^2b^2\sin^2\theta}\cos\theta\cos\phi \end{aligned}\right\} \tag{15.39}$$

where

$$A = -\frac{2\omega\mu C\pi b^3 J_1(\delta)}{\lambda\delta^2}$$

and C is a constant.

Substituting these expressions in eqn (15.38) and changing ϕ in eqn (15.39) into $(-\phi')$ gives

$$\mathbf{E}_a(\rho', \varepsilon') = \frac{Ae^{-jk(f+z_0)}}{f\sec^2\frac{\theta}{2}}\left\{\mathbf{a}_x\left[\frac{\sin 2\phi'}{2}\left(\frac{J_1(kb\sin\theta')}{kb\sin\theta'}\right.\right.\right.$$

$$\left.\left.-\frac{\delta^2 J_1'(kb\sin\theta')}{\delta^2 - k^2b^2\sin^2\theta'}\cos\theta\right)\right] + \mathbf{a}_y\left[\sin^2\phi'\frac{J_1(kb\sin\theta')}{kb\sin\theta'}\right.$$

$$\left.\left.+\frac{\delta^2 J_1'(kb\sin\theta')}{\delta^2 - k^2b^2\sin^2\theta'}\cos\theta\cos^2\phi'\right]\right\}$$

The x-component of the aperture field in the secondary coordinate system is

$$E_{xa}(\rho', \phi') = \frac{Ae^{-jk(f+z_0)}}{p\left(1 + \frac{\rho^{12}}{p^2}\right)}\sin 2\phi^1$$

$$\times\left[\frac{J_1\left(kb\frac{2p\rho^1}{p^2+\rho^{12}}\right)}{kb\frac{2p\rho^1}{p^2+\rho^{12}}} - \frac{\delta^2 J_1'\left(kb\frac{2p\rho'}{p^2+\rho^{12}}\right)}{\delta^2 - k^2b^2\frac{2p\rho'}{p^2+\rho^{12}}}\frac{p^2-\rho^{12}}{p^2+\rho^{12}}\right]$$

Similarly, the y-component is

$$E_{ya}(\rho', \phi') = \frac{Ae^{-jk(f+z_0)}}{p\left(1 + \frac{\rho^{12}}{p^2}\right)}\left[\frac{p^2-\rho^{12}}{p^2+\rho^{12}}\frac{\delta^2 J_1'\left(\frac{2kbp\rho'}{p^2+\rho^{12}}\right)}{\left(\delta^2 - \frac{2k^2b^2p\rho'}{p^2+\rho^{12}}\right)}\right.$$

$$\left.\times(1+\cos 2\phi') + \frac{J_1\left(\frac{2kbp\rho'}{p^2+\rho^{12}}\right)}{\frac{2kbp\rho'}{p^2+\rho^{12}}}(1-\cos 2\phi')\right]$$

These expressions can be written more succinctly as

$$E_{xa}(\rho', \phi') = -Af_1(\rho')e^{-jk(f+z_0)}\sin 2\phi'$$

$$E_{ya}(\rho', \phi') = A[f_1(\rho')\cos 2\phi' + f_2(\rho')]e^{-jk(f+z_0)}$$

where

$$f_{\frac{1}{2}}(\rho')$$

$$= \frac{p}{(p^2+\rho^{12})}\left[\left(\frac{p^2-\rho^{12}}{p^2+\rho^{12}}\right)\frac{\delta^2 J_1'\left(\frac{2kbp\rho'}{p^2+\rho^{12}}\right)}{\delta^2 - \left(\frac{2kb^2b^2p\rho'}{p^2+\rho^{12}}\right)} \mp \frac{J_1\left(\frac{2kbp\rho'}{p^2+\rho^{12}}\right)}{\frac{2kbp\rho'}{p^2+\rho^{12}}}\right]$$

Consequently, the radiated E_θ-field from the circular aperture of the paraboloid is

$$E_\theta(\theta,\phi)=\frac{j}{\lambda}\frac{e^{-jkr}}{r}\left\{-\cos\phi\int_0^a\int_0^{2\pi}A\sin 2\phi' f_1(\rho')\right.$$

$$\times e^{jk\rho'\sin\theta\cos(\phi-\phi')}\rho'\,d\rho'\,d\phi'$$

$$+\sin\phi\int_0^a\int_0^{2\pi}A[\cos 2\phi' f_1(\rho')+f_2(\rho')]$$

$$\left.\times e^{jk\rho'\sin\theta\cos(\phi-\phi')}\rho'\,d\rho'\,d\phi'\right\}e^{-jk(f+z_0)} \tag{15.40}$$

The integrations on ϕ' are standard, leading to

$$E_\theta(\theta,\phi)=j\frac{A2\pi}{\lambda r}e^{-jk(f+z_0+r)}$$

$$\times\int_0^a\{[\cos\phi\sin 2\phi-\sin\phi\cos 2\phi]J_2(k\rho'\sin\theta)$$

$$\times f_1(\rho')+\sin\phi J_0(k\rho'\sin\theta)f_2(\rho')\}\rho'\,d\rho'$$

$$=jkA\frac{e^{-jk(f+z_0+r)}}{r}\left\{\int_0^a J_0(k\rho'\sin\theta)f_2(\rho')\rho'\,d\rho'\right.$$

$$\left.+\int_0^a J_2(k\rho'\sin\theta)f_1(\rho')\rho'\,d\rho'\right\}\sin\phi \tag{15.41}$$

The integrations on ρ' may alternatively be written in terms of ψ, originally the angle θ in the primary coordinate system, through the equation

$$\rho'=p\tan\frac{\psi}{2}$$

so that $d\rho'=f\sec^2(\psi/2)d\psi$ and $(p^2+\rho^{12})=p^2\sec^2(\psi/2)$. Hence

$$E_\theta(\theta,\phi)=j2kf^2A\frac{e^{-jk(f+z_0+r)}}{r}$$

$$\times\sin\phi\left\{\int_0^{\psi_0}J_0\left(2kf\sin\theta\tan\frac{\psi}{2}\right)f_2(\psi)\tan\frac{\psi}{2}\sec^2\frac{\psi}{2}\,d\psi\right.$$

$$\left.+\int_0^{\psi_0}J_2\left(2kf\sin\theta\tan\frac{\psi}{2}\right)f_1(\psi)\tan\frac{\psi}{2}\sec^2\frac{\psi}{2}\,d\psi\right\} \tag{15.42}$$

where

$$f_{\frac{1}{2}}(\psi)=\frac{1}{p\sec^2\frac{\psi}{2}}\left[\left(\frac{1-\tan^2\frac{\psi}{2}}{\sec^2\frac{\psi}{2}}\right)\frac{\delta^2 J'_1\left(\dfrac{2kbp^2\tan\frac{\psi}{2}}{p^2\sec^2\frac{\psi}{2}}\right)}{\delta^2-\left(\dfrac{2k^2b^2p^2\tan\frac{\psi}{2}}{p^2\sec^2\frac{\psi}{2}}\right)}\mp\frac{J_1\left(\dfrac{2kbp^2\tan\frac{\psi}{2}}{p^2\sec^2\frac{\psi}{2}}\right)}{\left(\dfrac{2kbp^2\tan\frac{\psi}{2}}{p^2\sec^2\frac{\psi}{2}}\right)}\right]$$

Similarly the orthogonal field is given in terms of ρ' by

$$E_\phi(\theta,\phi)=-jk\frac{Ae^{-jk(f+z_0+r)}}{r}\cos\theta\cos\phi\left\{\int_0^a[J_2(k\rho'\sin\theta)\right.$$
$$\left.\times f_1(\rho')-J_0(k\rho'\sin\theta)\cdot f_2(\rho')]\rho'\,d\rho'\right\}\tag{15.43}$$

which may also be expressed in terms of an integral on ψ.

15.7 Power gain of paraboloid excited by TE_{11} mode cylindrical waveguide feed

In the forward direction with a predominantly y-directed electric field in the waveguide feed, the total forward field radiated by the paraboloid is

$$E_\theta\left(0,\frac{\pi}{2}\right)=j2kf^2A\frac{e^{-jk(f+z_0+r)}}{r}\int_0^{\psi_0}f_2(\psi)\tan\frac{\psi}{2}\sec^2\frac{\psi}{2}\,d\psi$$

$$=j\frac{4\pi f^2A}{\lambda 2f}\frac{e^{-jk(f+z_0+r)}}{r}\int_0^{\psi_0}\left[\left(\frac{1-\tan^2\frac{\psi}{2}}{\sec^2\frac{\psi}{2}}\right)\delta^2\right.$$

$$\times \frac{J_1'\left(\dfrac{2kbp^2\tan\dfrac{\psi}{2}}{p^2\sec^2\dfrac{\psi}{2}}\right)}{\delta^2-\left(\dfrac{2k^2b^2p^2\tan\dfrac{\psi}{2}}{p^2\sec^2\dfrac{\psi}{2}}\right)}+\frac{J_1\left(\dfrac{2kbp^2\tan\dfrac{\psi}{2}}{p^2\sec^2\dfrac{\psi}{2}}\right)}{\left(\dfrac{2kbp^2\tan\dfrac{\psi}{2}}{p^2\sec^2\dfrac{\psi}{2}}\right)}\Bigg]$$

$$\times \tan\frac{\psi}{2}\,d\psi = jkfA\frac{e^{-jk(f+z_0+r)}}{r}\int_0^{\psi_0} I\,d\psi$$

Hence the power gain of the combined paraboloid plus feed is

$$G=\frac{\left|E_\theta\left(0,\dfrac{\pi}{2}\right)\right|^2}{2Z_0}\frac{4\pi r^2}{P_F}$$

where P_F is the power radiated by the feed. This is related to the power gain G_F of the feed from eqn (15.39) by

$$G_F=\frac{A^2}{r^2}\frac{1}{8Z_0}\frac{4\pi r^2}{P_F}$$

so that the combined power gain of the paraboloid plus feed can be written as

$$\begin{aligned} G &= \frac{4\pi^2 f^2}{\lambda^2}\frac{1}{2Z_0}8Z_0G_F\left|\int_0^{\psi_0} I\,d\psi\right|^2 \\ &= \frac{4\pi^2}{\lambda^2}p^2G_F\left|\int_0^{\psi_0} I\,d\psi\right|^2 \end{aligned}$$

Hence the aperture efficiency can be expressed as

$$\begin{aligned} \eta &= G\frac{\lambda^2}{4\pi}\frac{4}{\pi D^2} \\ &= \frac{4}{D^2}a^2\cot^2\frac{\psi_0}{2}G_F\left|\int_0^{\psi_0} I\,d\psi\right|^2 \\ &= G_F\cot^2\frac{\psi_0}{2}\left|\int_0^{\psi_0} I\,d\psi\right|^2 \end{aligned} \tag{15.44}$$

Typical values for this particular feed would vary between 0.5 and 0.7.

Further reading

P.K. Chum: Aperture radiation: PhD thesis, University of Birmingham, 1977.

A.W. Rudge, K. Milne, A.D. Olver, P. Knight (Ed.): *The Handbook of Antenna Design*: Peter Peregrinus, 1982.

W.V.T. Rusch and P.D. Potter: *Analysis of Reflector Antennas*: Academic Press, New York, 1970.

S. Silver: *Microwave Antenna Theory and Design*: McGraw-Hill, 1949.

16

+−+

Cassegrain and offset reflector analysis

16.1 Cassegrain antennas

The front fed parabolic reflector with its feed placed at or close to the geometric focus forms a system of converting an approximately spherical wave from the source into a planar wave at the reflector surface. An alternative way of producing the same result would be to use the Cassegrain system shown in Fig. 16.1, where the use of a hyperboloidal subreflector allows the feed to be placed much nearer the vertex, and thus avoids the need for a long waveguide run with its attendant attenuation and associated increase of noise temperature in the receiving situation. Moreover, if the main beam of the paraboloid is directed towards the zenith the system noise properties will be improved significantly because feed spillover will be directed towards the cold sky rather than to the warm ground as in the front fed case. The relatively large subreflector also provides a sharper cut off in the excitation of the main reflector than can be

Fig. 16.1. Cassegrain antenna with feed horn and hyperboloidal subreflector.

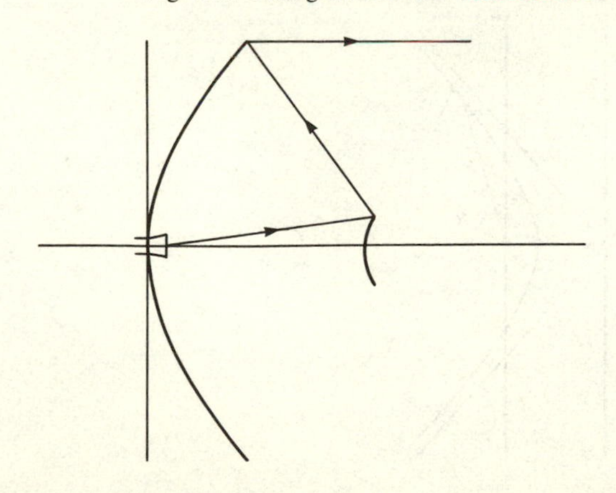

achieved with the smaller front feed, so that the spillover from the subreflector to ground does not cancel this improvement.

On the other hand, there are disadvantages in the cost of the subreflector and the increased cost of the larger feed required to irradiate the smaller dish. The presence of the subreflector also blocks out part of the direct illumination of the main paraboloid, and it turns out in practice that the positioning and alignment of the Cassegrain system is much more critical than for the front fed paraboloid.

The analysis of the Cassegrain system is also more involved analytically, because of the combined action of the dual reflectors, than that of the feed plus single reflector, and numerical integrations are necessary to obtain a complete solution. But, as for the paraboloid in isolation, useful results can be obtained from an analysis of the hyperboloidal reflector and the required geometry for this will next be developed.

16.2 Hyperboloidal reflector

A hyperboloidal reflector is the locus of a point which moves so that its distance from a fixed point, the focus, is equal to a constant greater than unity, times its distance from a fixed plane. The geometrical properties of the hyperboloid follow in general from its generating hyperbola.

16.2.1 *Equation of hyperbola*

Referring to Fig. 16.2, let the foci of the two hyperbolae shown be fixed at $F(\pm c, 0)$, and let the corresponding vertices be at $(\pm a, 0)$. The

Fig. 16.2. Hyperbolae with foci at $F(\pm c, 0)$.

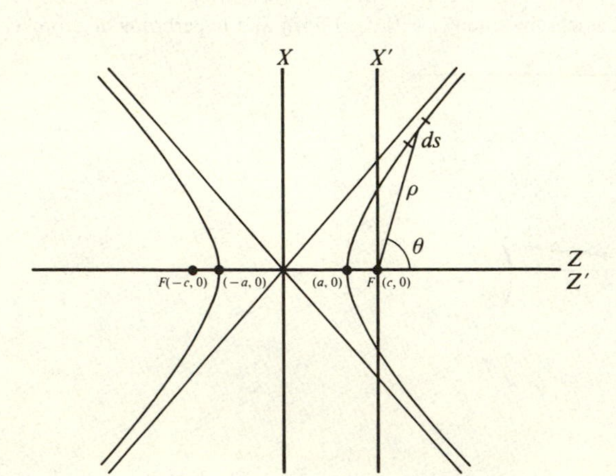

standard equation for these hyperbolae is given by

$$\frac{z^2}{a^2}-\frac{x^2}{(c^2-a^2)}=1 \tag{16.1}$$

Now let the origin of the coordinate system by moved to the right-hand focus F, so that the new coordinates (x', z') of any point are related to its previous coordinates (x, z) by the equations

$$x = x'$$
$$z = z' + c$$

Then the equation of the hyperbolae referred to this new origin at F are

$$\frac{(z'+c)^2}{a^2}+\frac{x'^2}{(c^2-a^2)}=1 \tag{16.2}$$

Transforming to polar coordinates through the equations

$$z' = \rho\cos\theta$$
$$x' = \rho\sin\theta$$

where θ is measured counter-clockwise from the z'-axis, gives for the equation of the hyperbolae

$$\frac{(\rho\cos\theta+c)^2}{a^2}-\frac{\rho^2\sin^2\theta}{(c^2-a^2)}=1$$

This simplifies to the quadratic equation in ρ

$$(c^2\cos^2\theta-a^2)\rho^2+2c(c^2-a^2)\cos\theta\cdot\rho+(c^2-a^2)^2=0$$

for which the solution is

$$\rho=-\frac{(c^2-a^2)}{(c\cos\theta-a)} \quad \textit{or} \quad \rho=-\frac{(c^2-a^2)}{(c\cos\theta+a)} \tag{16.3}$$

It is more usual to write these equations in terms of the eccentricity e and the distance p between the focus and directix, rather than in terms of the constants a and c. To bring this about note that from eqn (16.3) for the angles θ equal to $\pi/2$ and π respectively, from the first equation for ρ,

$$\rho_1=\left(\frac{c^2}{a}-a\right); \quad \rho_2=(c-a),$$

But by definition of a hyperbola $\rho_1 = ep$, and $\rho_2 = e(p-c+a)$. Hence, combining these equations gives

$$c-a=\frac{c^2}{a}-a-e(c-a)$$

or

$$e=\frac{c}{a} \tag{16.4}$$

Alternatively, using the equations for ρ_2 alone gives

$$e = \frac{c-a}{p-c+a} \tag{16.5}$$

From eqns (16.4) and (16.5), $(c^2 - a^2) = ape$, so that eqn (16.3) can be written in the desired form,

$$\rho = -\frac{pe}{(e\cos\theta - 1)} \quad or \quad -\frac{pe}{(e\cos\theta + 1)} \tag{16.6}$$

By considering, for example, the case of θ equal to π, it is readily seen that the first of these equations refers to the right-hand hyperbola in Fig. 16.2 and the second to the left-hand.

16.2.2 *Equation of hyperboloid*

The polar equations of the hyperboloids shown in cross-section in Fig. 16.2 are

$$\rho = -\frac{pe}{(e\cos\theta + 1)} \quad \rho = -\frac{pe}{(e\cos\theta - 1)} \tag{16.7}$$

for the left- and right-hand figures respectively.

16.2.3 *Element of area for hyperboloid*

Referring to Fig. 16.2 an element of area dS on the hyperboloid is equal to the product of the element of length ds on the generating hyperbola multiplied by the orthogonal element of length $\rho \sin\theta \, d\phi$, i.e.

$$dS = \rho \sin\theta \, d\phi \, ds$$

Since, however, in Fig. 16.3

$$ds^2 = dx^2 + dz^2$$

where

$$x = \rho \sin\theta = -\frac{pe\sin\theta}{1 + e\cos\theta}$$

and

$$z = \rho\cos\theta = -\frac{pe\cos\theta}{1 + e\cos\theta}$$

so that

$$\frac{dx}{d\theta} = -\frac{pe(e + \cos\theta)}{(1 + e\cos\theta)^2}$$

and

$$\frac{dz}{d\theta} = \frac{pe\sin\theta}{(1 + e\cos\theta)^2}$$

then

$$\frac{ds}{d\theta} = \left[\left(\frac{dx}{d\theta}\right)^2 + \left(\frac{dz}{d\theta}\right)^2\right]^{\frac{1}{2}} = \frac{pe[(1 + e\cos\theta)^2 + e^2\sin^2\theta]^{\frac{1}{2}}}{(1 + e\cos\theta)^2}$$

Hence the element of area dS on the paraboloid is given by

$$dS = -\frac{\rho^2 \sin\theta[(1 + e\cos\theta)^2 + e^2\sin^2\theta]^{\frac{1}{2}}}{(1 + e\cos\theta)} d\theta\, d\phi \tag{16.8}$$

16.2.4 *Equation of outward unit normal to hyperboloid*

The equation of the hyperboloid in Fig. 16.3 is

$$\rho = -\frac{pe}{1 + e\cos\theta}$$

Alternatively this may be written as

$$S = \rho + \frac{pe}{1 + e\cos\theta} = 0$$

so that

$$\begin{aligned}\nabla S &= \mathbf{a}_\rho \frac{\partial S}{\partial \rho} + \mathbf{a}_\theta \frac{1}{\rho}\frac{\partial S}{\partial \theta} + \mathbf{a}_\phi \frac{1}{\rho\sin\theta}\frac{\partial S}{\partial \phi} \\ &= \mathbf{a}_\rho + \mathbf{a}_\theta \frac{1}{\rho}\frac{pe^2\sin\theta}{(1 + e\cos\theta)^2} \\ &= \mathbf{a}_\rho - \mathbf{a}_\theta \frac{e\sin\theta}{(1 + e\cos\theta)}\end{aligned} \tag{16.9}$$

Hence

$$|\nabla S| = \left[1 + \frac{e^2\sin^2\theta}{(1 + e\cos\theta)^2}\right]^{\frac{1}{2}} \tag{16.10}$$

Fig. 16.3. Outward normal to hyperboloid.

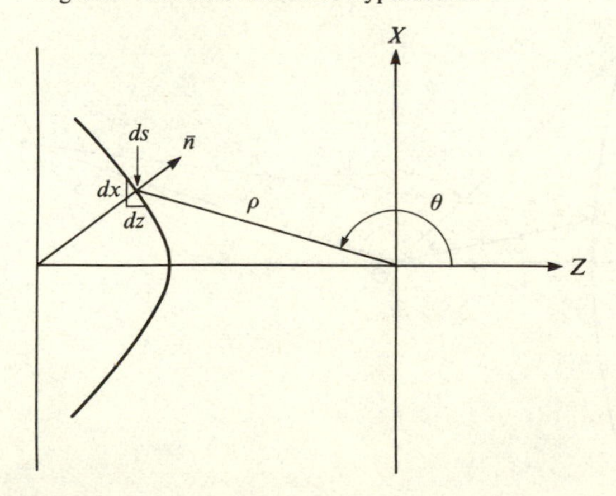

Since the unit normal is defined as

$$\mathbf{n} = \frac{\nabla S}{|\nabla S|}$$

This gives

$$\mathbf{n} = \frac{\mathbf{a}_\rho(1 + e\cos\theta) - \mathbf{a}_\theta e\sin\theta}{|(1 + 2e\cos\theta + e^2)^{\frac{1}{2}}|} \tag{16.11}$$

Since for θ equal to π this gives a negative value for $\mathbf{n}$, it represents the outward unit normal.

Transforming to rectangular coordinates gives

$$\mathbf{n} = \frac{(\hat{x}\sin\theta\cos\phi + \hat{y}\sin\theta\sin\phi + \hat{z}\cos\theta)(1 + e\cos\theta)}{|(1 + 2e\cos\theta + e^2)^{\frac{1}{2}}|} - \frac{(\hat{x}\cos\theta\cos\phi + \hat{y}\cos\theta\sin\phi - \hat{z}\sin\theta)(e\sin\theta)}{|(1 + 2e\cos\theta + e^2)^{\frac{1}{2}}|}$$

This reduces to

$$\mathbf{n} = \hat{\mathbf{x}}\frac{\sin\theta\cos\phi}{u} + \hat{\mathbf{y}}\frac{\sin\theta\sin\phi}{u} + \hat{\mathbf{z}}\frac{(e + \cos\theta)}{u} \tag{16.12}$$

where

$$u = |(1 + 2e\cos\theta + e^2)^{\frac{1}{2}}|$$

16.3 Radiation pattern of hyperboloid with current element feed – current distribution approach

Consider the current element Idx aligned along the x-axis of the coordinate system shown in Fig. 16.4, and positioned at the origin which is also the focus of the paraboloid. The only component of vector potential

Fig. 16.4. Current element feed at common focus of paraboloid and hyperboloid.

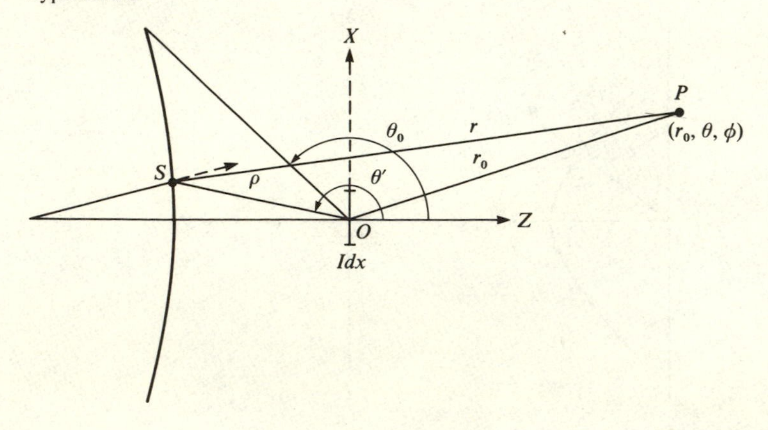

produced by the current element is given by

$$dA_x = \frac{\mu I dx}{4\pi} \frac{e^{-jkr}}{r} \tag{16.13}$$

Hence

$$dH_x = 0$$

$$dH_y = \frac{1}{\mu} \frac{\partial A_x}{\partial z}$$

$$dH_z = -\frac{1}{\mu} \frac{\partial A_x}{\partial y}$$

Now

$$r = (x^2 + y^2 + z^2)^{\frac{1}{2}}$$

so that

$$\frac{\partial}{\partial z}\left(\frac{e^{-jkr}}{r}\right) = -\frac{z}{r^2}\left(jk + \frac{1}{r}\right)e^{-jkr}$$

$$\frac{\partial}{\partial y}\left(\frac{e^{-jkr}}{r}\right) = -\frac{y}{r^2}\left(jk + \frac{1}{r}\right)e^{-jkr}$$

and hence

$$dH_y = -\frac{I dx}{4\pi} \frac{e^{-jkr}}{r}\left(\frac{1}{r} + jk\right)\cos\theta' \tag{16.14}$$

$$dH_z = \frac{I dx}{4\pi} \frac{e^{-jkr}}{r}\left(\frac{1}{r} + jk\right)\sin\theta' \sin\phi' \tag{16.15}$$

The current distribution on the reflector is then obtained from

$$\mathbf{J} = 2(\mathbf{n} \times \mathbf{H}) \tag{16.16}$$

with $\mathbf{n}$ given by

$$\mathbf{n} = \hat{\mathbf{x}}\frac{\sin\theta' \cos\phi'}{u} + \hat{\mathbf{y}}\frac{\sin\theta' \sin\phi'}{u} + \hat{\mathbf{z}}\frac{e + \cos\theta'}{u} \tag{16.17}$$

Thus, using eqns (16.14), (16.15) and (16.17) with ρ substituted for r in equations (16.14) and (16.15), gives

$$J_x = 2(n_y dH_z - n_z dH_y)$$

$$= \frac{I dx e^{-jk\rho}}{2\pi\rho^2}(1 + jk\rho)\left[\frac{\sin^2\theta' \sin^2\phi'}{u} + \frac{\cos\theta'(e + \cos\theta')}{u}\right] \tag{16.18}$$

Similarly,

$$J_y = -2n_x dH_z$$
$$= -\frac{Idx}{2\pi}\frac{e^{-jk\rho}}{\rho^2}(1+jk\rho)\frac{\sin^2\theta'\sin\phi'\cos\phi'}{u} \tag{16.19}$$

and

$$J_z = 2n_x dH_y$$
$$= -\frac{Idx}{2\pi}\frac{e^{-jk\rho}}{\rho^2}(1+jk\rho)\frac{\sin\theta'\cos\theta'\cos\phi'}{u} \tag{16.20}$$

These current densities are for the z-axis along the direction of maximum radiation from the paraboloid.

The corresponding vector potentials are

$$\left.\begin{aligned} dA_x &= \frac{\mu Idx}{8\pi^2}\frac{e^{-jk\rho}}{\rho^2}(1+jk\rho) \\ &\quad\times\frac{\sin^2\theta'\sin^2\phi' + (e+\cos\theta')\cos\theta'}{|(1+2e\cos\theta'+e^2)^{\frac{1}{2}}|}\frac{e^{-jkr}}{r}dS \\ dA_y &= -\frac{\mu Idx}{8\pi^2}\frac{e^{-jk\rho}}{\rho^2}(1+jk\rho)\frac{\sin^2\theta'\sin\phi'\cos\phi'}{|(1+2e\cos\theta'+e^2)^{\frac{1}{2}}|}\frac{e^{-jkr}}{r}dS \\ dA_z &= -\frac{\mu Idx}{8\pi^2}\frac{e^{-jk\rho}}{\rho^2}(1+jk\rho)\frac{\sin\theta'\cos\theta'\cos\phi'}{|(1+2e\cos\theta'+e^2)^{\frac{1}{2}}|}\frac{e^{-jkr}}{r}dS \end{aligned}\right\} \tag{16.21}$$

The contribution to the magnetic field at an observation point (x, y, z) from an element of current over the area dS is given by the components

$$\left.\begin{aligned} dH_x &= \frac{1}{\mu}\left[\frac{\partial}{\partial y}(dA_z) - \frac{\partial}{\partial z}(dA_y)\right] \\ dH_y &= \frac{1}{\mu}\left[\frac{\partial}{\partial z}(dA_x) - \frac{\partial}{\partial x}(dA_z)\right] \\ dH_z &= \frac{1}{\mu}\left[\frac{\partial}{\partial x}(dA_y) - \frac{\partial}{\partial y}(dA_x)\right] \end{aligned}\right\} \tag{16.22}$$

Since the derivatives are taken at the point of observation they involve the term $\exp(-jkr)/r$ only, for which in the radiation field,

$$\frac{\partial}{\partial x}\left(\frac{e^{-jkr}}{r}\right) = -\frac{e^{-jkr}}{r^3}(1+jkr)$$
$$\times\left(x + \frac{pe}{1+e\cos\theta'}\sin\theta'\cos\phi'\right) \doteqdot -j\frac{ke^{-jkr}}{r^2}x$$

Similarly,

$$\left.\begin{aligned}\frac{\partial}{\partial y}\left(\frac{e^{-jkr}}{r}\right) &= -\frac{e^{-jkr}}{r^3}(1+jkr)\\ &\quad\times\left(y+\frac{pe}{1+e\cos\theta'}\sin\theta'\sin\phi'\right) \doteqdot -j\frac{ke^{-jkr}}{r^2}y\\ \frac{\partial}{\partial z}\left(\frac{e^{-jkr}}{r}\right) &= -\frac{e^{-jkr}}{r^3}(1+jkr)\\ &\quad\times\left(z+\frac{pe}{1+e\cos\theta'}\cos\theta'\right) \doteqdot -j\frac{ke^{-jkr}}{r^2}z\end{aligned}\right\} \tag{16.23}$$

The distance r from the source point to the point of observation (r_0, θ, ϕ) in the far field is

$$\begin{aligned} r &\doteqdot r_0 + \rho\cos(\pi - S\hat{O}P)\\ &= r_0 + \frac{pe}{1+e\cos\theta'}(\cos\theta\cos\theta' + \sin\theta\sin\theta'\cos(\phi-\phi'))\end{aligned} \tag{16.24}$$

so that

$$\begin{aligned} dH_x &= j\frac{kIdx}{8\pi^2}\frac{e^{-jk(\rho+r)}}{\rho^2 r^2}(1+jk\rho)\left[y\frac{\sin\theta'\cos\theta'\cos\phi'}{|(1+2e\cos\theta'+e^2)^{\frac{1}{2}}|}\right.\\ &\quad\left. - z\frac{\sin^2\theta'\sin\phi'\cos\phi'}{|(1+2e\cos\theta'+e^2)^{\frac{1}{2}}|}\right]dS\\ &= jk\frac{Idx}{8\pi^2}\frac{e^{-jkr_0}}{r_0}\left\{\frac{(1+e\cos\theta')^2}{p^2e^2}\left(1 - jk\frac{pe}{(1+e\cos\theta')}\right)\right.\\ &\quad\times e^{+jkpe/(1+e\cos\theta')}\\ &\quad\times e^{-jkpe(\cos\theta\cos\theta'+\sin\theta\sin\theta'\cos(\phi-\phi'))/(1+e\cos\theta)}\\ &\quad\times(\sin\theta\sin\phi\sin\theta'\cos\theta'\cos\phi'\\ &\quad-\cos\theta\sin^2\theta'\sin\phi'\cos\phi')(-)\\ &\quad\times\frac{p^2e^2\sin\theta'}{(1+e\cos\theta')^3}d\theta'\,d\phi'\end{aligned}$$

Integrating over the surface of the hyperboloid gives

$$\begin{aligned} H_x &= -j\frac{kIdx}{8\pi^2}\frac{e^{-jkr_0}}{r_0}\int_{\theta_0}^{\pi} e^{jkpe(1-\cos\theta\cos\theta')/(1+e\cos\theta')}\\ &\quad\times\frac{\sin\theta'}{(1+e\cos\theta')}\left(1-jk\frac{pe}{(1+e\cos\theta')}\right)\\ &\quad\times\left\{\left(\sin\theta\sin\phi\sin\theta'\cos\theta'\int_0^{2\pi}\cos\phi'\right.\right.\end{aligned}$$

$$\times e^{-jkpe\sin\theta\sin\theta'\cos(\phi-\phi')/(1+e\cos\theta')}\,d\phi'\,d\theta'$$

$$-\cos\theta\sin^2\theta'\int_0^{2\pi}\sin\phi'\cos\phi'$$

$$\times e^{-jkpe\sin\theta\sin\theta'\cos(\phi-\phi')/(1+e\cos\theta')}\,d\phi'\,d\theta'\bigg)\bigg\}$$

Defining

$$\psi_1=\frac{kpe(1-\cos\theta\cos\theta')}{1+e\cos\theta'}\quad\text{and}$$

$$\psi_2=-\frac{kpe\sin\theta\sin\theta'\cos(\phi-\phi')}{(1+e\cos\theta')}$$

gives, for the total magnetic field in the x-direction,

$$H_x=-j\frac{kI\,dx}{8\pi^2}\frac{e^{-jkr_0}}{r_0}\int_{\theta_0}^{\pi}e^{j\psi_1}\frac{\sin\theta'}{(1+e\cos\theta')}$$

$$\times\left(1-jk\frac{pe}{(1+e\cos\theta')}\right)\bigg\{\bigg(\sin\theta\sin\phi\sin\theta'\cos\theta'$$

$$\times\int_0^{2\pi}e^{j\psi_2}\cos\phi'\,d\phi'\,d\theta'-\cos\theta\sin^2\theta'\int_0^{2\pi}e^{j\psi_2}$$

$$\times\sin\phi'\cos\phi'\,d\phi'\,d\theta'\bigg)\bigg\}\tag{16.25}$$

Similarly the component of magnetic field in the y-direction due to an element of current over the area dS is given by

$$dH_y=-j\frac{kI\,dx}{8\pi^2}\frac{e^{-jk(\rho+r)}}{\rho^2r^2}(1+jk\rho)$$

$$\times\left[z\frac{\sin^2\theta'\sin^2\phi'+(e+\cos\theta')\cos\theta'}{|(1+2e\cos\theta'+e^2)^{\frac{1}{2}}|}\right.$$

$$\left.+x\frac{\sin\theta'\cos\theta'\cos\phi'}{|(1+2e\cos\theta'+e^2)^{\frac{1}{2}}|}\right]dS$$

$$=-j\frac{kI\,dx}{8\pi^2}\frac{e^{-jkr_0}}{r_0}\bigg\{\frac{(1+e\cos\theta')^2}{p^2e^2}\left(1-jk\frac{pe}{(1+e\cos\theta')}\right)$$

$$\times e^{jkpe/(1+e\cos\theta')}$$

$$\times e^{-jkpe(\cos\theta\cos\theta'+\sin\theta\sin\theta'\cos(\phi-\phi'))/(1+e\cos\theta')}$$

$$\times[\cos\theta(\sin^2\theta'\sin^2\phi'+(e+\cos\theta')\cos\theta')$$

$$+\sin\theta\cos\phi\sin\theta'\cos\theta'\cos\phi']$$

$$\times(-)\frac{p^2e^2\sin\theta'}{(1+e\cos\theta')^3}\,d\theta'\,d\phi'\bigg\}$$

Integrating over the surface of the hyperboloid gives, for the total magnetic field in the y-direction,

$$H_y = j\frac{kIdx}{8\pi^2}\frac{e^{-jkr_0}}{r_0}\int_{\theta_0}^{\pi} e^{j\psi_1}\frac{\sin\theta'}{(1+e\cos\theta')}$$

$$\times\left(1 - jk\frac{pe}{(1+e\cos\theta')}\right)\Bigg\{\Bigg(\cos\theta\sin^2\theta'$$

$$\times\int_0^{2\pi} e^{j\psi_2}\sin^2\phi'\,d\phi'\,d\theta' + \cos\theta\cos\theta'(e+\cos\theta')$$

$$\times\int_0^{2\pi} e^{j\psi_2}\,d\phi'\,d\theta' + \sin\theta\cos\phi\sin\theta'\cos\theta'$$

$$\times\int_0^{2\pi} e^{j\psi_2}\cos\phi'\,d\phi'\,d\theta'\Bigg)\Bigg\} \tag{16.26}$$

The component of magnetic field in the z-direction due to an element of current over the area dS is given by

$$dH_z = j\frac{kIdx}{8\pi^2}\frac{e^{-jk(\rho+r)}}{\rho^2 r^2}(1+jk\rho)\left[x\frac{\sin^2\theta'\sin\phi'\cos\phi'}{|(1+2e\cos\theta'+e^2)^{\frac{1}{2}}|}\right.$$

$$\left. + y\frac{\sin^2\theta'\sin^2\phi' + (e+\cos\theta')\cos\theta'}{|(1+2e\cos\theta'+e^2)^{\frac{1}{2}}|}\right]dS$$

$$= j\frac{kIdx}{8\pi^2}\frac{e^{-jkr_0}}{r_0}\Bigg\{\frac{(1+e\cos\theta')^2}{p^2e^2}$$

$$\times\left(1 - jk\frac{pe}{(1+e\cos\theta')}\right)e^{jkpe/(1+e\cos\theta')}$$

$$\times e^{-jkpe(\cos\theta\cos\theta' + \sin\theta\sin\theta'\cos(\phi-\phi'))/(1+e\cos\theta')}$$

$$\times[\sin\theta\cos\phi\sin^2\theta'\sin\phi'\cos\phi' + \sin\theta\sin\phi(\sin^2\theta'\sin^2\phi'$$

$$+ (e+\cos\theta')\cos\theta')](-)\frac{p^2e^2\sin\theta'}{(1+e\cos\theta')^3}d\theta'\,d\phi'\Bigg\}$$

Integrating over the surface of the hyperboloid gives

$$H_z = j\frac{kIdx}{8\pi^2}\frac{e^{-jkr_0}}{r_0}\int_{\theta_0}^{\pi} e^{j\psi_1}\left(\frac{1}{\rho'} + jk\right)\frac{ep\sin\theta'}{(1+e\cos\theta')^2}$$

$$\times\Bigg\{\Bigg(\sin\theta\cos\phi\sin^2\theta'\int_0^{2\pi} e^{j\psi_2}\sin\phi'\cos\phi'\,d\phi'\,d\theta'$$

$$+ \sin\theta\sin\phi\sin^2\theta'\int_0^{2\pi} e^{j\psi_2}\sin^2\phi'\,d\phi'\,d\theta'$$

$$+ \sin\theta\sin\phi(e+\cos\theta')\cos\theta'\int_0^{2\pi} e^{j\psi_2}\,d\phi'\,d\theta'\Bigg\} \tag{16.27}$$

Transforming to spherical coordinates using

$$H_\theta = H_x \cos\theta \cos\phi + H_y \cos\theta \sin\phi - H_z \sin\theta$$
$$H_\phi = -H_x \sin\phi + H_y \cos\phi$$

leads to the far field result

$$H_\theta = \frac{Idx}{2\lambda^2} \frac{e^{-jkr_0}}{r_0} \int_{\theta_0}^{\pi} e^{j\psi_1} \frac{ep \sin\theta'}{(1 + e\cos\theta')^2} \times \left\{ \sin\phi(1 + e\cos\theta') \int_0^{2\pi} e^{j\psi_2}\, d\phi'\, d\theta' + \sin^2\theta' \int_0^{2\pi} e^{j\psi_2} \cos\phi' \sin(\phi' - \phi)\, d\phi'\, d\theta' \right\} \tag{16.28}$$

and

$$H_\phi = \frac{Idx}{2\lambda^2} \frac{e^{-jkr_0}}{r_0} \int_{\theta_0}^{\pi} e^{j\psi_1} \frac{ep \sin\theta'}{(1 + e\cos\theta')^2} \times \left\{ \sin\theta \sin\theta' \cos\theta' \int_0^{2\pi} e^{j\psi_2} \cos\phi'\, d\phi'\, d\theta' - \cos\theta \sin^2\theta' \int_0^{2\pi} e^{j\psi_2} \cos\phi' \cos(\phi - \phi')\, d\phi'\, d\theta' + \cos\theta \cos\phi(1 + e\cos\theta') \int_0^{2\pi} e^{j\psi_2}\, d\phi'\, d\theta' \right\} \tag{16.29}$$

16.3.1 *Reduction of field components to single integral*

In eqns (16.28) and (16.29) the following integrals arise:

$$\left.\begin{aligned} I_1(\phi) &= \int_0^{2\pi} e^{j\psi_2}\, d\phi' \\ I_2(\phi) &= \int_0^{2\pi} \cos\phi' e^{j\psi_2}\, d\phi' \\ I_3(\phi) &= \int_0^{2\pi} \cos\phi' \cos(\phi' - \phi) e^{j\psi_2}\, d\phi' \\ I_4(\phi) &= \int_0^{2\pi} \cos\phi' \sin(\phi' - \phi) e^{j\psi_2}\, d\phi' \end{aligned}\right\} \tag{16.30}$$

where

$$\psi_2 = -\frac{kpe \sin\theta \sin\theta' \cos(\phi - \phi')}{(1 + e\cos\theta')} = \psi \cos(\phi - \phi')$$

Then since

$$\begin{aligned} e^{j\psi\cos(\phi-\phi')} &= \cos(\psi\cos(\phi-\phi')) + j\sin(\psi\cos(\phi-\phi')) \\ &= J_0(\psi) - 2\{J_2(\psi)\cos 2(\phi-\phi') \\ &\quad - J_4(\psi)\cos 4(\phi-\phi') + \ldots\} + j2\{J_1(\psi)\cos(\phi-\phi') \\ &\quad - J_3(\psi)\cos 3(\phi-\phi') + \ldots\} \\ &= J_0(\psi) + 2\sum_1^\infty j^n J_n(\psi)\cos n(\phi-\phi') \end{aligned}$$

it follows that

$$\begin{aligned} I_1(\phi) &= \int_0^{2\pi} J_0(\psi)\,d\phi' + j\int_0^{2\pi} 2J_1(\psi)\cos(\phi-\phi')\,d\phi' \\ &\quad - \int_0^{2\pi} 2J_2(\psi)\cos 2(\phi-\phi')\,d\phi' + \ldots \\ &= 2\pi J_0(\psi) \end{aligned}$$

Similarly it may be shown that

$$\left.\begin{aligned} I_2(\phi) &= j2\pi\cos\phi J_1(\psi) \\ I_3(\phi) &= \pi\cos\phi[J_0(\psi) - J_2(\psi)] \\ \text{and}\quad I_4(\phi) &= -\pi\sin\phi[J_0(\psi) + J_2(\psi)] \end{aligned}\right\} \tag{16.31}$$

Hence the scattered electric field from the paraboloid has a component E_θ, given by

$$\begin{aligned} E_\theta^s = Z_0 H_\phi &= \frac{I\,dx}{2\lambda^2}\sqrt{\left(\frac{\mu}{\varepsilon}\right)}\frac{e^{-jkr_0}}{r_0}\int_{\theta_0}^{\pi} e^{j\psi_1}\frac{ep\sin\theta'}{(1+e\cos\theta')^2} \\ &\quad \times \{j\sin\theta\sin\theta'\cos\theta'\cdot 2\pi J_1(\psi)\cos\phi \\ &\quad - \cos\theta\sin^2\theta'\pi\cos\phi[J_0(\psi) - J_2(\psi)] \\ &\quad + \cos\theta\cos\phi(1+e\cos\theta')2\pi J_0(\psi)\}\,d\theta' \end{aligned} \tag{16.32}$$

Now the direct field radiated from the x-directed dipole has a θ-component

$$\begin{aligned} E_\theta^d &= E_x\cos\theta\cos\phi + E_y\cos\theta\sin\phi - E_z\sin\theta \\ &= -jkZ_0\frac{I\,dx}{4\pi}\frac{e^{-jkr}}{r}\cos\theta\cos\phi \end{aligned} \tag{16.33}$$

The total θ-component of electric field is then

$$\begin{aligned} E_\theta^T &= \frac{Z_0 I\,dx}{2\lambda}\frac{e^{-jkr_0}}{r_0}\cos\phi\left\{-j\cos\theta + k\int_{\theta_0}^{\pi}\frac{ep\sin\theta' e^{j\psi_1}}{(1+e\cos\theta')^2}\right. \\ &\quad \times\left[\cos\theta\left(\frac{\sin^2\theta'}{2} - (1+e\cos\theta')\right)J_0(\psi) + \frac{\cos\theta\sin^2\theta'}{2}\right. \\ &\quad \left.\left.\times J_2(\psi) + j\sin\theta\sin\theta'\cos\theta' J_1(\psi)\right]d\theta'\right\} \end{aligned} \tag{16.34}$$

Similarly it may be shown that the total ϕ-directed component of electric field is

$$E_\phi^T = \frac{Z_0 I dx}{2\lambda}\frac{e^{-jkr_0}}{r_0}\sin\phi\left\{ j\cos^2\theta - k\int_{\theta_0}^{\pi}\frac{ep\sin\theta' e^{j\psi_1}}{(1+e\cos\theta')^2}\right.$$
$$\left.\times\left[(1+e\cos\theta') - \frac{\sin^2\theta'}{2}J_0(\psi) - \frac{\sin^2\theta'}{2}J_2(\psi)\right]d\theta'\right\} \quad (16.35)$$

16.4 Forward scattered field of hyperboloid with cylindrical waveguide feed

Because a current element would not be used in practice as a feed for a reflector antenna, it is worthwhile to demonstrate how the forward field scattered by a hyperboloid from a cylindrical waveguide feed with a y-directed electric field can be calculated. From eqns (12.56)–(12.58) the radiated fields from such a feed can be written as

$$E_\theta = f_1(\theta)\sin\phi\frac{e^{-jk\rho}}{\rho}$$

$$E_\phi = f_2(\theta)\cos\phi\frac{e^{-jk\rho}}{\rho}$$

Converting to rectangular coordinates gives, for the far field,

$$dH_x = -\frac{1}{Z_0}[f_2(\theta)\cos\theta\cos^2\phi + f_1(\theta)\sin^2\phi]\frac{e^{-jk\rho}}{\rho}$$

$$dH_y = -\frac{1}{Z_0}[f_2(\theta)\cos\theta - f_2(\theta)]\sin\phi\cos\phi\frac{e^{-jk\rho}}{\rho}$$

$$dH_z = \frac{1}{Z_0}f_2(\theta)\sin\theta\cos\phi\frac{e^{-jk\rho}}{\rho}$$

Then the standard approximation

$$\mathbf{J} = 2\mathbf{n}\times\mathbf{H}$$

with $\mathbf{n}$ given by eqn (16.17), gives for the dominant y-directed current density

$$J_y = -\frac{2}{Z_0}\frac{e^{-jk\rho}}{\rho}\left[f_1(\theta)\frac{e+\cos\theta}{u}\sin^2\phi\right.$$
$$\left.+ f_2(\theta)\frac{(1+e\cos\theta)\cos^2\phi}{u}\right]$$

In the forward scattered direction the contribution to the total vector potential A_y from an elemental area dS on the hyperboloid is given by

$$dA_y = \frac{\mu}{2\pi Z_0}\frac{e^{-jk\rho}}{\rho}\left[f_1(\theta)\frac{e+\cos\theta}{u}\sin^2\phi + f_2(\theta)\right.$$
$$\left. \times \frac{(1+e\cos\theta)}{u}\cos^2\phi \right]\frac{e^{-jk(r_0-\cos\theta)}}{r}dS \tag{16.36}$$

where

$$dS = -\frac{\rho^2\sin\theta\, u}{(1+e\cos\theta)}d\theta\, d\phi$$

Hence, integrating over ϕ using the standard integrals,

$$\int_0^{2\pi}\frac{\sin^2 x}{\cos^2 x}dx = \pi$$

Fig. 16.5. Comparison of experimental and theoretical H-plane radiation pattern from 7.8 wavelength hyperboloid reflector with horn illumination. (After Rusch, 1963.)

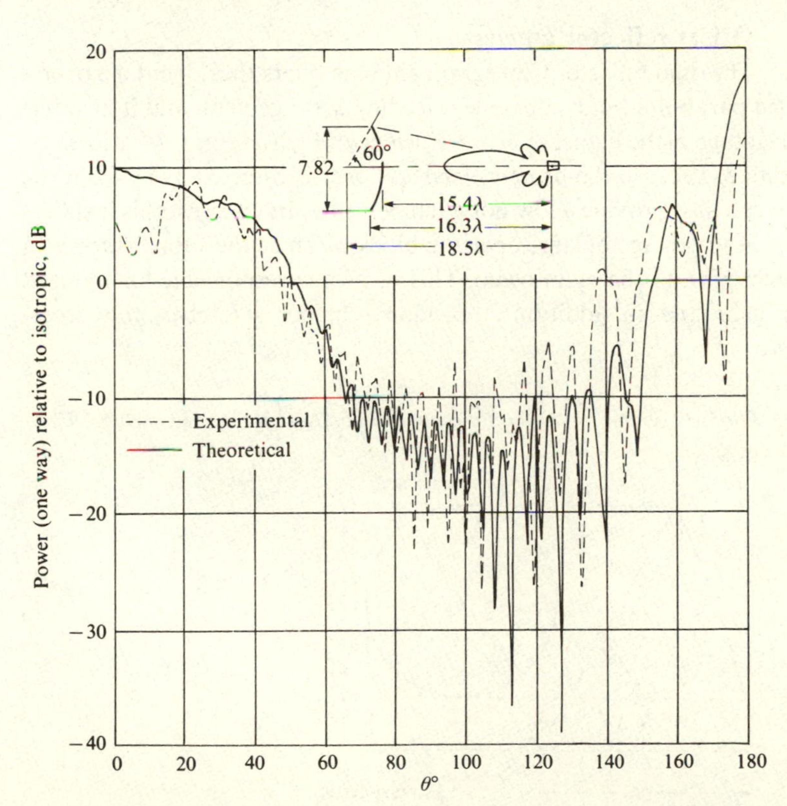

gives for this total vector potential, in the form of an integrand over θ only,

$$A_y = -\frac{\mu pe}{2Z_0}\frac{e^{-jkr_0}}{r_0}\int_0^{\theta_0}[f_1(\theta)(e+\cos\theta)+f_2(\theta)(1+e\cos\theta)]$$

$$\times\frac{\sin\theta}{(1+e\cos\theta)}\cdot\exp jke(1-\cos\theta)/(1+e\cos\theta)\cdot d\theta \qquad (16.37)$$

This function can be readily evaluated by computer, and in the far field the associated magnetic intensity H_x is given by

$$H_x = \frac{j\omega A_y}{Z_0}$$

16.4.1 *Scattered radiation pattern from hyperboloidal reflector with horn illumination*

The results obtained in Section 16.4 can clearly be generalised to include observation points other than in the forward direction, and results from such an analysis have been compared with experiment, as illustrated in Fig. 16.5. The agreement obtained can be seen to be very satisfactory.

16.5 **Offset reflector antennas**

The dual reflector Cassegrain antenna offers the advantage over a front fed paraboloid of a convenient feeding arrangement, and it also has the advantage of the higher aperture efficiency which a large f/D ratio gives. In addition, because the primary feed spillover is directed away from the earth it can also provide a low noise temperature. But against this it suffers the big disadvantage of large aperture blocking since the subreflector is in the direct path of the main beam. This in turn is responsible for reduced power gain and in addition it produces higher sidelobes and cross-polarisation.

Fig. 16.6. Simple offset reflector configuration. (After Rudge and Adatia, 1978.)

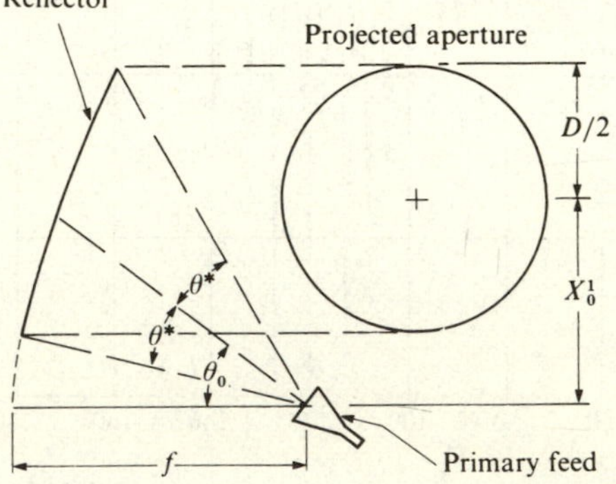

A solution to the problem of aperture blocking is offered by the offset reflector antenna, as shown in Fig. 16.6. This design also effectively eliminates the reflection from the main reflector into the primary waveguide feed, but has the disadvantage of generating a cross-polarised field when it is fed from a linearly polarised feed.

The offset parabolic reflector can also be designed as a dual reflector system, as shown in Fig. 16.7, where the main reflector is illuminated by the combination of a subreflector and a primary feed. The location of this primary feed may be either below the dual reflectors, or it may be within the main reflector as shown in Fig. 16.7, an arrangement which is usually referred to as an open Cassegrain system.

16.5.1 *Geometry of offset reflector*

Referring to Fig. 16.8 the convention is adopted that unprimed coordinates (ρ, θ, ϕ), (x, y, z) are used for the offset paraboloid, and primed

Fig. 16.7. Dual offset reflector – open Cassegrain. (After Rudge and Adatia, 1978.)

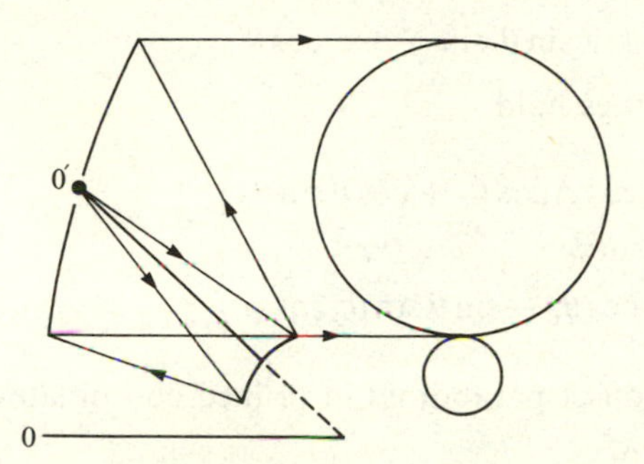

Fig. 16.8. Geometry and coordinate system for offset paraboloid. (After Rudge and Adatia, 1978.)

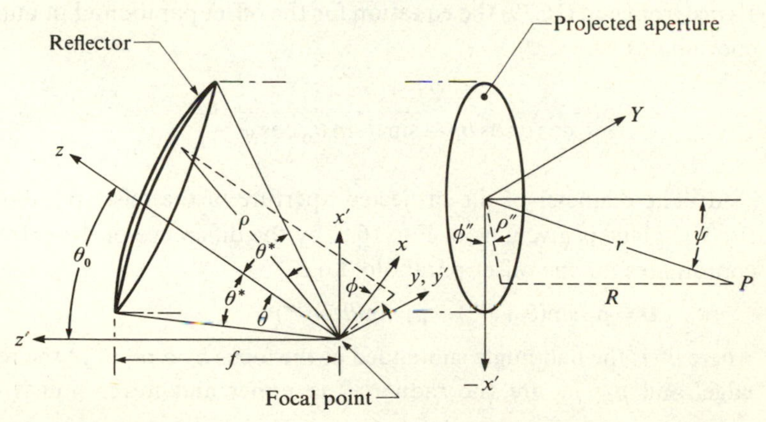

coordinates (ρ', θ', ϕ'), (x', y', z') for the parent paraboloid from which the offset is cut. The following relationships hold between these coordinates:

(a) $\rho = \rho'$ (16.38)

This follows since the radius from the focus to any point on the reflector surface is identical in both coordinate systems.

(b) In the unprimed coordinate system the spherical-rectangular transformation is given by

$$\rho = x \sin\theta \cos\phi + y \sin\theta \sin\phi + z \cos\theta$$

Since the primed-to-unprimed conversions are

$$x = x' \cos\theta_0 - z' \sin\theta_0$$
$$y = y'$$
$$z = x' \sin\theta_0 + z' \cos\theta_0$$

it follows that the primed spherical–rectangular transformation is given by

$$\begin{aligned}\rho' &= (x' \cos\theta_0 - z' \sin\theta_0) \sin\theta \cos\phi + y' \sin\theta \sin\phi \\ &\quad + (x' \sin\theta_0 + z' \cos\theta_0) \cos\theta \\ &= x' \sin\theta' \cos\phi' + y' \sin\theta' \sin\phi' + z' \cos\theta'\end{aligned}$$

Hence the following equalities hold:

$$\left.\begin{aligned}\sin\theta' \cos\phi' &= \sin\theta \cos\phi \cos\theta_0 + \cos\theta \sin\theta_0 \\ \sin\theta' \sin\phi' &= \sin\theta \sin\phi \\ \cos\theta' &= \cos\theta \cos\theta_0 - \sin\theta \sin\theta_0 \cos\phi\end{aligned}\right\} \quad (16.39)$$

(c) The equation of the offset paraboloid in primed coordinates is

$$\rho' = f \sec^2 \frac{\theta'}{2} = \frac{2f}{1 + \cos\theta'}$$

Hence from eqn (16.39) the equation for the offset paraboloid in unprimed coordinates is

$$\rho = \frac{2f}{1 + \cos\theta \cos\theta_0 - \sin\theta \sin\theta_0 \cos\phi} \quad (16.40)$$

(d) The diameter of the projected aperture of the offset paraboloid in the $x'y'$-plane is given, from Fig. 16.8, by the difference of the extreme x'-coordinates for the offset paraboloid, i.e.

$$D = \rho_U \sin(\theta + \theta^*) - \rho_L \sin(\theta - \theta^*)$$

where θ^* is the half-angle subtended at the focus by a point at the reflector edge, and ρ_U, ρ_L are the radii to the upper and lower points on the paraboloid. These radii are defined by the equations

$$\rho_U = \frac{2f}{1 + \cos(\theta_0 + \theta^*)}$$

$$\rho_L = \frac{2f}{1 + \cos(\theta_0 - \theta^*)}$$

Substituting in the equation for the diameter, and simplifying, leads to the result

$$D = \frac{4f \sin\theta^*}{(\cos\theta_0 + \cos\theta^*)} \tag{16.41}$$

(e) When radiation is considered to take place from the projected diameter D it is convenient to regard the centre of this circular aperture as a new reference point for all the other radiating elements in the aperture. The distance of this centre from the origin, with both these reference points being on the x' axis, is given by

$$x'_0 = \frac{D}{2} + \frac{2f \sin\theta^*}{(1 + \cos(\theta_0 - \theta^*))}$$

By substituting from eqn (16.41) this expression may be simplified to give

$$x'_0 = \frac{2f \sin\theta_0}{(\cos\theta_0 + \cos\theta^*)} \tag{16.42}$$

Hence the path length from a radiating element in the circular aperture of radius $D/2$ located at (ρ'', ϕ''), as shown in Fig. 16.8, to a point of observation $P(r, \psi, \Phi)$ in the far field, is given by

$$R \approx r - \rho'' \sin\psi \cos(\phi'' - \Phi)$$

where r is the distance from the centre of the aperture to P. Since ρ'' is the length of the radius to the source point from the centre of the aperture, this equation may be written as

$$R \approx r - (x'_0 - x') \sin\psi \cos\Phi - y \sin\psi \sin\Phi \tag{16.43}$$

where (x', y) are the rectangular coordinates of the source point, given by

$$\left.\begin{aligned} x' &= \rho' \sin\theta' \cos\phi' \\ &= \rho(\sin\theta \cos\phi \cos\theta_0 + \cos\theta \sin\theta_0) \\ y &= \rho \sin\theta \sin\phi \end{aligned}\right\} \tag{16.44}$$

(f) The elemental area $dx'\,dy$ of a source element in the radiating aperture may be transformed to polar coordinates (θ', ϕ') through the

Jacobian

$$J = \begin{vmatrix} \dfrac{\partial x}{\partial \theta} & \dfrac{\partial x}{\partial \phi} \\ \dfrac{\partial y}{\partial \theta} & \dfrac{\partial y}{\partial \phi} \end{vmatrix}$$

In applying this relationship it must be remembered that in eqn (16.44) the symbol ρ is itself a function of (θ, ϕ) as given by eqn (16.40). The evaluation of the determinant is therefore laborious, but the result simplifies to

$$dx'\,dy = \rho^2 \sin\theta\, d\theta\, d\phi \tag{16.45}$$

(g) For the purpose of finding the reflected field from the surface of the offset paraboloid it is necessary to know the equation of the unit normal at the point of reflection. For the parent paraboloid this has previously been given as

$$\mathbf{a}_n = -\mathbf{a}_{x'} \sin\frac{\theta'}{2}\cos\phi' - \mathbf{a}_{y'} \sin\frac{\theta'}{2}\sin\phi' - \mathbf{a}_{z'}\cos\frac{\theta'}{2}$$

which may also be written as

$$\mathbf{a}_n = -\frac{1}{2\cos\dfrac{\theta'}{2}}\left(\mathbf{a}_{x'}\sin\theta'\cos\phi' + \mathbf{a}_{y'}\sin\theta'\sin\phi' + \mathbf{a}_{z'}2\cos^2\frac{\theta'}{2}\right) \tag{16.46}$$

Using the rectangular conversion formulae

$$\mathbf{a}_x = \mathbf{a}_{x'}\cos\theta_0 - \mathbf{a}_{z'}\sin\theta_0$$
$$\mathbf{a}_z = \mathbf{a}_{x'}\sin\theta_0 + \mathbf{a}_{z'}\cos\theta_0$$

this simplifies to

$$\mathbf{a}_n = -\frac{1}{2\cos\dfrac{\theta'}{2}}[(\sin\theta\cos\phi - \sin\theta_0)\mathbf{a}_x + \sin\theta\sin\phi\mathbf{a}_y + (\cos\theta + \cos\theta_0)\mathbf{a}_z]$$

The final dependence on $\theta'/2$ may then by removed by using the relationship

$$\rho = \frac{2f}{\cos^2\dfrac{\theta'}{2}} = \frac{2f}{1 + \cos\theta\cos\theta_0 - \sin\theta\sin\theta_0\cos\phi}$$

so that the equation for the normal becomes

$$\mathbf{a}_n = -\sqrt{\left(\frac{\rho}{4f}\right)}((\sin\theta\cos\phi - \sin\theta_0)\mathbf{a}_x + \sin\theta\sin\phi\mathbf{a}_y + (\cos\theta + \cos\theta_0)\mathbf{a}_z) \quad (16.47)$$

16.6 Tangential electric fields in focal aperture plane of offset paraboloid

Assuming that the phase centre of the primary feed is located at the geometric focus, and that optical type reflection occurs at the surface of the offset paraboloid, the tangential electric field in the focal aperture plane will be given by

$$E_a = E^r e^{-jk\rho\cos\theta'} \quad (16.48)$$

where E^r is the reflected field at the surface of the paraboloid. This may also be written as

$$E_a = E^r e^{jk(\rho - 2f)} \quad (16.49)$$

To find E^r use is made of eqn (15.8)

$$\mathbf{E}^r = 2(\mathbf{a}_n \cdot \mathbf{E}^i)\mathbf{a}_n - \mathbf{E}^i$$

where the incident field from the primary feed, when expressed in spherical coordinates, is

$$\mathbf{E}^i = E_\theta \mathbf{a}_\theta + E_\phi \mathbf{a}_\phi$$

Transforming this to rectangular coordinates gives

$$\mathbf{E}^i = \mathbf{a}_x(E_\theta\cos\theta\cos\phi - E_\phi\sin\phi) + \mathbf{a}_y(E_\theta\cos\theta\sin\phi + E_\phi\cos\phi) - \mathbf{a}_z E_\theta\sin\theta \quad (16.50)$$

Hence by substituting eqns (16.46) and (16.49) into (15.8) and then using eqn (16.48) the tangential aperture electric field becomes

$$\begin{aligned}\mathbf{E}_a = \frac{\rho}{2f}\{&\mathbf{a}_x[E_\theta\cos\theta(\sin\theta\sin\theta_0 - \cos\phi(1 + \cos\theta\cos\theta_0)) \\ &+ E_\phi\cos\theta_0\sin\phi(\cos\theta + \cos\theta_0)] \\ &- \mathbf{a}_y[E_\theta\sin\phi(\cos\theta + \cos\theta_0) \\ &- E_\phi(\sin\theta\sin\theta_0 - \cos\phi(1 + \cos\theta\cos\theta_0))]\}E^{jk(\rho - 2f)}\end{aligned} \quad (16.51)$$

Using this equation the calculation of the radiation field from the offset paraboloid for any specified primary feed radiation pattern may now be dealt with, following the method of Section 10.6.

16.7 Offset paraboloid radiation pattern

Eqn (16.51) gives the tangential electric field in the circular aperture, of which each element of area is $dx'\,dy$. From eqn (16.45) the variables of integration may be changed to the angles $d\theta, d\phi$ associated directly with the offset paraboloid, in which the x, y-components of electric field in equation (16.51) have already been expressed. Hence, referring to Fig. 16.8, the radiation field has components E_ψ, E_Φ which from eqns (10.32), (10.33) and (16.43) may be expressed as

$$E_\psi = \frac{j}{\lambda}\frac{e^{-jkr_0}}{r_0}\int_0^{2\pi}\int_0^{\theta^*}(E_{ax}\cos\Phi + E_{ay}\sin\Phi) \times e^{jkR'}\rho^2\sin\theta\,d\theta\,d\phi$$

Fig. 16.9. Measured radiation pattern of offset fed paraboloid antenna linearly polarised at 18.5 GHz. Aperture diameter is 12″; $f/D = 0.25$. (After Chu and, Turrin, 1973.)

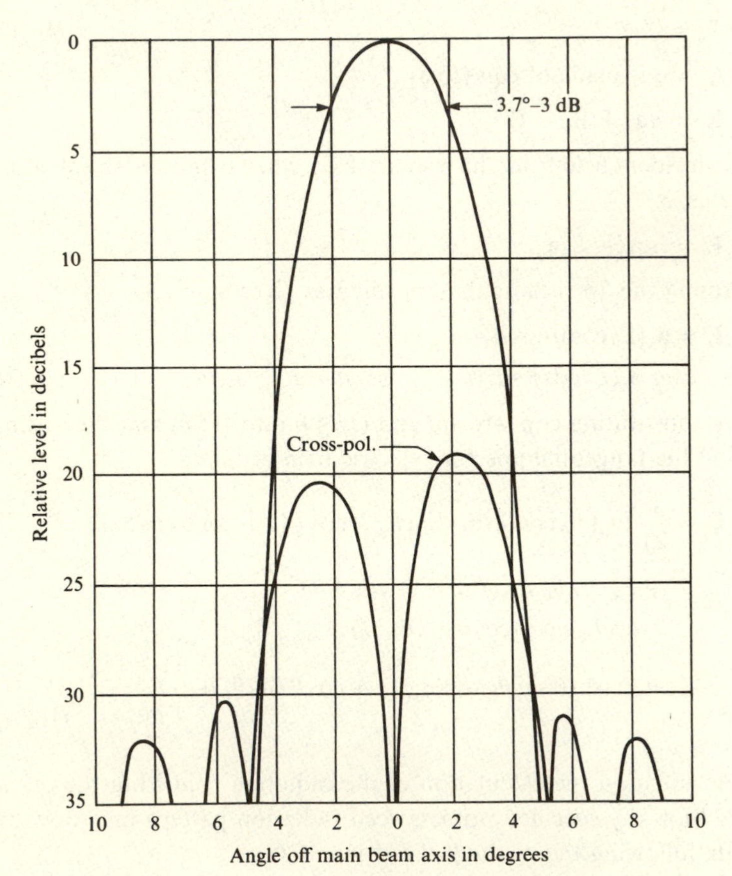

$$E_\Phi = \frac{j}{\lambda}\frac{e^{-jkr_0}}{r_0}\int_0^{2\pi}\int_0^{\theta^*}(E_{ax}\sin\Phi - E_{ay}\cos\Phi) \times e^{jkR'}\rho^2\sin\theta\, d\theta\, d\phi \quad (16.52)$$

where

$$R' = (x'_0 - x')\sin\psi\cos\Phi + y\sin\psi\sin\Phi$$

and x'_0, x', y are given in eqns (16.42) and (16.44).

From eqn (16.52) the pattern may be computed by double numerical integration, once the offset paraboloid parameters f, θ_0 and θ^*, and the primary feed radiation fields E_θ, E_ϕ at the reflector surface have been specified. A measured radiation pattern is shown in Fig. 16.9 for an offset paraboloid of 12″ diameter at 18.5 GHz, with $\theta_0 = \theta^x = 45°$. The feed used was a circular symmetric dual mode feed which provided a pattern taper of − 10 dB at the edge of the offset paraboloid section.

Further reading

T.S. Chu and R.H. Turrin: 'Depolarisation properties of offset reflector antennas': *IEEE Trans. Antennas and Propagation*, **AP-21**, 1973, 339–45. (Copyright C, 1973, *IEEE*.)

V. Galindo: 'Design of dual-reflector, antennas with arbitrary phase and amplitude distributions': *IEEE Trans. Antennas and Propagation*, **AP-12**, 1964, 403–8.

A.W. Rudge and N. Adatia: 'Offset-parabolic-reflector antennas: a review': *Proc. IEEE*, **66**, 1978, 1592–1618. (Copyright C, 1978, *IEEE*.)

A.W. Rudge, K. Milne, A.D. Olver, P. Knight (Ed.): *The Handbook of Antenna Design*: Peter Peregrinus, 1982.

W.V.T. Rusch: 'Scattering, from hyperboloidal reflector in a Cassegrainian feed system': *Trans. IEEE* (*APG*), **AP-21**, 1963. (Copyright C, 1963, *IEEE*.)

Index

+ – +

admittance matrix, 65
Ampères law, 2
analysis of infinite sheath helix, 126
  transmitting paraboloid, 312
angular spectrum of plane waves, 210
  for two dimensional aperture, 223
aperture antennas, 189
  efficiency for paraboloid with cylindrical waveguide feed, 305; corrugated waveguide feed, 309; infinitesimal dipole feed, 297
areal current density, 15
array of director elements, 145
axial line current approximation, 15

Bessel function, 102
Bifin antenna, 158

Cassegrain antenna, 333
centre-fed transmitting dipole, 39
charge, point, 9, 13
  uniformly distributed, 18
characteristic impedance, 158
circular aperture in planar conducting sheet, 229
comparison of aperture field techniques, 201
complementary structure, 159
computer solutions of dipoles, receiving operation, 68
  transmitting operation, 66
computer solutions of monopoles, with infinite ground, 70
  with finite ground plane, input impedance, 71; radiation pattern, 75
corner reflector antenna, 62
  with half-wave dipole, 78
contours, electric field for corner reflector, 81
  paraboloid, 285
  magnetic field for paraboloid, 291
coupling between apertures, 300
coupling between paraboloid and circular waveguide feed, 303
  and corrugated waveguide feed, 307
cross spectral density, 182
  between two elements of linear dipole, 183
  in terms of radiation resistance, 186
curl, rectangular coordinates, 11
  spherical coordinates, 27
current density, areal, 15
  linear, 3
current distribution of linear dipole, 41, 47, 67, 68
  corner reflector, 80
  hyperboloidal reflector, 339
  paraboloidal reflector, 272
current element feed for hyperboloid, 338
current elements along orthogonal axes, 27
current element sources, 9
cylindrical current element, 15
  waveguide-derivation of wave equation, 254; solution of wave equation for TE modes, 255; tangential electric and magnetic fields in aperture, 257; radiation pattern for TE_{11} mode, 259

dipole antennas, 39
  centre-fed transmitting, 39
  computer solution of, 65; in corner reflector, 78; transmitting operation, 66; receiving operation, 68
  current distribution when transmitting, 41
  filamentary, 34
  general expression for fields, 56
  half-wave dipole in corner reflector, 62
  Hertzian, 34; input impedance of, 36

impedance loaded, 50
impedance transformation using Smith chart, 52
open-circuited, 47
output impedance of, 49
radiation pattern with standing wave of current, 54
receiving power gain, 58
short-circuited reflecting, 43; current distribution on, 47
short, of finite radius, 36; input impedance of, 37
transmitting power gain, 58
travelling wave of current on, 53
director currents, 154
director elements, 145
optimisation of, 153

E-plane sectoral horn, 253
effective length of short dipole, 49
half-wave dipole, 49
electric current, 2
density, 5, 6
electric field, antiphase, 12
electric flux, 3
electric line current element, 9
electric vector potential, 30
of latitudinal magnetic current on sphere, 196
equiangular spiral antenna, 159
equivalence principle, 190
Love's form, 191
application to aperture antennas, 199
equivalent currents, 191
evanescent electric field components, 219, 232, 234
experimental approach to Yagi–Uda antenna design, 151

Faraday's law, 4
field equivalence theorem, 190
fields of linear dipole, 56
filamentary dipole, 34, 39
Fourier transform relationship, 212, 214
frequency independent antenna, 157
linearly polarised, 164
unidirectional, 166

generating sheet, 8

H-plane sectoral horn, 248
half-wave dipole, 49, 78
Hertzian dipole, 34
input impedance, 36
helical antenna, 118
axial mode, 123
linearly polarised, 138
normal mode, 118; closed, 122; open, 120; radiation resistance, 122
sheath helix model, 126
hyperboloidal reflector, 334
element of area, 336
equation of outward normal, 337
radiation pattern with current feed, 338

Impedance loaded dipole, 50
matrix, 66
mutual collinear, 21; orthogonal, 23; coplanar, 25
self, 20
to waves from circular aperture, 236
infinitesimal dipole, 34
feed for paraboloidal reflector, 296
infinite slot in conducting plane, 214
input resistance of Hertzian dipole, 235
isotropic noise field, 176, 178

linear current density, 3
line current element, 9
logarithmically periodic antenna, 157
with circumferential teeth, 164; trapezoidal teeth, 165; triangular teeth, 166
logarithmically periodic dipole array, 168
long Yagi–Uda array, 145
loop antenna, 84
balanced screened, 108
direction finding, 106
electric field approach, 87
flux cutting approach, 86
flux linking approach, 85
input impedance, 96, 105
loss resistance, 97
magnetic field probe, 105
o.c. broadside incidence, 90
o.c. endfire incidence, 93, 104
s.c. broadside incidence, 87, 99
s.c. endfire incidence, 92, 102
transmitting, 111
unabalanced screened, 109
Love's form of equivalence principle, 191
magnetic current density, 7
element, 30
sheet, 6
magnetic field, 11
magnetic flux, 4
magnetic force-antiphase, 32
magnetic pole strength, 32
magnetic vector potential, 10
of longitudinal current on spherical surface, 193
measured radiation patterns, 242
monopole antenna, 39, 69
computer solution of, 65, 70
input impedance on finite ground plane, 71; infinite ground plane, 70
power gain on infinite ground plane, 58, 60, 61

radiation pattern on finite ground plane, 75
mutual impedance between current elements collinear, 21
orthogonal, 23
coplanar, 25
mutual impedance between parallel half-wave dipoles, 63

network analysis of uniformly loaded periodic structure, 171
expression for noise power, 184
representation for noise power, 181
noise power delivered to any antenna, 178
to infinitesimal dipole, 176

offset paraboloid-tangential electric fields in aperture plane, 353
radiation pattern, 354
offset reflector antenna, 348
geometry, 349
open-circuited linear dipole, 47
optimisation of feed reflector combination in Yagi–Uda antenna, 153
director elements in Yagi–Uda antenna, 153
orthogonal current elements, 27
output impedance of linear dipole, 49

paraboloidal reflector, 266
asymmetry of scattered fields on axis, 276
current distribution on, 272
element of area, 267
focal plane magnetic fields, 279
focal plane electric fields, 286
inward normal, 268
power across focal plane, 291
power gain with current element feed, 325
power gain with cylindrical waveguide feed, 330
scattered magnetic field on axis, 273
scattered electric field on axis, 275
with infinitesimal dipole feed, 296
with cylindrical waveguide feed, 303
with corrugated waveguide feed, 307
radiation pattern with infinitesimal dipole feed 298, 312; cylindrical waveguide feed, 327
parasitic element, 141
periodic structure, 171
application to periodic dipole array, 175
phase switching, 169
planar equiangular spiral, 159
power coupling between apertures, 300
power flow across focal plane of paraboloid, 291
from current element, 29
from magnetic current element, 33
power gain, 58
of *E*-plane sectoral horn, 253
of *H*-plane sectoral horn, 253
of half-wave dipole, 60
of half-wave dipole in corner reflector, 78
of monopole, 58, 60, 61
of pyramidal horn, 253
of paraboloid with current element feed, 324
of paraboloid with cylindrical waveguide feed, 330
of uniformly excited circular aperture, 238
of uniformly excited rectangular aperture, 241
of Yagi–Uda antenna, 149
pyramidal horn, 253

radiating electric field component, 219, 232, 234, 235
radiation fields from infinite slot in conducting plane, 214
two dimensional aperture in conducting plane, 227
circular aperture, 237
rectangular aperture, 239
Radiation from combined electric and magnetic currents, 207
electric current in infinite magnetic ground plane, 205
magnetic current in infinite ground plane, 202
Radiation intensity, 58
radiation pattern of cylindrical waveguide, 259
H-plane sectoral horn, 248
hyperboloid with current element feed, 338
large rectangular loop, 113
linear dipole, 54
paraboloid with infinitesimal dipole feed, 298, 312, 318; cylindrical waveguide feed 327; single wire helix, 135; sheath helix, 129; rectangular waveguide, 245
radiation resistance of Hertzian dipole, 35
loop antenna, 113
receiving power gain, 58
of half-wave dipole, 61
rectangular waveguide, TE_{01} mode, 245
reflecting dipole, 43

scalar potential, 13
scattered field of hyperboloid with cylindrical waveguide feed, 346
scattered fields on axis of paraboloid, 273, 275, 277

scattered focal plane fields of paraboloid, 279, 286
self complementary structure, 158
self impedance, 20
short current element, 17
short dipole of finite radius, 36
signal to noise ratio of paraboloid plus feed, 309
switching of successive dipole feeds, 169
standing wave of current, 54

transmitting power gain, 58
  of half-wave dipole, 60
trapezoidal teeth on logarithmic periodic antenna, 165
travelling wave current on filamentary dipole, 53
triangular logarithmic periodic anetnna, 166
two dipole array, 141
  computer solution, 144

uniform plane wave, 1, 5, 6
uniformly distributed charge, 18
unidirectional frequency independent antenna, 166
uniqueness theorem, 199

waveguide cylindrical-derivation of wave equation, 254
  solution of wave equation for TE modes, 255
  tangential electric and magnetic fields in aperture, 257
  radiation pattern for TE_{11} modes, 259
waveguide radiators, 244
waveguide rectangular, TE_{01} mode, 245

Yagi–Uda antenna, 140
  continuous array of short director elements, 145
  director currents, 154
  experimental approach, 151
  near field amplitude and phase plots, 152
  power gain, 153